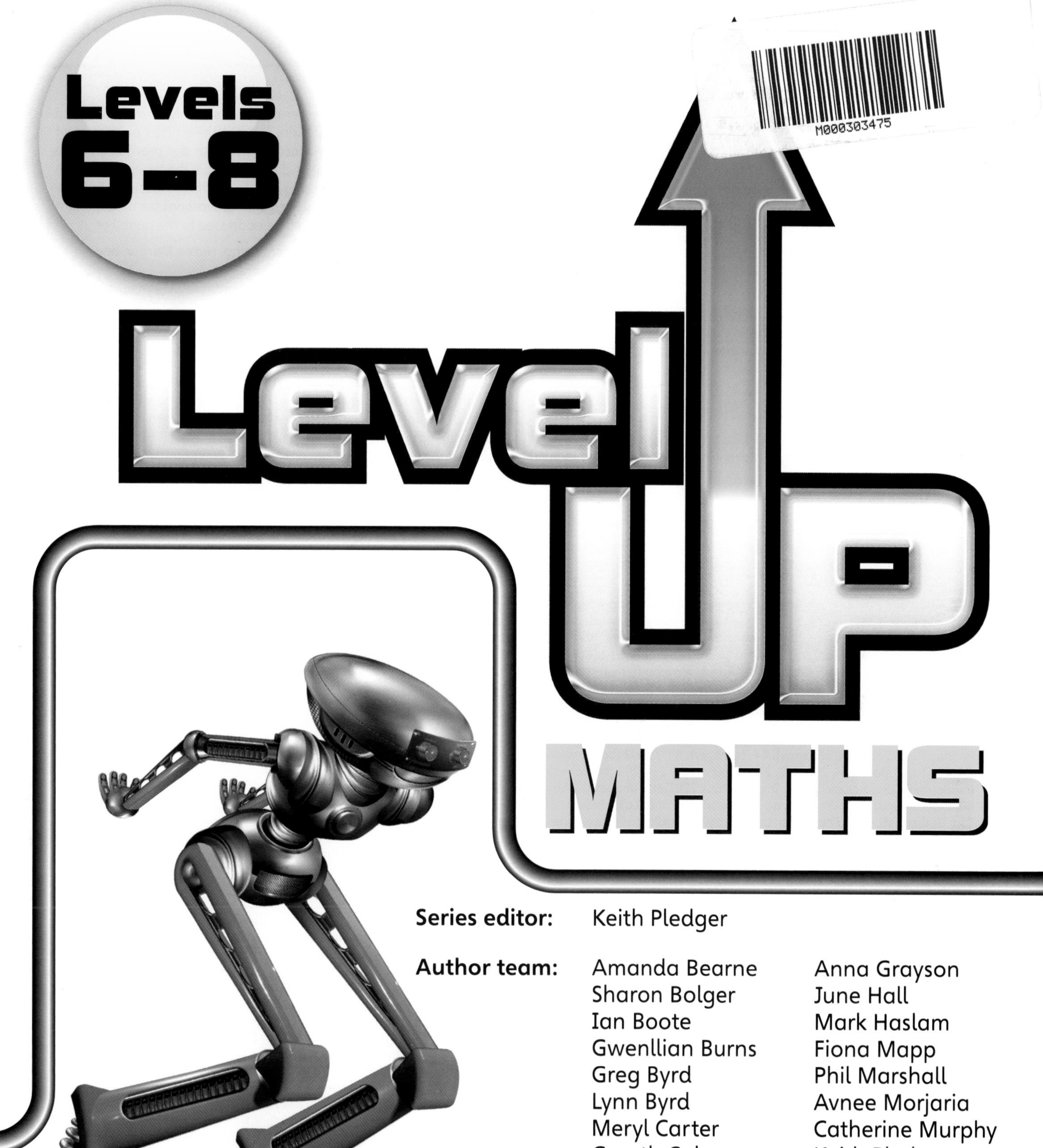

Series editor: Keith Pledger

Author team: Amanda Bearne, Sharon Bolger, Ian Boote, Gwenllian Burns, Greg Byrd, Lynn Byrd, Meryl Carter, Gareth Cole, Crawford Craig, Jackie Fairchild, Freda Gardiner, Anna Grayson, June Hall, Mark Haslam, Fiona Mapp, Phil Marshall, Avnee Morjaria, Catherine Murphy, Keith Pledger, Harry Smith, Robert Ward-Penny, Angela Wheeler

LiveText

Heinemann

Heinemann is an imprint of Pearson Education Limited, a company incorporated in England and Wales, having its registered office at Edinburgh Gate, Harlow, Essex, CM20 2JE. Registered company number: 872828

www.heinemann.co.uk

Heinemann is a registered trademark of Pearson Education Ltd

First published 2009

13 12 11 10
10 9 8 7 6 5 4 3 2

British Library Cataloguing in Publication Data is available from the British Library on request.

ISBN 978 0 435537 33 3

Edited by Gwenllian Burns, Jane Glendening, Nicola Morgan and Jim Newall
Designed by Bigtop Design, Tom Cole (Seamonster design) and Debbie Oatley (room9design)
Typeset by Tech-Set Ltd, Gateshead
Original illustrations © Pearson Education Limited, 2008
Illustrated by Tech-Set Ltd, Matt Buckley (Chrome Dome Design), Richard Duckett (NB Illustration), Alisdair Bright (NB illustration), Jonathan Edwards, Andy Hammond, Tom Cole (Seamonster design)
Cover design by Tom Cole (Seamonster design), Stephen Holmes
Cover illustration © Pearson Education Limited
Printed in Malaysia (CTP-VP)

Acknowledgements

We would like to thank all of those schools who provided invaluable help in the development and trialling of this course.

We would also like to thank the following individuals and organisations for their help in providing information: James Allison, Renault p214; Mike Fossum, NASA p102; Kishore Patel, Fiona Cairns Ltd p2; Amaia Ruiz de Alegria, Plymouth University p 232; Marcus du Sautoy, University of Oxford p140; Anne Vorley, Standard Life p266, Matt Cooper p248.

The author and publisher would like to thank the following individuals and organisations for permission to reproduce photographs and images: Alamy/Inmagine p166; Alamy/ICP-Traffic p48; Alamy/Ian Miles-Flashpoint p132; Alamy/Wolfgang Kuehler p110 (mosaic); Alamy/Sean Clarkson p110 (mosaic); Alamy/Blinckwinkel p297 (dog beside pond); Alamy/Trevor Smith p274; Alamy/David Coleman p14 (background); Alamy/Amana Images p36; Alamy/Images & Stories p196 (background); Alamy/Keith Morris p258; Alamy/Ace Stock Ltd p122 (background); Alamy/Jim Betty p164; Alamy/William Caram p179; Alamy/Steve Allen (fractal image); Alamy/Chris Howes, Wild Places p120 (background); Alamy/Doug Web p246; Alamy/Charles Polidano/Touch the Skies p118; Alamy/Jupiter Images/Polka Dot p277 (girls playing the flute); Alamy/David R Frazier p100; Amaia Ruiz de Aegria p232; Brand X Pictures p133; Ancient Art and Architecture Collection p196 (inset); BBC p186; Cormstock Images p64; Corbis/Bettman p82; Corbis/Michael Hall Photography Ltd p38 (girl); Corbis/Karen Kasmauski p140 (cicadas), Corbis/Robert Michael p234; Corbis/Digital Art p184 (UFO); Corbis/Araldo de Luca p69; Corbis/Juice Images p208; Corbis/Alan Copson/JAI p244 (Las Palmas); Digital Stock pp66, 130, 136 (Eiffel Tower), 218; Digital Vision pp40, 136 (forest), 225, 284, 286; Eyewire pp134, 171; Getty Images/PhotoDisc pp34, 38 (x-ray), 44 (teenager), 49, 73, 102 (background), 108, 114, 129, 138, 144, 146, 168 (basketball), 176, 210 (background), 211 (spacecraft) 219, 224 (saucepan), 244, 250, 262, 277 (apples), 285, 294, 299 (blood vessels), 301 (typhoon); Getty Images pp102 (astronaut), 276 (school register); Getty Images/AFP p15 (solar car); Getty Images/Stone 221; Getty Images/Ulf Anderson p140 (Marcus du Sautoy); Getty Images/Gallo Images/Wolfgang Kaehier p18 (background); iStockPhoto/Long Ha p84 ; iStockPhoto/Lajos Repasi p86 ; iStockPhoto/Andy Cook p96; iStockPhoto/Levente Janos p170; iStockPhoto/Tom Guther p46; iStockPhoto/Alan Lagadu p50; iStockPhoto/Nancy Honeycutt p52; iStockPhoto pp20, 28; iStockPhoto/Niels Laan p30; iStockPhoto/Yakv Stavchansky p220; iStockPhoto/Javier Fontanella p6; iStockPhoto/Nina Shannon p8; iStockPhoto/Janne Ahvo p200; iStockPhoto/Gareth Jago p204; iStockPhoto/Imagestock p42; iStockPhoto/George Peters p211 (distorted clocks); iStockPhoto/Rich Legg p268; iStockPhoto/Donald Johansson p270; iStockPhoto/Martin Vegh p70; iStockPhoto/Oleksandr Koval p72; iStockPhoto/Jeff Chevrier p180; iStockPhoto/Bart Coenders p188; iStockPhoto/Appletat (buoy in ocean) p287; iStockPhoto/William Walsh pp 284, 288; iStockPhoto/Joery Reimann pp 285, 292; iStockPhoto/Marcin Laskowski p299 (Menger sponge); iStockPhoto/Dra Schwartz p148; iStockPhoto/Geotrac p152; iStockPhoto/Technotr p154; iStockPhoto/Pali Rad p158; iStockPhoto/Chang p160; iStockPhoto/David Steffes p253; iStockPhoto/Kelvin Wakefield p252; iStockPhoto pp80, 98; iStockPhoto/Nigel Silcock p212; iStockPhoto/Chris Scredon p156 (tennis ball); iStockPhoto/Andres Peiro Palmer p301 (fists); iStockPhoto/Cathleen Abers-Kimball p156 (car headlight); iStockPhoto/Stanislave Mikhalev p56; iStockPhoto/FotoVoyager p60 (background); Jackie Fairchild pp18, 22, 24; Jupiter Images/Photos.com p51; Katherine Pate p248 (joinery workshop); Kishore Patel p2 (inset); LAT Photography/Glenn Dunbar p214; M C Escher's 'Ascending and Descending' © The M C Escher Company Holland. All rights reserved p192; Mary Evans Picture Library p184 (UFO); NASA p260; National Geophysical Data Centre p289; Nature PL/Tony Heald p285 (bird of prey), 296 (bird of prey); Pearson Education Ltd/Rob Judges pp126, 168, 172; Pearson Education Ltd/Jules Seimes pp44 (woman with cake), 55, 142, 216, 276 (canteen food); Pearson Education Ltd/Debbie Rowe pp12, 104, 226, 287 (ferris wheel); Pearson Education Ltd/Clark Wiseman, Studio 8 pp 209, 212 (earrings supplied by Jackie Fairchild); Pearson Education Ltd/Tudor Photography p62; Pearson Education Ltd/Gareth Boden p242; Pearson Education Ltd/Ian Wedgewood p199; Pearson Education Ltd/Chris Parker p89; Peter Evans p41 (Ancient Egyptian statues); Reed International Books Australia PTY LTD/Lindsey Evans Photography p131; Rex Features p280; Rex Features/Chris Capstick p120; Richard Smith p277 (hospital sign); Science Photo Library pp150, 178 (inset and background), 210 (Einstein); Science Photo Library/Henry Groskinsky, Peter Arnold Inc p156 (background); Science Photo Library/Emilio Serge Visual Archives/American Institute for Physics p210 (Schwarzschild); Science Photo Library/Pasieka p26 (background), p27; Shutterstock/Walter 71 p88; Shutterstock/Phase4Photography p90; Shutterstock/8781118005 p94; Shutterstock/Morgan Lane p41; Shutterstock/Monkey Business Images p54; Shutterstock/Jenny Horne pp32, 291; Shutterstock/Nat Ulrich p124; Shutterstock/ALBA p128; Shutterstock/Chee-Ohn Leong p222; Shutterstock/C pp4, 224; Shutterstock/Christofoto p10; Shutterstock/Rachaeil Grazias p198; Shutterstock/Vivian Fung p202; Shutterstock/Captured by JJ p206; Shutterstock/BestWeb p92 (background); Shutterstock/WitR p110 (background); Shutterstock/Rostislav Glinsky p157 (Golden Gate Bridge); Shutterstock/Katherine Wittfeld p156 (telecommunications satellite); Shutterstock/Krkr p156 (jet of water); Shutterstock/Kulish Viktroila p278 (background); Shutterstock/Mike Fleming p279 (party hat); Shutterstock/Elnur p296 (boat on curve); Shutterstock/Lim Yong Hian p272; Shutterstock/Digital File p276 (sun and clouds); Shutterstock/Mike Flippo p276 (tennis racket and ball); Shutterstock/Christopher Elwell p 277 (countryside); Shutterstock/Thomas M Perkins p68; Shutterstock/Ben Smith p76; Shutterstock/Adrian Baras p106; Shutterstock/Maxymuss p182; Shutterstock/Morekuliasz p190; Shutterstock/Robyn Mackenzie p236; Shutterstock/Stephen Finn p240; Shutterstock/Kurt Tutschek p2 (background); Shutterstock/Angelo Gilardelli p248 (wood shavings); Shutterstock/Joshua Havlv p266; Shutterstock/Christopher Waters p287 (tsunami hazard warning sign); Shutterstock/Beerkof pp 284, 290; Shutterstock/Harris Shiffman pp 285, 298 (Romanesco broccoli); Shutterstock/Marten Czamanske p127; Shutterstock/Andre Maritz p256; Shutterstock/Suzanne Tucker p116; Shutterstock/Cindy Hughes p16; Shutterstock/prism_68 p264; Shutterstock/Lim Yong Hien (computer coding) pp285, 300; Shutterstock/Meliksetyan p300 (knot); Shutterstock/Muldix p300 (calculator); Shutterstock/Jaimie Duplass p301 (vaccination); Shutterstock/Ben Smith p230; Shutterstock/Andrey Armyagov p194; Shutterstock/Artkot p112; Shutterstock/David Davis p58; Shutterstock/Stephen Gibson p74.

Every effort has been made to contact copyright holders of material reproduced in this book. Any omissions will be rectified in subsequent printings if notice is given to the publishers.

Contents

Expanded contents			iv
Introduction			viii
Unit 1	An opening sequence	Algebra 1/2	2
Unit 2	Keep your balance	Algebra 3	18
Unit 3	Share and share alike	Number 1	36
Unit 4	Be constructive	Geometry and measures 1	60
Unit 5	Stat's entertainment	Statistics 1	82
Revision 1			100
Unit 6	Extreme measures	Geometry and measures 2	102
Unit 7	Power up	Number 2	120
Unit 8	Graphic detail	Algebra 4	140
Unit 9	Strictly come chancing	Statistics 2	164
Unit 10	Shape shifter	Geometry and measures 3	178
Revision 2			194
Unit 11	Magic formula	Algebra 5	196
Unit 12	No problem?	Solving problems	214
Unit 13	Data day statistics	Statistics 3	232
Unit 14	Trig or treat?	Geometry and measures 4	248
Unit 15	A likely story	Statistics 4	266
Revision 3			282
Unit 16	Dramatic mathematics	GCSE transition	284
Index			302

Expanded contents

Unit 1 An opening sequence – Algebra 1/2

Introduction 2

1.1 Generating terms in a sequence 4

1.2 Linear sequences 6

1.3 Quadratic and fraction sequences 8

1.4 Sequences from patterns 10

1.5 Functions and mappings 12

1.6 World's Greatest Maths! World solar challenge 14

Plenary Stacking fairy cakes 16

Unit 2 Keep your balance – Algebra 3

Introduction 18

2.1 Using letter symbols in algebra 20

2.2 Constructing and solving linear equations 22

2.3 Solving linear equations with x and brackets on both sides 24

maths! Top profit 26

2.4 Using trial and improvement to solve equations 28

2.5 Solving simultaneous equations graphically 30

2.6 Solving simultaneous equations algebraically 32

Plenary Travelling equations 34

Unit 3 Share and share alike – Number 1

Introduction 36

3.1 Decimals and fractions in order 38

3.2 Adding and subtracting fractions 40

maths! Joining the resistance 42

3.3 Multiplying and dividing fractions 44

3.4 Using percentages 46

3.5 Get things in proportion 48

3.6 Ratio and proportion 50

3.7 Brain power 52

3.8 More calculation strategies 54

3.9 Efficient calculation 56

Plenary Towards a better lifestyle 58

Unit 4 Be constructive – Geometry and measures 1

Introduction 60

4.1 Quadrilaterals, angles and proof 62

4.2 Interior and exterior angles in polygons 64

4.3 Angles in triangles and quadrilaterals 66

4.4 Pythagoras' theorem 68

4.5 Using Pythagoras' theorem to solve problems 70

4.6 Constructions 72

4.7 Construction problems 74

4.8 Circles and tangents 76

4.9 World's Greatest Maths! Perilous paper round 78

Plenary Engineering calculations 80

Unit 5 Stat's entertainment – Statistics 1

Introduction 82

5.1 Planning an investigation 84

5.2 Organising data 86

5.3 Calculating and using statistics 88

5.4 Pie charts 90

maths! Power for the future 92

5.5 Line graphs 94

5.6 Frequency diagrams and frequency polygons 96

Plenary Data on disease 98

Revision 1 100

Unit 6 Extreme measures – Geometry and measures 2

Introduction 102

6.1 Converting between measures 104

6.2 Compound measures and bounds 106

6.3 Circles and perimeter 108

maths! Roman mosaics 110

6.4 Circles, sectors and area 112

6.5 Area and volume 114

6.6 Volume and surface area 116

Plenary All wrapped up 118

Unit 7 Power up – Number 2

Introduction 120

7.1 World's Greatest Maths Labyrinth 122

7.2 To round or not to round? 124

7.3 Using a calculator 126

7.4 More rounding 128

7.5 Roots and standard form on a calculator 130

7.6 Written methods 132

7.7 Using a calculator efficiently 134

7.8 Problem solving 136

Plenary Decimal Day! 138

Unit 8 Graphic detail – Algebra 4

Introduction 140

8.1 Prime factors, HCF and LCM 142

8.2 Using factors and multiples 144

8.3 Multiplying and dividing with indices 146

8.4 Index laws with negative and fractional powers 148

8.5 Reviewing straight-line graphs 150

8.6 Investigating the properties of straight-line graphs 152

8.7 Graphs of quadratic and cubic functions 154

maths! Parabolas 156

8.8 Functions and graphs in real life 158

8.9 Investigating patterns 160

Plenary Primed for action 162

Unit 9 Strictly come chancing – Statistics 2

Introduction 164
9.1 Fair play 166
9.2 It's exclusive! 168
9.3 Tree diagrams 170
9.4 Relative frequency 172
maths! Pollyopoly 174
Plenary I've got the 'skill' factor 176

Unit 10 Shape shifter – Geometry and measures 3

Introduction 178
10.1 Testing for congruence 180
10.2 Transforming shapes 182
10.3 World's Greatest Maths UFO hunt 184
10.4 Similarity 186
10.5 Introducing trigonometry 188
10.6 Using trigonometry 1 190
Plenary The shape illusions 192
Revision 2 194

Unit 11 Magic formula – Algebra 5

Introduction 196
11.1 Simplifying algebraic expressions 198
11.2 Working with double brackets 200
11.3 Factorising quadratics 202
11.4 Using and writing formulae 204
11.5 Working with formulae 206
11.6 Inequalities 208
maths! Algebra from the edge 210
Plenary Accessorised with algebra 212

Unit 12 No problem? – Solving problems

Introduction 214
12.1 Data problem solving 216
12.2 Number problem solving 218
12.3 Algebra problem solving 220
12.4 Geometrical reasoning: lines, angles and shapes 222
12.5 Percentage and proportion problems 224
12.6 Functions and graphs 226
maths! Unsolved problems 228
Plenary Cars, cars, cars! 230

Unit 13 Data day statistics – Statistics 3

Introduction 232
13.1 Questionnaires and samples 234
13.2 Averages from grouped data 236
13.3 World's Greatest Maths Classroom challenge 238
13.4 Scatter graphs and correlation 240
13.5 Misleading graphs and charts 242
13.6 Comparing distributions 244
Plenary There's a storm coming 246

Unit 14 Trig or treat? – Geometry and measures 4

Introduction 248
14.1 Solving geometrical problems 250
14.2 More geometrical problems 252
maths! The story of Pythagoras 254
14.3 3-D shapes 256
14.4 Prisms and cylinders 258
14.5 Using trigonometry 2 260
14.6 Solving problems in trigonometry 262
Plenary Accessibility problems 264

Unit 15 A likely story – Statistics 4

Introduction 266
15.1 This will probably be familiar! 268
15.2 Growing trees 270
15.3 Experiment! 272
15.4 What's the problem? 274
15.5 Back to the future 276
maths! The birthday problem 278
Plenary Understanding probabilities 280
Revision 3 282

Unit 16 Dramatic mathematics – GCSE transition

Introduction 284
16.1 Making waves 286
16.2 The mathematics of earthquakes 288
16.3 Medical statistics 290
16.4 Made to measure 292
16.5 Stellar mathematics 294
maths! Curves of pursuit 296
16.6 Fractals 298
Plenary Mathematics at work 300
Index 302

Welcome to Level Up Maths!

Level Up Maths is an inspirational new course for today's classroom. With stunning textbooks and amazing software, Key Stage 3 Maths has simply never looked this good!

This textbook is divided into 16 units. Here are the main features.

Why learn this?

You can represent pr problems using inequ for example, decidin to sell to maximise p

Why are you learning this? For each topic there's a reason why it is useful.

Unit introduction

Each unit begins with an introduction. These include a striking background image to set the scene and short activities to check what you know.

Here's a quick quiz to see how much you already know. The arrows tell you where to look for help.

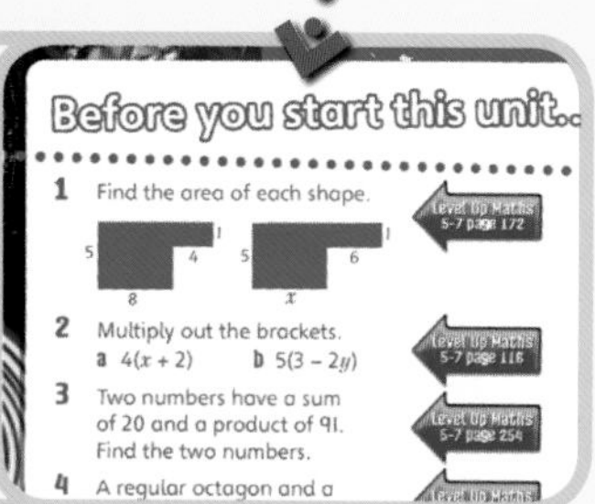
Before you start this unit...

1 Find the area of each shape.

2 Multiply out the brackets.
a $4(x + 2)$ b $5(3 - 2y)$

3 Two numbers have a sum of 20 and a product of 91. Find the two numbers.

4 A regular octagon and a

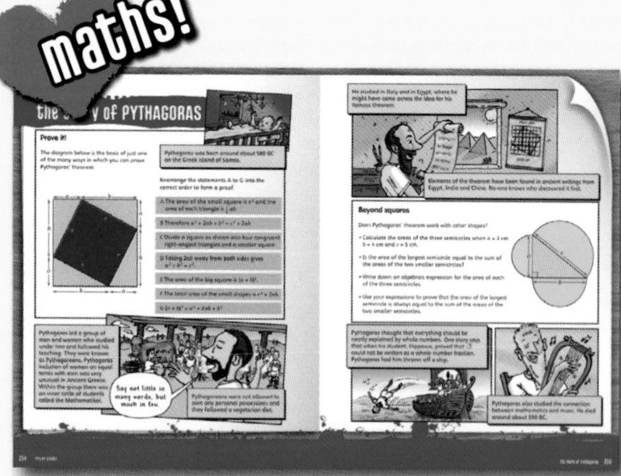

Eleven sets of special activities highlight some intriguing **applications and implications of maths**. Where in history and culture does today's maths come from? How does it affect our lives? Why is it so important to get it right?

Main content

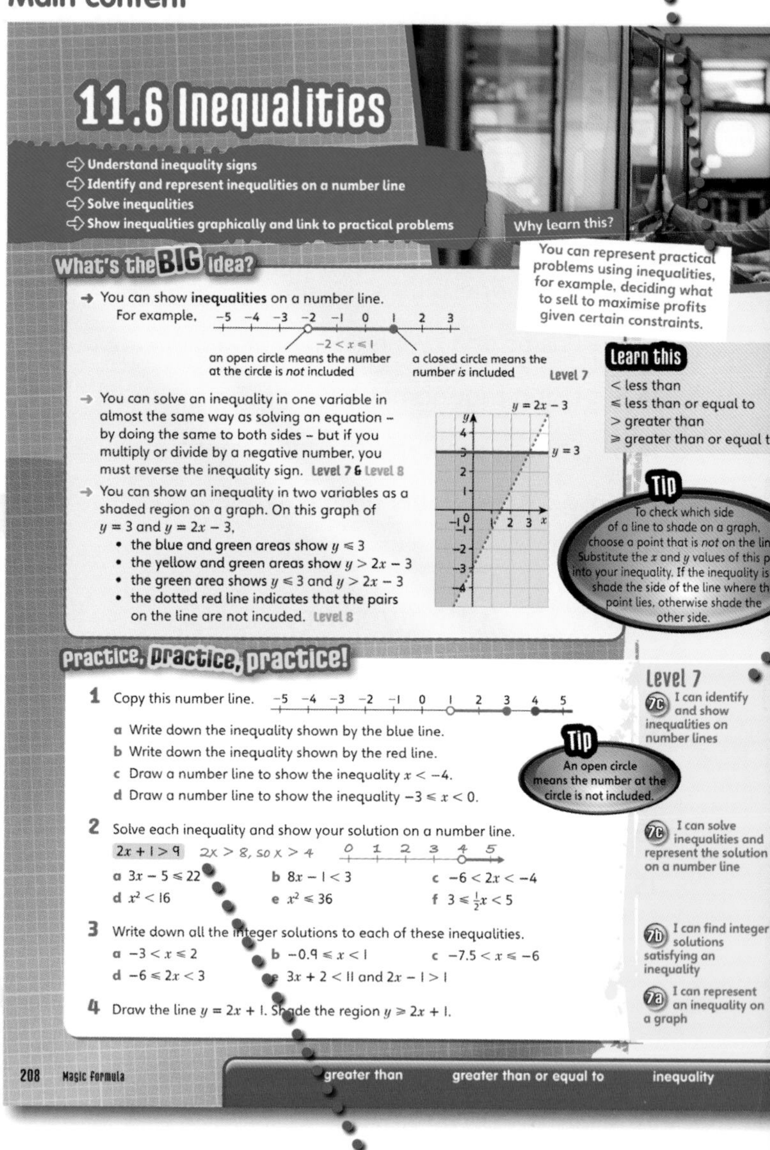
11.6 Inequalities

- Understand inequality signs
- Identify and represent inequalities on a number line
- Solve inequalities
- Show inequalities graphically and link to practical problems

Why learn this?

You can represent practical problems using inequalities, for example, deciding what to sell to maximise profits given certain constraints.

What's the BIG idea?

- You can show **inequalities** on a number line.
For example, $-2 < x \leqslant 1$
an open circle means the number at the circle is *not* included; a closed circle means the number *is* included. Level 7
- You can solve an inequality in one variable in almost the same way as solving an equation – by doing the same to both sides – but if you multiply or divide by a negative number, you must reverse the inequality sign. Level 7 & Level 8
- You can show an inequality in two variables as a shaded region on a graph. On this graph of $y = 3$ and $y = 2x - 3$,
  - the blue and green areas show $y \leqslant 3$
  - the yellow and green areas show $y > 2x - 3$
  - the green area shows $y \leqslant 3$ and $y > 2x - 3$
  - the dotted red line indicates that the pairs on the line are not included. Level 8

Learn this

$<$ less than
$\leqslant$ less than or equal to
$>$ greater than
$\geqslant$ greater than or equal to

Tip: To check which side of a line to shade on a graph, choose a point that is *not* on the line. Substitute the x and y values of this point into your inequality. If the inequality is true, shade the side of the line where the point lies, otherwise shade the other side.

Practice, practice, practice!

1 Copy this number line.
a Write down the inequality shown by the blue line.
b Write down the inequality shown by the red line.
c Draw a number line to show the inequality $x < -4$.
d Draw a number line to show the inequality $-3 \leqslant x < 0$.

Tip: An open circle means the number at the circle is not included.

2 Solve each inequality and show your solution on a number line.
$2x + 1 > 9$ $2x > 8$, so $x > 4$
a $3x - 5 \leqslant 22$ b $8x - 1 < 3$ c $-6 < 2x < -4$
d $x^2 < 16$ e $x^2 \leqslant 36$ f $3 \leqslant \frac{1}{2}x < 5$

3 Write down all the integer solutions to each of these inequalities.
a $-3 < x \leqslant 2$ b $-0.9 \leqslant x < 1$ c $-7.5 < x \leqslant -6$
d $-6 \leqslant 2x < 3$ e $3x + 2 < 11$ and $2x - 1 > 1$

4 Draw the line $y = 2x + 1$. Shade the region $y \geqslant 2x + 1$.

Level 7
7b I can identify and show inequalities on number lines
7c I can solve inequalities and represent the solution on a number line
7b I can find integer solutions satisfying an inequality
7a I can represent an inequality on a graph

208 Magic Formula greater than greater than or equal to inequality

There are plenty of questions for you to practise on. All the questions have been **levelled** and **sublevelled**, so you can see how well you are doing. The **'I can ...'** statements confirm what you can do.

$2x + 1 > 9$ $2x > 8$,

Highlighted sample questions show you how to set out your working.

This shows that the questions are at Level 7.

- 'c' questions are easiest.
- 'b' questions are of medium difficulty.
- 'a' questions are hardest.

Level 7

7c I can identify and show inequalities on number lines

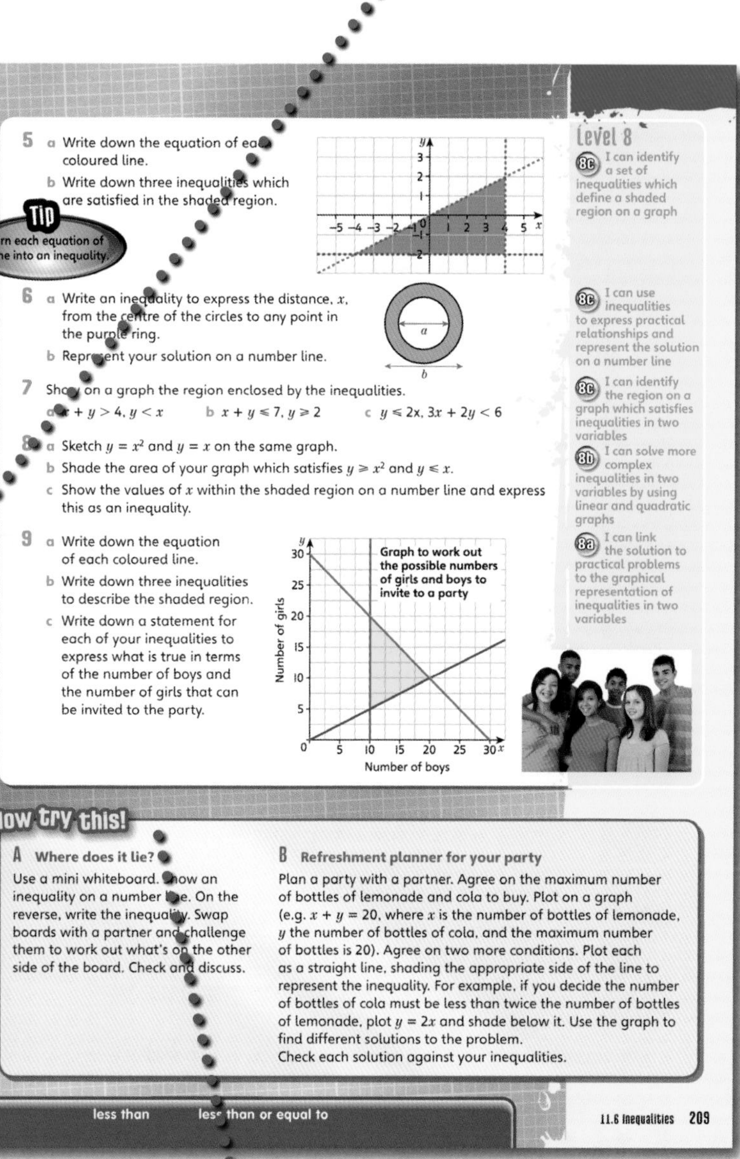

5 a Write down the equation of each coloured line.
b Write down three inequalities which are satisfied in the shaded region.

Tip: Turn each equation of a line into an inequality.

6 a Write an inequality to express the distance, x, from the centre of the circles to any point in the purple ring.
b Represent your solution on a number line.

7 Show on a graph the region enclosed by the inequalities.
a $x + y > 4$, $y < x$ b $x + y \leqslant 7$, $y \geqslant 2$ c $y \leqslant 2x$, $3x + 2y < 6$

8 a Sketch $y = x^2$ and $y = x$ on the same graph.
b Shade the area of your graph which satisfies $y \geqslant x^2$ and $y \leqslant x$.
c Show the values of x within the shaded region on a number line and express this as an inequality.

9 a Write down the equation of each coloured line.
b Write down three inequalities to describe the shaded region.
c Write down a statement for each of your inequalities to express what is true in terms of the number of boys and the number of girls that can be invited to the party.

Level 8

8c I can identify a set of inequalities which define a shaded region on a graph

8c I can use inequalities to express practical relationships and represent the solution on a number line

8c I can identify the region on a graph which satisfies inequalities in two variables

8b I can solve more complex inequalities in two variables by using linear and quadratic graphs

8a I can link the solution to practical problems to the graphical representation of inequalities in two variables

Now try this!

A Where does it lie?
Use a mini whiteboard. Show an inequality on a number line. On the reverse, write the inequality. Swap boards with a partner and challenge them to work out what's on the other side of the board. Check and discuss.

B Refreshment planner for your party
Plan a party with a partner. Agree on the maximum number of bottles of lemonade and cola to buy. Plot on a graph (e.g. $x + y = 20$, where x is the number of bottles of lemonade, y the number of bottles of cola, and the maximum number of bottles is 20). Agree on two more conditions. Plot each as a straight line, shading the appropriate side of the line to represent the inequality. For example, if you decide the number of bottles of cola must be less than twice the number of bottles of lemonade, plot $y = 2x$ and shade below it. Use the graph to find different solutions to the problem.
Check each solution against your inequalities.

less than less than or equal to

11.6 Inequalities 209

Now try this!

A Where does it lie?

Use a mini whiteboard. Show an inequality on a number line. On the

Two '**Now try this!**' activities give you the chance to **solve problems** and explore the maths further, sometimes with a partner. Activity B is generally more challenging than Activity A.

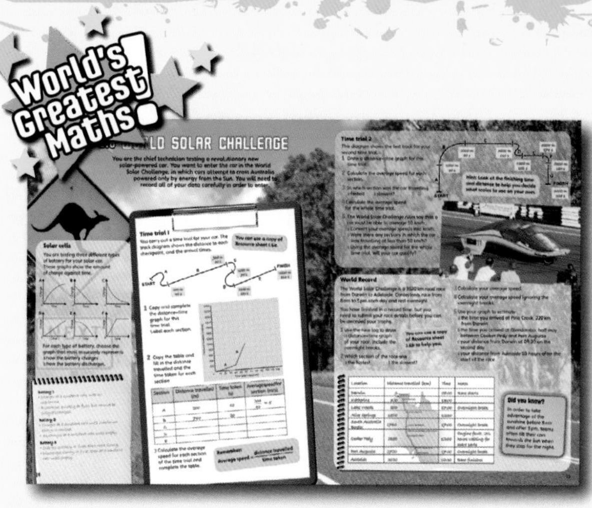

We've scoured the planet to find examples of the World's greatest maths for you to try. They include a driving test, enlarging images, cumulative frequency and some barking dogs!

Unit plenary

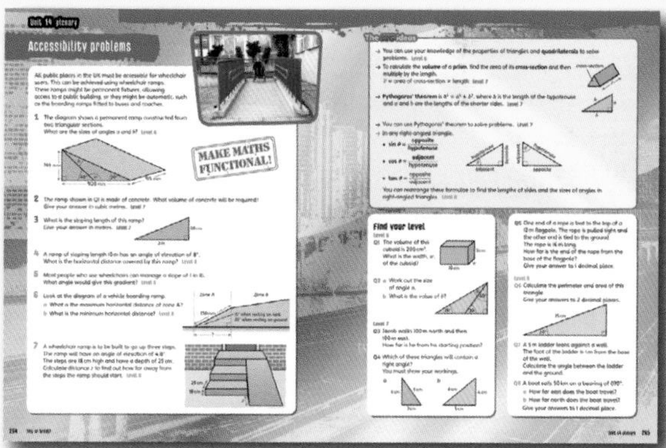

Each unit ends with an extended, levelled activity to help you practise **functional maths** and demonstrate your understanding. There are also **Find your level** questions to try.

Revision activities

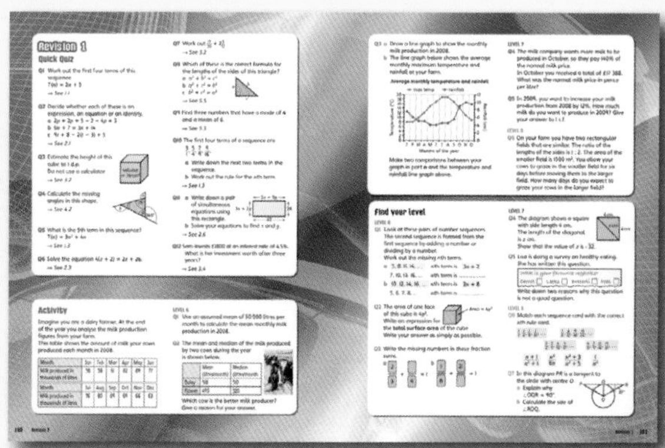

Three revision sections (one for each term) give you the chance to double-check your understanding of previous units. Each consists of a quick quiz, an extended activity and **Find your level** questions.

LiveText software

The LiveText software gives you a wide range of additional materials – interactive explanations, games, activities and extra questions.

Simply turn the pages of the electronic book to the page you need, and explore!

Interactive explanations

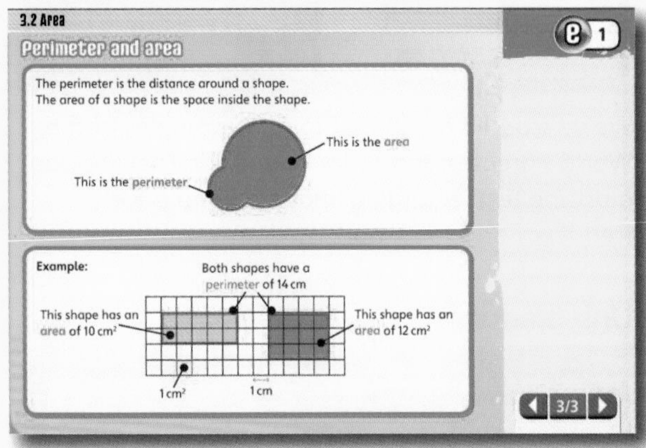

Interactive

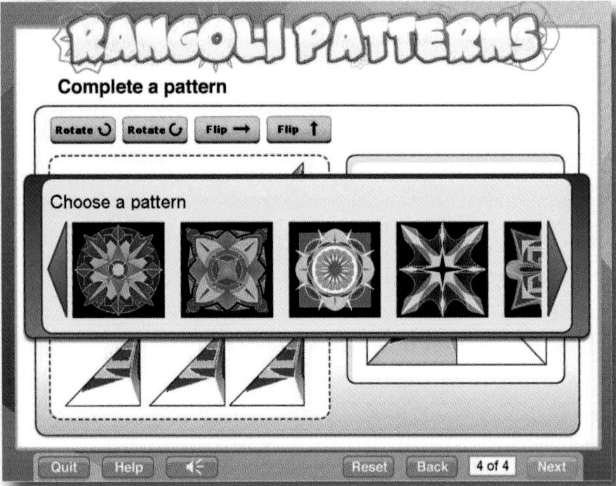

Competitive maths games

- Use linear expressions to describe the nth term in an arithmetic sequence
- Use the pattern in the first differences to help find the nth term in an arithmetic sequence
- Justify the form of the nth term by looking at the pattern that generated it

Zoom in on any part of the page with a single click.

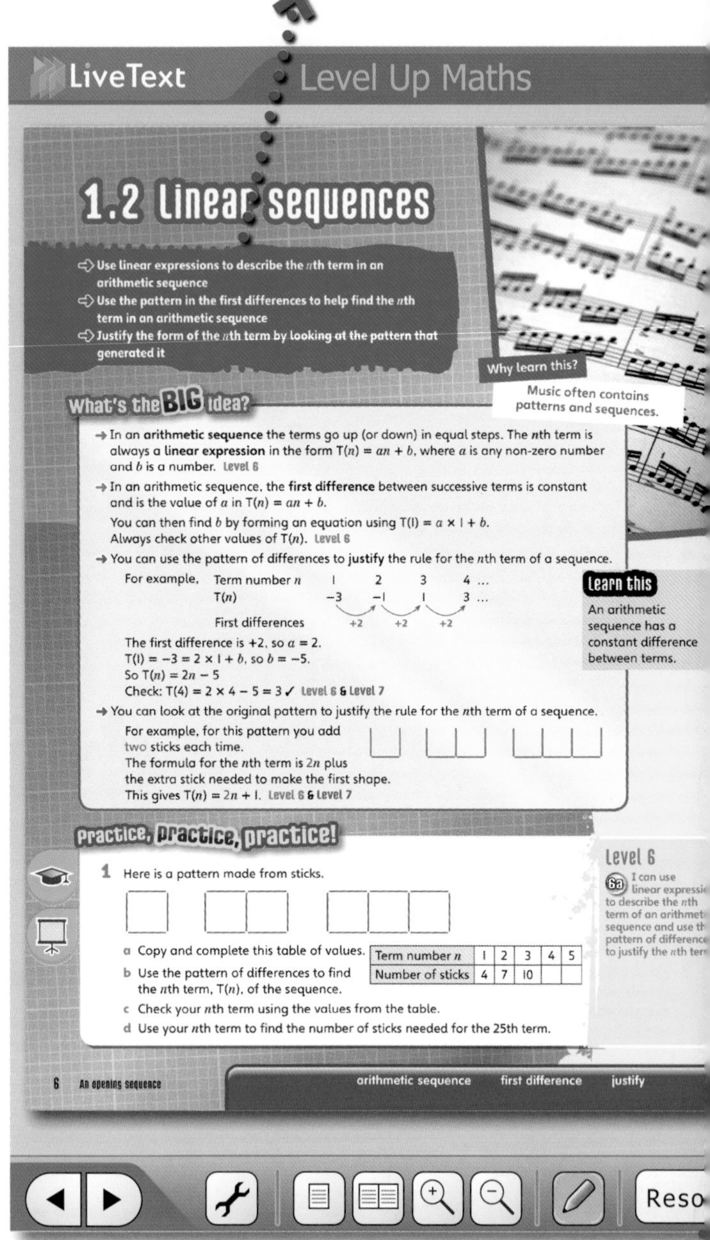

Resources – a comprehensive list of all relevant resources plus lesson plans.

Glossary – contains definitions of key terms. Play audio to hear translations in Bengali, Gujarati, Punjabi, Turkish and Urdu.

Click on the icons to access the extra resources.

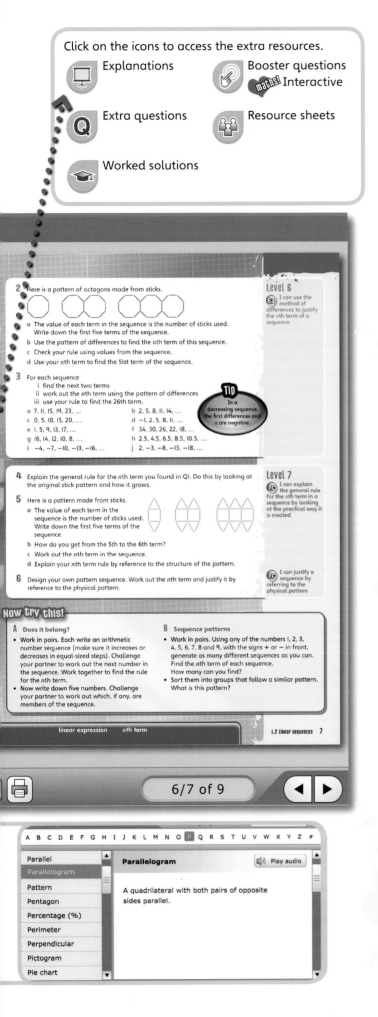

- Explanations
- Booster questions
- maths! Interactive
- Extra questions
- Resource sheets
- Worked solutions

2 Here is a pattern of octagons made from sticks.

a The value of each term in the sequence is the number of sticks used. Write down the first five terms of the sequence.
b Use the pattern of differences to find the nth term of this sequence.
c Check your rule using values from the sequence.
d Use your nth term to find the 51st term of the sequence.

3 For each sequence
i find the next two terms
ii work out the nth term using the pattern of differences
iii use your rule to find the 26th term.

a 7, 11, 15, 19, 23, …
b 2, 5, 8, 11, 14, …
c 0, 5, 10, 15, 20, …
d −1, 2, 5, 8, 11, …
e 1, 5, 9, 13, 17, …
f 34, 30, 26, 22, 18, …
g 16, 14, 12, 10, 8, …
h 2.5, 4.5, 6.5, 8.5, 10.5, …
i −4, −7, −10, −13, −16, …
j 2, −3, −8, −13, −18, …

Tip In a decreasing sequence, the first differences and a are negative.

Level 6
6a I can use the method of differences to justify the nth term of a sequence

4 Explain the general rule for the nth term you found in Q1. Do this by looking at the original stick pattern and how it grows.

5 Here is a pattern made from sticks.
a The value of each term in the sequence is the number of sticks used. Write down the first five terms of the sequence.
b How do you get from the 5th to the 6th term?
c Work out the nth term in the sequence.
d Explain your nth term rule by reference to the structure of the pattern.

6 Design your own pattern sequence. Work out the nth term and justify it by reference to the physical pattern.

Level 7
7c I can explain the general rule for the nth term in a sequence by looking at the practical way it is created
7c I can justify a sequence by referring to the physical pattern

Now try this!

A Does it belong?
- Work in pairs. Each write an arithmetic number sequence (make sure it increases or decreases in equal-sized steps). Challenge your partner to work out the next number in the sequence. Work together to find the rule for the nth term.
- Now write down five numbers. Challenge your partner to work out which, if any, are members of the sequence.

B Sequence patterns
- Work in pairs. Using any of the numbers 1, 2, 3, 4, 5, 6, 7, 8 and 9, with the signs + or − in front, generate as many different sequences as you can. Find the nth term of each sequence. How many can you find?
- Sort them into groups that follow a similar pattern. What is this pattern?

linear expression nth term

1.2 Linear sequences 7

6/7 of 9

A B C D E F G H I J K L M N O P Q R S T U V W X Y Z #

Parallel
Parallelogram
Pattern
Pentagon
Percentage (%)
Perimeter
Perpendicular
Pictogram
Pie chart

Parallelogram Play audio

A quadrilateral with both pairs of opposite sides parallel.

World's Greatest Maths!

Interactive

BREAK THE CODE, OR...

Boosters

To make one cake a chef uses 45g of dried fruit.
How much dried fruit will he need to make 10 cakes?
Answer: g
Reset
Reload
Submit

Extra practice questions

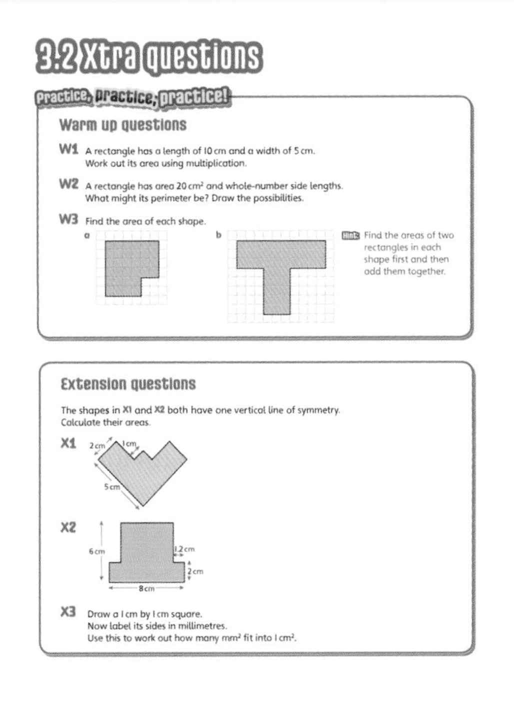

3.2 Xtra questions

Practice, practice, practice!

Warm up questions

W1 A rectangle has a length of 10 cm and a width of 5 cm. Work out its area using multiplication.

W2 A rectangle has area 20 cm² and whole-number side lengths. What might its perimeter be? Draw the possibilities.

W3 Find the area of each shape.

Hint: Find the areas of two rectangles in each shape first and then add them together.

Extension questions

The shapes in X1 and X2 both have one vertical line of symmetry. Calculate their areas.

X1

X2

X3 Draw a 1 cm by 1 cm square. Now label its sides in millimetres. Use this to work out how many mm² fit into 1 cm².

1 An opening sequence

This unit is about looking for patterns in sequences of numbers. Mathematicians spend a lot of time analysing patterns to find the rule behind each pattern.

Husband and wife team Kishore Patel and Fiona Cairns run a business supplying wedding cakes to Waitrose, Selfridges and Harrods. Kishore is the managing director, and Fiona makes and decorates the cakes. Kishore has always loved maths. 'There is a surprising amount of maths in making wedding cakes,' he says. 'Take a wedding cake with square layers, for example. If you know the volume of the top layer, and all the layers are the same height, is there a quick way of working out the volume of the bottom layer?'

Activities

A The volumes of the layers in a wedding cake form a sequence. The layers are all the same height, but each has a different side length.

Layer	Height × length × width	Volume
1	10 cm × 10 cm × 10 cm	1000 cm^3
2	10 cm × 20 cm × 20 cm	4000 cm^3
3	10 cm × 30 cm × 30 cm	9000 cm^3
4	10 cm × 40 cm × 40 cm	16 000 cm^3

Can you see a pattern in the volumes?

Predict the volume of a fifth layer.

B You can use the pattern to work out the quantities of ingredients for each layer. For example, the top layer uses 100 g of flour, so layer 2 uses 400 g, layer 3 uses 900 g and so on.

- The top layer uses 2 eggs. How many eggs do you need for layers 2, 3 and 4?
- Write down an expression that can be used to work out how many eggs are needed for layer n.

Before you start this unit...

1 Find the next two terms in each of these sequences.

a 6, 12, 18, 24, ...

b 4, 9, 14, 19, ...

c 46, 39, 32, 25, ...

d 1, 4, 9, 16, 25, ...

e 1, 3, 6, 10, 15, ...

f 1, 8, 27, 64, ...

g 17, 12, 7, 2, ...

h 1, 1, 2, 3, 5, 8, ...

2 Some of the sequences in Q1 have special names. What are they?

1.1 Generating terms in a sequence
1.2 Linear sequences
1.3 Quadratic and fraction sequences
1.4 Sequences from patterns
1.5 Functions and mappings
World's Greatest Maths! 1.6 World solar challenge
Unit plenary: Stacking fairy cakes
Plus digital resources

Did you know?

A hotel in India claims to have made a Christmas cake 8.9 m tall and weighing 2500 kg.

The record for the largest wedding cake is one with seven layers and weighing 6818 kg. It was made in the USA in 2004.

1.1 Generating terms in a sequence

- Generate terms of a sequence using term-to-term rules
- Generate terms of a sequence using position-to-term definitions
- Generate the terms in a quadratic sequence given a rule for finding each term from the position

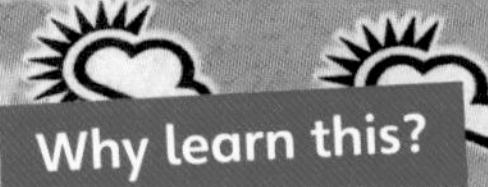

You can use sequences to predict some future events. Many events follow sequences, for example weather patterns.

What's the BIG idea?

→ A **term-to-term rule** tells you what to do to each **term** to obtain the next term in the **sequence**.
For example, the term-to-term rule 'start at 5, add 3, and multiply by −1' **generates** the sequence 5, −8, 5, −8, ... Level 6

→ A **position-to-term rule** tells you what to do to the term number to find that term in the sequence.
For example, the position-to-term rule 'multiply the term number by 4 and subtract 7' generates the sequence −3, 1, 5, 9, ... Level 6

→ Looking at the pattern of **first differences** can help you generate terms in a sequence.
For example, Term −3 1 5 9 ...
First differences

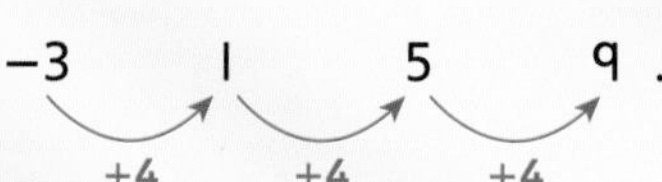

To find the next term, add 4. Level 6 & Level 7

→ You can use the notation T(1) for the 1st term, T(2) for the 2nd term, ..., T(n) for the **nth term**.
For example, T(1) = −3, T(2) = 1, T(3) = 5, ..., T(n) = $4n - 7$ Level 6 & Level 7

→ A **quadratic sequence** has n^2 as the highest power of n in T(n). Level 7

Practice, practice, practice!

Level 6

6C I can find a term in a sequence using term-to-term rules with positive and negative numbers

6C I can find a term given its position and a position-to-term rule using positive and negative numbers

1 Use these term-to-term rules to find the first five terms of each sequence.

- **a** Start at −3, add 2, then multiply by 3.
- **b** Start at 4, multiply by −2, then add 5.
- **c** Start at 96, divide by 2, then add 1.
- **d** Start at −100, multiply by 0.2, then add 3.

2 Use these position-to-term rules to find the 1st, 2nd, 3rd and 10th terms of each sequence.

- **a** Multiply the term number by 3, then subtract 2.
- **b** Multiply the term number by −4.
- **c** Subtract 2 from the term number, then multiply by 5.
- **d** Multiply the term number by 10, then add 6.

first difference generate nth term position-to-term rule

3 Match each term-to-term definition to the correct position-to-term definition.

Term-to-term	Position-to-term
a Start at 2, add 5 each time.	Multiply the term number by 3, then subtract 3.
b Start at 5, add 3 each time.	Multiply the term number by 5, then subtract 3.
c Start at 0, add 3 each time.	Multiply the term number by 3, then add 2.

4 Write down the first four terms of each sequence.

a $T(n) = 3n + 2$ **b** $T(n) = 50 - 4n$

c $T(n) = 5n - 0.4$ **d** $T(n) = 3n + \frac{1}{2}$

e $T(n) = \frac{n}{2} + 1$ **f** $T(n) = 6 - 5n$

Tip

If the number in front of n (its coefficient) is negative, the sequence decreases.

5 Here is a rule for a sequence.

To find the next term of the sequence add ☐

Choose a 1st term for the sequence and a number to go in the box so that

a every term is a positive integer

b every third term is a multiple of 3

c every fifth term is an integer.

6 Use the pattern of first differences to generate the next term in each sequence.

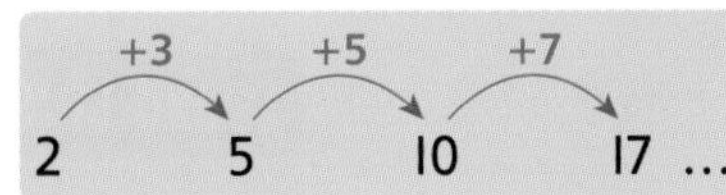

The next term is 17 + 9 = 26

a 4, 7, 12, 19, … **b** 3, 9, 19, 33, … **c** −2, 7, 22, 43, … **d** 2, 6, 12, 20, …

7 Generate the first six terms of each sequence.

a $T(n) = n^2 - 1$ **b** $T(n) = 4n^2 + 3$ **c** $T(n) = n(2n + 1)$ **d** $T(n) = \frac{n^2 + 1}{2}$

Level 6

6c I can match term-to-term definitions to their position-to-term definitions

6b I can find terms in a sequence using the nth term or position-to-term rule

6b I can generate sequences to satisfy defined rules

6b I can find the next term in a sequence by looking at the pattern of differences

6a I can generate a quadratic sequence

8 The nth term of a sequence is given by $T(n) = 3n + (n - 1)(n + 2)(n - 3)$. Write down the first five terms.

9 Use a spreadsheet to generate the first ten terms of each sequence.

a $T(n) = 2n^2 + 3n + 1$ **b** $T(n) = n(n - 1)(2n + 3)$

c $T(n) = \frac{n}{2}(2n + 3)$ **d** $T(n) = 4 - 5n - 2n^2$

Watch out!

Make sure you use the correct order of operations.
1st brackets
2nd indices (or powers)
3rd ÷, ×
4th +, −

Level 7

7c I can find terms using the nth term rule for a complex sequence

7b I can use ICT to generate non-linear sequences

Now try this!

A What comes next? ...and next?

Use a spreadsheet or calculator to find all the multiples of 29 up to the 100th multiple.

B Multiples

Explore with a calculator or spreadsheet to find a rule for a sequence where every second term is a multiple of 5. How many different rules can you find?

quadratic sequence   sequence   term   term-to-term rule

1.2 Linear sequences

- Use linear expressions to describe the nth term in an arithmetic sequence
- Use the pattern in the first differences to help find the nth term in an arithmetic sequence
- Justify the form of the nth term by looking at the pattern that generated it

Music often contains patterns and sequences.

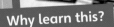

What's the BIG idea?

- In an **arithmetic sequence** the terms go up (or down) in equal steps. The nth term is always a **linear expression** in the form $T(n) = an + b$, where a is any non-zero number and b is a number. **Level 6**
- In an arithmetic sequence, the **first difference** between successive terms is constant and is the value of a in $T(n) = an + b$.

  You can then find b by forming an equation using $T(1) = a \times 1 + b$.
  Always check other values of $T(n)$. **Level 6**
- You can use the pattern of differences to **justify** the rule for the nth term of a sequence.

  For example,

Term number n	1	2	3	4 …
$T(n)$	−3	−1	1	3 …
First differences	+2	+2	+2	

  The first difference is +2, so $a = 2$.
  $T(1) = -3 = 2 \times 1 + b$, so $b = -5$.
  So $T(n) = 2n - 5$
  Check: $T(4) = 2 \times 4 - 5 = 3$ ✓ **Level 6 & Level 7**
- You can look at the original pattern to justify the rule for the nth term of a sequence.

  For example, for this pattern you add two sticks each time.
  The formula for the nth term is $2n$ plus the extra stick needed to make the first shape.
  This gives $T(n) = 2n + 1$. **Level 6 & Level 7**

Learn this

An arithmetic sequence has a constant difference between terms.

Practice, practice, practice!

1 Here is a pattern made from sticks.

a Copy and complete this table of values.

Term number n	1	2	3	4	5
Number of sticks	4	7	10		

b Use the pattern of differences to find the nth term, $T(n)$, of the sequence.

c Check your nth term using the values from the table.

d Use your nth term to find the number of sticks needed for the 25th term.

Level 6

6a I can use linear expressions to describe the nth term of an arithmetic sequence and use the pattern of differences to justify the nth term

arithmetic sequence first difference justify

2 Here is a pattern of octagons made from sticks.

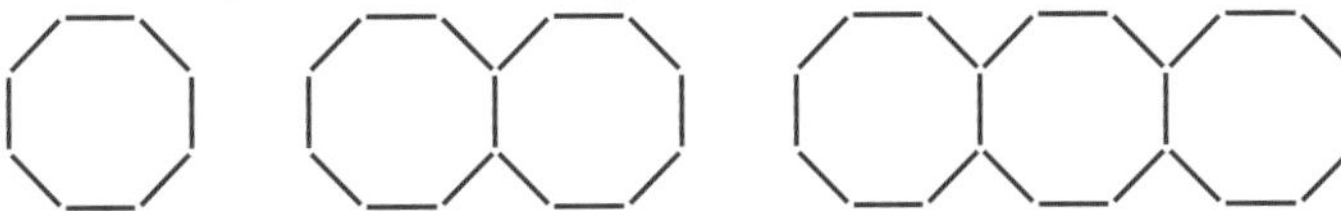

a The value of each term in the sequence is the number of sticks used. Write down the first five terms of the sequence.

b Use the pattern of differences to find the nth term of this sequence.

c Check your rule using values from the sequence.

d Use your nth term to find the 51st term of the sequence.

3 For each sequence

i find the next two terms

ii work out the nth term using the pattern of differences

iii use your rule to find the 26th term.

a 7, 11, 15, 19, 23, …
b 2, 5, 8, 11, 14, …
c 0, 5, 10, 15, 20, …
d −1, 2, 5, 8, 11, …
e 1, 5, 9, 13, 17, …
f 34, 30, 26, 22, 18, …
g 16, 14, 12, 10, 8, …
h 2.5, 4.5, 6.5, 8.5, 10.5, …
i −4, −7, −10, −13, −16, …
j 2, −3, −8, −13, −18, …

Tip
In a decreasing sequence, the first differences and a are negative.

Level 6

6a I can use the method of differences to justify the nth term of a sequence

4 Explain the general rule for the nth term you found in Q1. Do this by looking at the original stick pattern and how it grows.

5 Here is a pattern made from sticks.

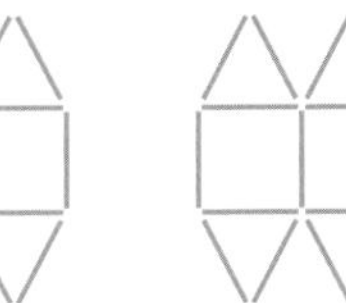

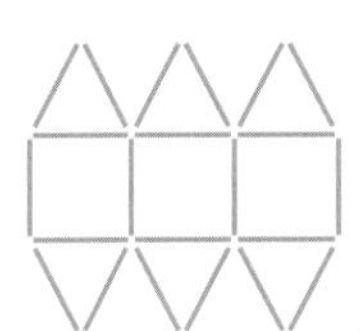

a The value of each term in the sequence is the number of sticks used. Write down the first five terms of the sequence.

b How do you get from the 5th to the 6th term?

c Work out the nth term in the sequence.

d Explain your nth term rule by reference to the structure of the pattern.

6 Design your own pattern sequence. Work out the nth term and justify it by reference to the physical pattern.

Level 7

7c I can explain the general rule for the nth term in a sequence by looking at the practical way it is created

7c I can justify a sequence by referring to the physical pattern

Now try this!

A Does it belong?

- Work in pairs. Each write an arithmetic number sequence (make sure it increases or decreases in equal-sized steps). Challenge your partner to work out the next number in the sequence. Work together to find the rule for the nth term.
- Now write down five numbers. Challenge your partner to work out which, if any, are members of the sequence.

B Sequence patterns

- Work in pairs. Using any of the numbers 1, 2, 3, 4, 5, 6, 7, 8 and 9, with the signs + or − in front, generate as many different sequences as you can. Find the nth term of each sequence. How many can you find?
- Sort them into groups that follow a similar pattern. What is this pattern?

linear expression **nth term**

1.3 Quadratic and fraction sequences

- Find any term in a quadratic sequence given the rule for the nth term
- Find the nth term of a quadratic sequence where the rule is of the form $T(n) = an^2$, $T(n) = an^2 \pm b$ or $T(n) = an^2 \pm bn \pm c$
- Justify the nth term of a quadratic sequence
- Explore fraction sequences

Why learn this?

A quadratic sequence lets you work out the area covered by a number of tiles.

What's the BIG idea?

- The nth term of a quadratic sequence is of the form $T(n) = an^2 + bn + c$, where a is any non-zero number and b and c are any numbers. **Level 7**
- You can use the rule for the nth term of a quadratic sequence to find any term in the sequence. **Level 7**
- For an n^2 sequence the **second differences** equal 2. You can use this to find the nth term.

For example,

Term number n	1		2		3		4		5 ...
$T(n)$	4		7		12		19		28 ...
First differences		+3		+5		+7		+9	
Second differences			+2		+2		+2		

Level 7

- For a quadratic sequence with nth term $T(n) = an^2 + bn + c$, the second differences equal $2a$. **Level 8**

Tip In all quadratic sequences the second difference is constant.

Practice, practice, practice!

1 For each of these expressions
- **i** find five consecutive terms
- **ii** work out the difference pattern.

a n^2 **b** $2n^2$
c $2n^2 + 3$ **d** $n^2 + 2n$
e $n(n + 1)$ **f** $6 - 5n^2$
g $n^2 + 2n + 1$ **h** $n^2 + 2n - 1$
i $n^2 + 6n - 4$ **j** $n(2 - n)$

Learn this

Make sure you use the correct order of operations.
1st brackets
2nd indices (or powers)
3rd ÷, ×
4th +, −

Level 7

7c I can find any term in a quadratic sequence given the nth term and examine term-to-term difference patterns

2 Each of these quadratic sequences has nth term in the form $T(n) = an^2$. Use the pattern of differences to find the nth term.

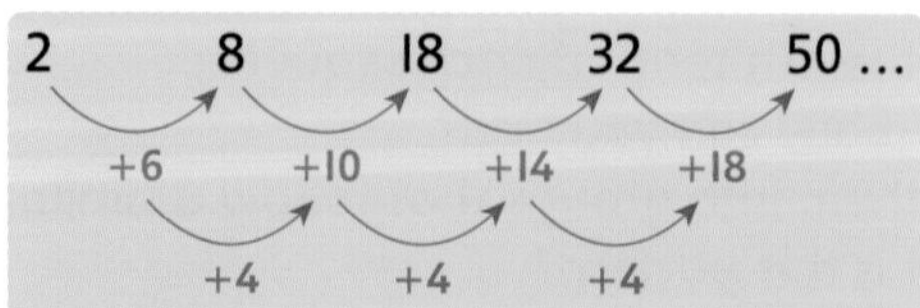

From the second differences,
$2a = 4$ so $a = 2$
So $T(n) = 2n^2$
Check: $T(3) = 2 \times 3^2 = 18$ ✓

a 3, 12, 27, 48, 75, ... **b** 5, 20, 45, 80, 125, ... **c** 8, 32, 72, 128, 200, ...

7a I can find the nth term of a quadratic sequence of the form $T(n) = an^2$ by using the pattern of second differences

Tip Remember to check for other values of n.

justify predict

3 Each of these quadratic sequences has nth term in the form $T(n) = an^2 + b$. Use the pattern of differences to find the nth term.

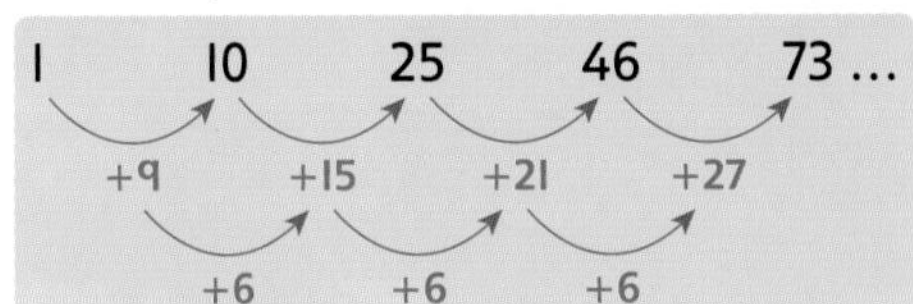

From the second differences,
$2a = 6$ so $a = 3$ and $T(n) = 3n^2 + b$
For $T(1)$, $1 = 3 \times 1^2 + b$ so $b = -2$
So $T(n) = 3n^2 - 2$
Check: $T(3) = 3 \times 9 - 2 = 25$ ✓

a 0, 3, 8, 15, 24, … b 5, 11, 21, 35, 53, …
c 3, 18, 43, 78, 123, … d 9, 15, 25, 39, 57, …

4 Each of these quadratic sequences has nth term in the form $T(n) = an^2 + bn + c$. Use the pattern of differences to find the nth term.

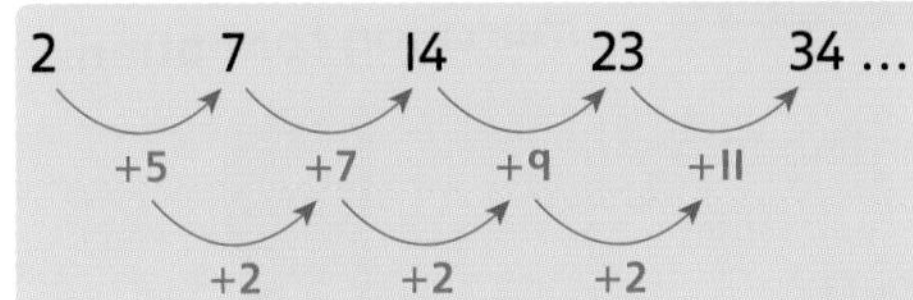

From the second differences, $2a = 2$, so $a = 1$ and $T(n) = n^2 + bn + c$
From $T(1)$, $2 = 1 + b + c$ so $c = 1 - b$ ①
From $T(2)$, $7 = 4 + 2b + c$ ②
Substituting in ②, $7 = 4 + 2b + 1 - b$
Rearranging, $7 = 5 + b$ so $b = 2$
Substituting in ①, $c = 1 - 2$ so $c = -1$
So $T(n) = n^2 + 2n - 1$
Check: $T(3) = 9 + 6 - 1 = 14$ ✓

a 1, 5, 11, 19, 29, … b 5, 10, 17, 26, 37, …
c 1, 3, 6, 10, 15, … d 4, 7, 14, 25, 40, …

5 For each of these fraction sequences

i predict other values

ii work out the rule for the nth term.

$\frac{2}{1}, \frac{3}{4}, \frac{4}{9}, \frac{5}{16}, \ldots$

i $T(5) = \frac{6}{25}$, $T(6) = \frac{7}{36}$, $T(10) = \frac{11}{100}$

ii $T(n) = \frac{n + 1}{n^2}$

a $\frac{1}{2}, \frac{1}{4}, \frac{1}{8}, \frac{1}{16}, \ldots$ b $\frac{1}{3}, \frac{1}{4}, \frac{1}{5}, \frac{1}{6}, \ldots$ c $\frac{1}{1}, \frac{2}{1}, \frac{3}{2}, \frac{5}{3}, \frac{8}{5}, \ldots$

Did you know?

In the sequence $\frac{1}{1}, \frac{2}{1}, \frac{3}{2}, \frac{5}{3}, \frac{8}{5}, \ldots$ both the numerators and the denominators follow the Fibonacci sequence.

Level 8

8c I can find the nth term of a quadratic sequence of the form $T(n) = 2a + b$ by using the pattern of second differences

8b I can find the nth term of a quadratic sequence of the form $T(n) = an^2 + bn + c$ using the pattern of second differences

8a I can explore fractional sequences

Now try this!

A Quadratic generation

Use ICT to generate quadratic sequences. Check that the second differences are constant.

B Working from the differences

Use ICT to generate quadratic sequences. Swap them with a partner. Work out the next three terms in each sequence by working backwards from the second difference. Find the nth term for each sequence.

quadratic sequence second difference

1.4 Sequences from patterns

- Explore spatial patterns
- Generate terms and sequences from spatial patterns
- Find the general terms for spatial patterns
- Find the sum of a series generated from spatial patterns

Why learn this?

Patterns of squares and triangles are used in marching formations.

What's the BIG idea?

→ A spatial **sequence** is a sequence of patterns that grow according to a mathematical **rule**.

For example, this is a spatial sequence of triangle numbers.

Level 7

→ A **series** is the **sum** of the terms in a sequence. S(n) is the sum of a series with n terms.

For example, 2, 4, 6, 8, ..., $2n$ is a sequence,
$2 + 4 + 6 + 8 + \ldots + 2n$ is a series. Level 7

→ You can use patterns of dots to help you work out the algebraic rule for the nth term of a sequence or series. Level 7 & Level 8

Practice, practice, practice!

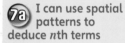

Level 7

7a I can use spatial patterns to deduce nth terms

1 Here are the first six terms of a sequence of square numbers.

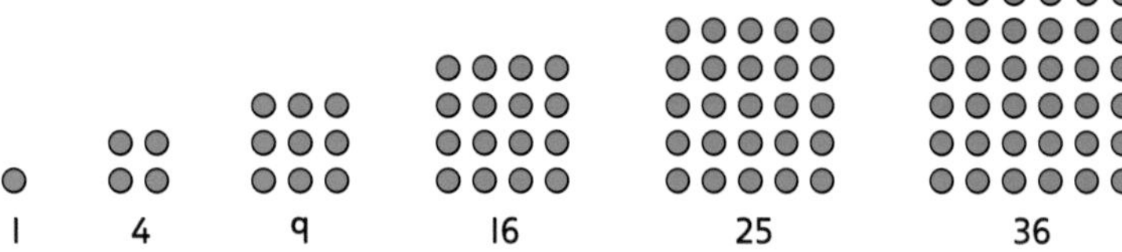

By looking at ways of dividing the square pattern of dots, deduce

a that $T(n) = n^2$

b that $T(n) = (n - 1)^2 + 2n - 1$

c other general rules for the nth term.

Learn this

T(n) is the nth term of the sequence.

7a I can use spatial patterns to express terms as a series

2 Here are the first six triangular numbers.

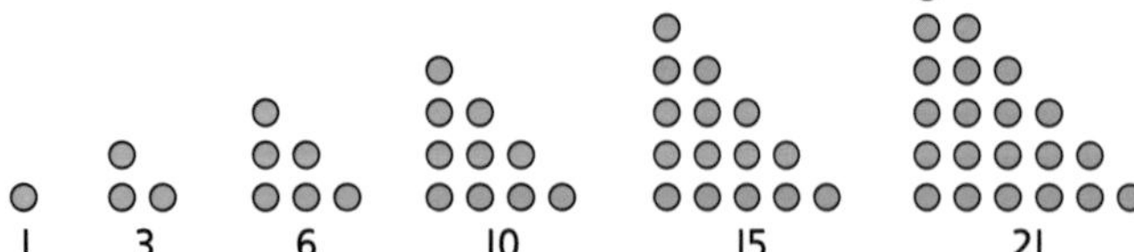

Hint: Look at the number of dots in each column.

a Write each of these terms as an arithmetic series.

b Write the nth triangle number, T(n), as an arithmetic series.

arithmetic sequence rule sequence

3 The drawing shows how you can investigate the relationship between square numbers and triangular numbers by splitting a square pattern of dots into two triangles.

T(5) T(4) T(4) is the 4th triangular number

a What do you notice?

b Repeat for four other square numbers.

c Deduce the general result.

d Test your general result by predicting a new result.

e Check and test your prediction.

Level 7

7a I can deduce results that link triangle numbers and square numbers together

4 The drawing shows another way of dividing a square pattern of dots. It shows that $1 + 3 + 5 + 7 + 9 + 11 = 6^2$

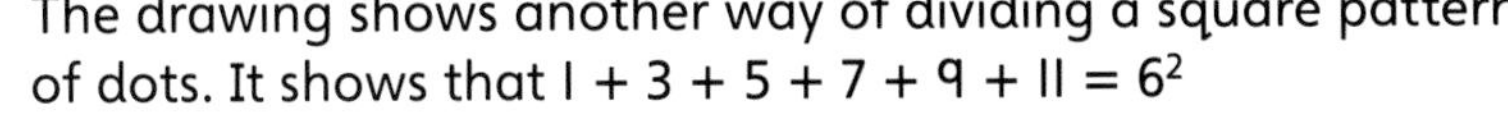

a Repeat for four other square patterns.

b Deduce the general formula for the sum of the first n odd numbers.

c Investigate other ways of dividing a square pattern of dots like this one.

Level 8

8c I can use spatial patterns to deduce general results

5 Here is a rectangle made up of two identical triangles.
Rectangle area = 5 × 6 = 30
T(5) = 15, so T(5) + T(5) = 30

T(5) T(5) T(5) is the 5th triangular number

a Draw diagrams for these.

i T(1) + T(1) **ii** T(2) + T(2) **iii** T(3) + T(3) **iv** T(4) + T(4)

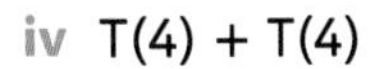

b Work out a general rule for T(n) + T(n).

c Use this rule to work out the sum of the first 100 whole numbers.

8a I can deduce the rule for nth triangular number

6 Each pair of lamp posts is connected by a separate strip of flags.

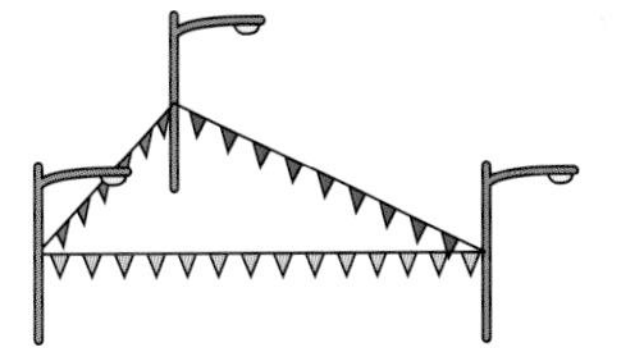

a How many strips of flags do you need for three lamp posts so that every post is connected to all the others?

b Work out how many strips you need for different numbers of lamp posts.

c Find a general rule for n lamp posts. Explain your rule.

8a I can find and explain the general rule for a practical problem

Now try this!

A Tram lines

The drawings show the maximum number of times lines cross for a given number of tram lines.

a Predict the next term and test your prediction.

b Work out the rule for the nth term, T(n).

c Justify the rule by looking at the line patterns.

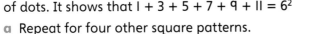

1 tram line
0 crossing

2 tram lines
4 crossings

3 tram lines
12 crossings

B An even challenge

Use a square pattern of dots to deduce a rule for finding the sum of the first n even numbers.

series sum

1.5 Functions and mappings

- Write functions in symbols
- Draw mapping diagrams
- Find the inverse of linear functions
- Draw straight-line graphs of linear functions
- Know the properties of graphs of quadratic functions

Why learn this?

Recipes are like functions. Recipes give you rules and instructions to make food. If different people follow the same recipe, all the results should be the same.

What's the BIG idea?

→ You can represent functions in different ways:

$x \rightarrow 2x + 8$ $\quad$ $y = 2x + 8$

- a **function machine**
- a **mapping**
- an equation Level 6

→ You can use a **mapping diagram** to show the inputs and outputs of a function. Level 6

→ An **inverse function** undoes what the original **function** did.
For example, the inverse of $x \rightarrow 2x$ is $x \rightarrow \frac{x}{2}$. Level 6

→ A function where the inverse is the same as the original function is a **self-inverse** function.
For example, $x \rightarrow 3 - x$ is a self-inverse function because its inverse is $x \rightarrow 3 - x$. Level 6

→ **Linear functions** have straight-line graphs. Level 6 & Level 7

→ A **quadratic function** is of the form $y = ax^2 + bx + c$, where a is any non-zero number and b and c are any numbers. Level 8

→ The graph of a quadratic function is a curve that is symmetrical about its turning point.
The y-coordinate at the turning point is either a maximum or a minimum.
$y = x^2 - 2$ is an example of a quadratic function. Level 8

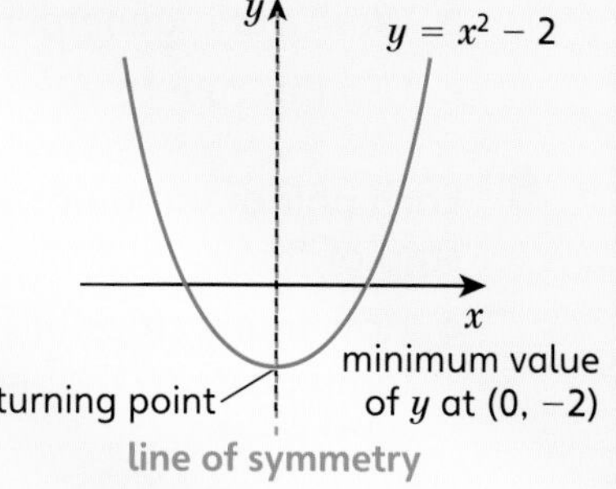

Practice, practice, practice!

Level 6

6c I can express simple functions using symbols

1 Write each of these function machines as a mapping and as an equation.

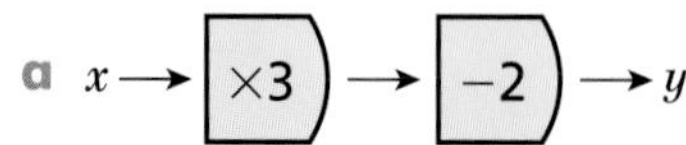

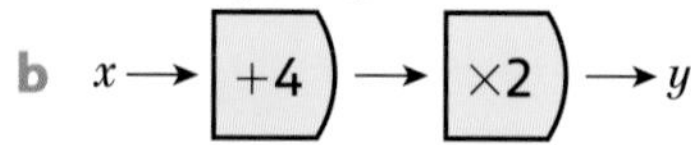

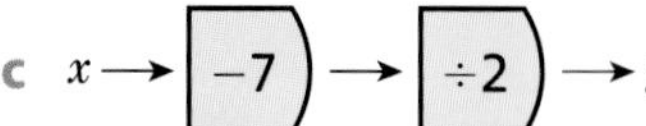

d $x \rightarrow$ +5 $\rightarrow$ ×3 $\rightarrow y$

6b I can draw mapping diagrams for simple functions for positive values of x

2 Draw a mapping diagram for each function for $x = 0$ to $x = 5$.

$x \rightarrow \frac{x}{2} + 1$

0 1 2 3 4 5

0 1 2 3 4 5

a $x \rightarrow 3x - 4$ **b** $x \rightarrow 5 - 2x$ **c** $x \rightarrow \frac{x}{2} + 3$ **d** $x \rightarrow x + 5$

function function machine inverse function linear function mapping

3 Using suitable positive and negative values of x, plot the graph of each function.

a $x \rightarrow 2x$ **b** $x \rightarrow 3x + 2$ **c** $x \rightarrow 4x - 5$ **d** $x \rightarrow \frac{x}{2} + 1$

e $x \rightarrow 5 - x$ **f** $x \rightarrow 10 - x$ **g** $x \rightarrow 3 - x$ **h** $x \rightarrow 20 - x$

4 Draw a mapping diagram for each function in Q2 for values of x from 0 to −5. Compare them with your diagrams from Q2. What do you notice?

5 Repeat Q3 using a graphical calculator or ICT.

6 Find the inverse of each linear function in parts **a** to **d** of Q3.

$x \rightarrow 2x + 1$

Function: $x \rightarrow$ [×2] $\rightarrow$ [+1] $\rightarrow 2x + 1$

Inverse: $\frac{x-1}{2} \leftarrow$ [÷2] $\leftarrow$ [−1] $\leftarrow x$

So the inverse function is $x \rightarrow \frac{x-1}{2}$.

Tip
Write out the inverse function at the end.

7 Find the inverse of each function in parts **e** to **h** of Q3. What do you notice?

8 Match each graph to its inverse.

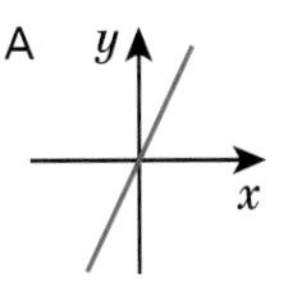

A $y = 2x$

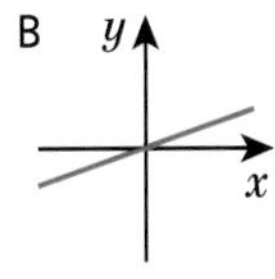

B $y = \frac{1}{3}x$

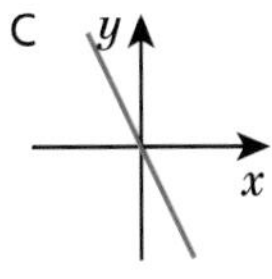

C $y = -2x$

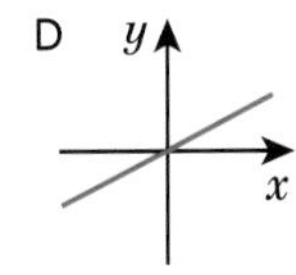

D $y = \frac{1}{2}x$

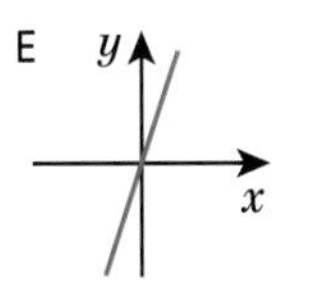

E $y = 3x$

F $y = -\frac{1}{2}x$

Level 6

6b I can plot linear functions in four quadrants.

6a I can draw mapping diagrams for simple functions for negative values of x

6a I can plot linear functions using ICT

6a I can find the inverse of linear functions

6a I can recognise the graphs of the inverse of simple linear functions

9 What do you notice about the graphs of $y = 2x$, $y = 3x$ and $y = -2x$ and their inverses in Q8?

Level 7

7a I can recognise the relationship between the graphs of a linear function and its inverse

10 These are graphs of quadratic functions.

i For each graph write down the equation of the line of symmetry.

ii Write the coordinates of the turning point, saying whether it is a maximum or a minimum.

a $y = x^2 - 3$

b $y = 2 - x^2$

Level 8

8c I can find the line of symmetry and write down the turning point of a quadratic graph

Now try this!

A Linear generalisations

Plot the graphs of $y = 3x$, $y = 3x + 2$, $y = 3x + 4$, $y = 3x - 2$ between $x = -5$ and +5. What do you notice? Generalise this for other straight lines that have the same value (coefficient) in front of x.

B Quadratic generalisations

Use ICT to generate quadratic graphs of the form $y = x^2 + bx + c$, where b and c are whole numbers. For each write down the equation of the vertical line of symmetry, the y-coordinate of the turning point and whether it is a maximum or a minimum.
Describe any patterns.

mapping diagram quadratic function self-inverse

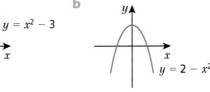

World's Greatest Maths!

1.6 WORLD SOLAR CHALLENGE

You are the chief technician testing a revolutionary new solar-powered car. You want to enter the car in the World Solar Challenge, in which cars attempt to cross Australia powered only by energy from the Sun. You will need to record all of your data carefully in order to enter.

Solar cells

You are testing three different types of battery for your solar car. These graphs show the amount of charge against time.

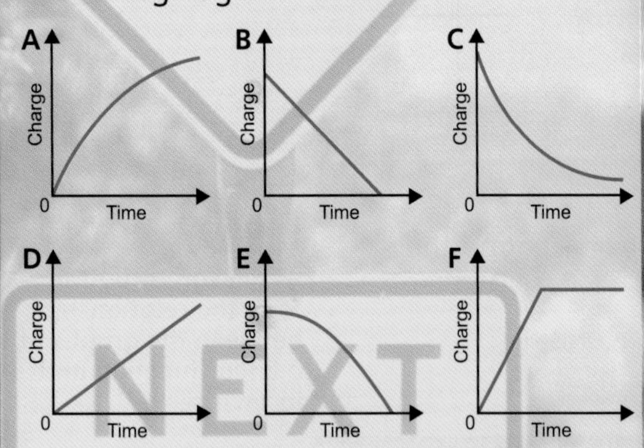

For each type of battery, choose the graph that most accurately represents

a how the battery charges

b how the battery discharges.

Battery 1
- *Charges at a constant rate, with no maximum.*
- *Discharges quickly at first, but cannot be fully discharged.*

Battery 2
- *Charges at a constant rate until maximum charge is reached.*
- *Discharges at a constant rate until empty.*

Battery 3
- *Charges quickly at first, then more slowly.*
- *Discharges slowly at first, then at a constant rate until empty.*

Time trial 1

You carry out a time trial for your car. The track diagram shows the distance to each checkpoint, and the arrival times.

You can use a copy of Resource sheet 1.6a.

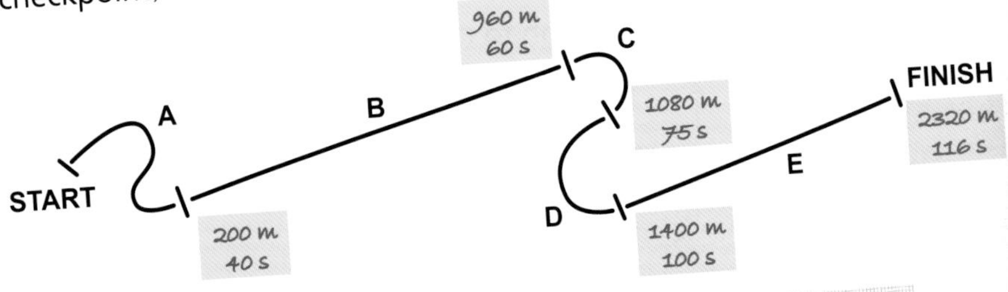

1 Copy and complete the distance–time graph for this time trial.
Label each section.

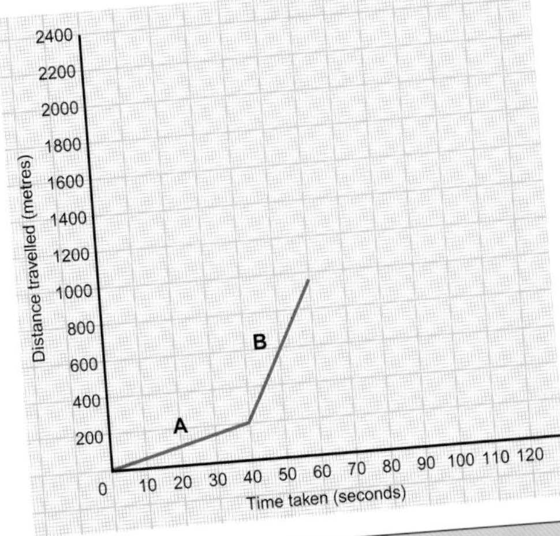

2 Copy the table and fill in the distance travelled and the time taken for each section.

Section	Distance travelled (m)	Time taken (s)	Averagespeedfor section (m/s)
A	200	40	$\frac{200}{40} = 5$
B	760	20	
C			
D			
E			

3 Calculate the average speed for each section of the time trial and complete the table.

Remember:
$$\text{Average speed} = \frac{\text{distance travelled}}{\text{time taken}}$$

Time trial 2

This diagram shows the test track for your second time trial.

1 Draw a distance–time graph for this time trial.

2 Calculate the average speed for each section.

3 In which section was the car travelling
   a fastest   b slowest?

4 Calculate the average speed for the whole time trial.

5 The World Solar Challenge rules say that a car must be able to average 50 km/h.
   a Convert your average speeds into km/h.
   b Were there any sections in which the car was travelling at less than 50 km/h?
   c Using the average speed for the whole time trial, will your car qualify?

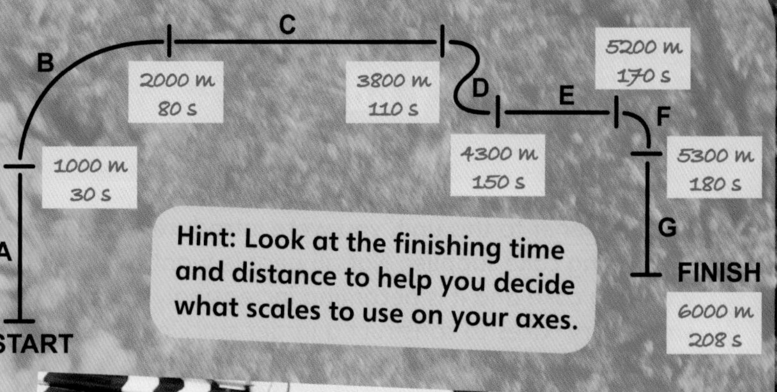

Hint: Look at the finishing time and distance to help you decide what scales to use on your axes.

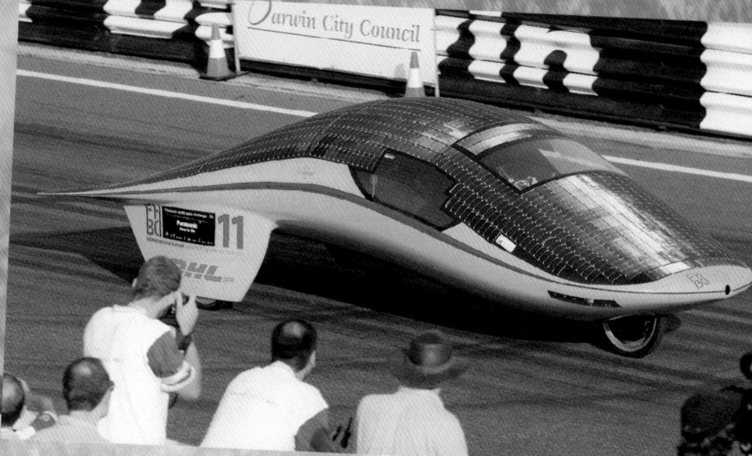

World Record

The World Solar Challenge is a 3020 km road race from Darwin to Adelaide. Contestants race from 8 am to 5 pm each day and rest overnight.

You have finished in a record time, but you need to submit your race details before you can be awarded your trophy.

1 Use the race log to draw a distance–time graph of your race. Include the overnight breaks.

You can use a copy of Resource sheet 1.6b to help you.

2 Which section of the race was
   a the fastest   b the slowest?

3 Calculate your average speed.

4 Calculate your average speed ignoring the overnight breaks.

5 Use your graph to estimate
   a the time you arrived at Pine Creek, 220 km from Darwin
   b the time you arrived at Glendambo, half way between Coober Pedy and Port Augusta
   c your distance from Darwin at 09:30 on the second day
   d your distance from Adelaide 50 hours after the start of the race.

Location	Distance travelled (km)	Time	Notes
Darwin	0	08:00	Race starts
Katherine	320	13:00	
Lake Woods	800	17:00	Overnight break
Alice Springs	1500	12:30	
South Australia Border	1760	17:00	Overnight break
Coober Pedy	2180	12:00	Engine fault. $2\frac{1}{2}$ hours waiting for spare parts
Port Augusta	2720	17:00	Overnight break
Adelaide	3020	10:30	Race finishes

Did you know?

In order to take advantage of the sunshine before 8 am and after 5 pm, teams often tilt their cars towards the Sun when they stop for the night.

Stacking fairy cakes

In the opener to this unit, you were introduced to Fiona Cairns, a wedding cake supplier. Not all people want a traditional wedding cake and it is becoming increasingly popular to use small fairy cakes to construct a larger stacking display. Fiona Cairns makes the fairy cakes and the clients arrange them themselves.

1 A client arranges the fairy cakes in a pyramid with a square base. The table shows how many cakes are in each layer.

Layer number	Number of cakes in the layer	Total number of fairy cakes
1	1	1
2	4	5
3	9	14
4	16	30

a Predict the number of cakes in the 5th and 6th layers of this type of arrangement.

b Look at the sequence 'Number of cakes in the layer'. What type of numbers does this sequence represent? **Level 6**

2 **a** Which layer would have 81 cakes?

b How many cakes would there be *in total* in a wedding cake with the number of layers you identified in part **a**? **Level 6**

3 **a** Work out the first three terms of the sequence with the general term $1 + 4(n - 1) + \frac{5}{2}(n - 1)(n - 2)$

b What sequence of numbers do you think this general term might represent?

c Does it work for the 4th term of the sequence you suggested in part **b**? **Level 7**

4 Look again at the sequence 'Number of cakes in the layer'.

a Comment on the first differences between successive terms.

b What is the second difference between terms?

c What type of sequence is this?

d What is the general term of this sequence?

e Use the pattern of differences to find the general term of each of these sequences.
i 6, 24, 54, 96, 150 **ii** 4, 10, 20, 34, 52 **Level 8**

5 Another client arranges the fairy cakes in triangular layers.

a The diagram shows the plan view of layers 1, 2 and 3. Construct and complete a table to give the layer number, the number of cakes in that layer and the total number of cakes. Include seven layers in your table.

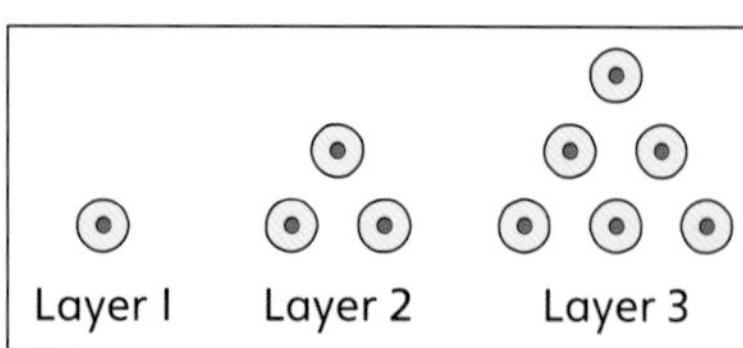

b What type of numbers does the sequence 'Number of cakes in the layer' in your table represent?

c Describe the shape of this wedding cake. **Level 6**

6 Look at your table from Q5a.
What is the general term for the sequence of the number of cakes in each layer? **Level 8**

The BIG ideas

- In an **arithmetic sequence**, the terms go up (or down) in equal steps. The ***n*th term** is always a **linear expression** in the form $T(n) = an + b$, where a is any non-zero number and b is any number. **Level 6**
- In an arithmetic sequence, the **first difference** between successive terms is constant and is the value of a in $T(n) = an + b$. You can find b by forming an equation using $T(1) = a \times 1 + b$. **Level 6**
- Looking at the pattern of first differences can help you generate terms in a sequence. **Level 6 & Level 7**
- You can use the pattern of differences to **justify** the rule for the nth term of a sequence. **Level 6 & Level 7**
- A **quadratic sequence** has n^2 as the highest power of n in $T(n)$. **Level 6 & Level 7**
- The nth term of a quadratic sequence is of the form $T(n) = an^2 + bn + c$, where a is any non-zero number and b and c are any numbers. **Level 7**
- In all quadratic sequences, the **second difference** is constant. **Level 7 & Level 8**
- You can use patterns of dots to help you work out the algebraic rule for the nth term of a sequence. **Level 7 & Level 8**
- For a quadratic sequence with nth term $T(n) = an^2 + bn + c$, the second difference is $2a$. **Level 8**

Find your level

Level 6

Q1 Copy and complete.

a 3, 5, 7, 9 — nth term is $2n + 1$
4, 6, 8, 10 — nth term is ____

b 1, 4, 7, 10 — nth term is $3n - 2$
2, 5, 8, 11 — nth term is ____

Q2 Copy and complete the mappings for these functions.

a $n \rightarrow n + 4$
$2 \rightarrow$ ____
____ $\rightarrow 13$

b $n \rightarrow 3n$
$6 \rightarrow$ ____
____ $\rightarrow 24$

Q3 Find the inverse of each function.

a $x \rightarrow 2x + 6$

b $x \rightarrow 3(x + 7)$

Q4 Write the next three terms of each sequence.

a Rule: add 7
2, 9, ___, ___, ___

b Rule: multiply by 2 and subtract 3
4, 5, ___, ___, ___

Level 7

Q5 **a** The nth term of a sequence is $2n + 3$.
What is the 9th term of this sequence?

b The nth term of another sequence is $(n + 7)(n - 5)$.
Work out the first three terms.

Q6 The nth term of a sequence is $n^3 + 2$.
Work out the first four terms.

Q7 Match each nth term rule to its number sequence.

nth term
$(n + 2)^2$
$7n$
$n(n - 1)$
$4 + n^2$

Number sequence
0, 2, 6, 12, …
9, 16, 25, 36, …
5, 8, 13, 20, …
7, 14, 21, 28, …

Level 8

Q8 Find the nth term of this sequence.
3, 8, 15, 24, …

Q9 Look at this sequence.
0, 1, 4, 9, …
The nth term of this sequence is in the form $an^2 + bn + c$.
Find the values of a, b and c.

2 Keep your balance

This unit is about expressing mathematical relationships and solving problems using letters and equations. We can use algebra to describe and model situations and to predict what will happen.

Early civilisations investigated mathematical problems before the alphabet was even invented. Instead of using letters, they used geometric diagrams to show relationships between lengths and areas. Patterns can give us useful ways of visualising, describing and understanding different aspects of mathematics.

This pattern is from a box with a traditional Lebanese design. It illustrates the relationships between many different shapes. What shapes can you see in it? Look for the pattern that shows how a triangle can be divided into four smaller congruent triangles.

Relationships between length and area in a pattern can be expressed algebraically, and then used to produce the design in different sizes.

Activities

A Here is a regular hexagon, split into six congruent triangles meeting at the centre.

What kind of triangles are these?

Rose makes this pattern out of wire. She predicts that she will need twice as much wire to make it as she would need to make a regular hexagon. Is Rose correct? Justify your answer.
Investigate this for other regular polygons.

B Make your own geometric pattern, using no more than three different shapes which tessellate.

Label one of the lengths x.
Describe the other lengths and areas in terms of x.

Before you start this unit...

1 Find the area of each shape.

Level Up Maths 5-7 page 172

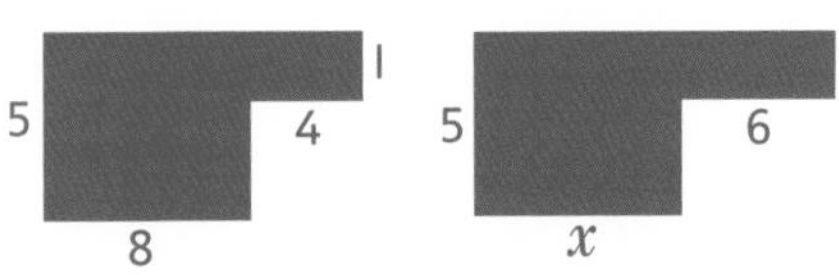

2 Multiply out the brackets.

a $4(x + 2)$ **b** $5(3 - 2y)$

Level Up Maths 5-7 page 116

3 Two numbers have a sum of 20 and a product of 91. Find the two numbers.

Level Up Maths 5-7 page 254

4 A regular octagon and a regular hexagon have the same perimeter and side lengths that are whole numbers.

Level Up Maths 5-7 page 254

a Write one set of possible values for the two side lengths.

b Write a relationship between the side length of the octagon, t, and the side length of the hexagon, h.

2.1 Using letter symbols in algebra
2.2 Constructing and solving linear equations
2.3 Solving linear equations with x and brackets on both sides
maths! Top profit
2.4 Using trial and improvement to solve equations
2.5 Solving simultaneous equations graphically
2.6 Solving simultaneous equations algebraically
Unit plenary: Travelling equations
Plus digital resources

Did you know?

Scientists use equations to predict how a disease like flu will spread over a period of time and to predict epidemics. The health service can use this information to plan ahead so that they are ready for any extra demand for medicines, medical staff and hospital beds.

2.1 Using letter symbols in algebra

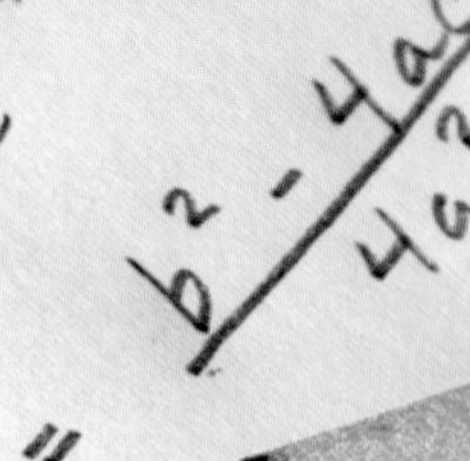

- Describe what letters stand for in formulae and functions
- Distinguish between expressions, equations and identities
- Expand brackets and simplify expressions
- Write expressions

Why learn this?

You can understand and use other people's mathematics if you understand the language of algebra.

What's the BIG idea?

- An **expression** has no equals sign. It is a way of expressing something general. **Level 6**
- **Equations** are true for particular values.
  For example, $3x + 7 = 13$ is only true for $x = 2$.
  $x^2 = 9$ is only true for $x = \pm 3$.
  $3x + 7 = 2x + 2$ is only true for $x = -5$. **Level 6**
- An **identity** shows the same thing in two different ways.
  For example, $3x + 2x + 4 \equiv 5x + 4$
  An identity is true for all values of the **variable** (x in this case). **Level 6**
- To **expand** a bracket, multiply every term inside the bracket by the term on the outside.
  For example, $4(3x - 2) = 4 \times 3x + 4 \times (-2) = 12x - 8$
  $2x(x + y^2) = 2x \times x + 2x \times y^2 = 2x^2 + 2xy^2$ **Level 6 & Level 7**

Tip

Sometimes the identity symbol, ≡, is used but more often the equals sign is used.

Practice, practice, practice!

Level 6

1 Explain what each letter represents in these.

- **a** This function converts from hours to seconds: $y = 3600x$
- **b** This function converts from pounds to euros: $y = kx$
- **c** This formula finds the perimeter of a kite: $P = 2(x + y)$
- **d** This formula finds the angle sum of a regular polygon: $S = 180(n - 2)$

6c I can explain what letters represent in functions and formulae

2 Copy these statements. Put = or ≡ in each empty box as appropriate.

- **a** $a + 8 \square 5$ **b** $x + x + x \square 3 \times x$ **c** $x \times x + x \times x \square 2x^2$
- **d** $5 \times d + 5 \square 5^2$ **e** $35 \square p^2 - 1$ **f** $2 + r^2 \square r \times r + 2$

6a I can use the identity sign correctly

3 Decide whether each of these is an expression, an equation or an identity.

- **a** $3(x + 4) + 4 = 3x + 16$ **b** $9x + 11$ **c** $7a - 3 = 3a + 7$
- **d** $7m - 3 - 3m - 7$ **e** $2(y + 7) = 2y + 14$ **f** $15 - 2x = 3x - 7$

6a I can distinguish between expressions, equations and identities

equation expand expression

4 Caroline is x years old.
Jacqueline is 3 years younger than Caroline.
Rosemary is 1 year older than Jacqueline.

a Write an expression for the sum of their ages. Simplify your expression.

b Their father is 10 years younger than twice Jacqueline's age. Write an expression for his age.

c Find an expression for the difference between the father's age and the sum of his daughters' ages.

Tip
Write an expression, function or formula using key words for each unknown. Then you can see clearly what each unknown represents.

Level 6
6a I can write and simplify algebraic expressions involving brackets

5 a Write an expression for the combined area of both yellow shapes using brackets.

b Expand and simplify your answer to part a.

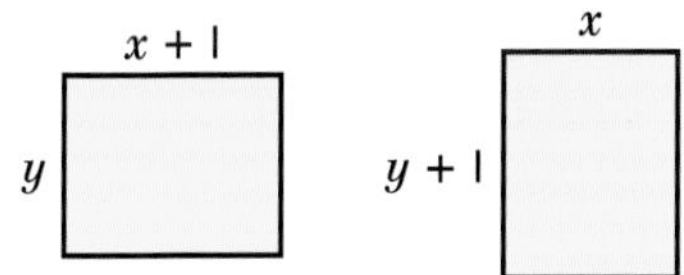

6

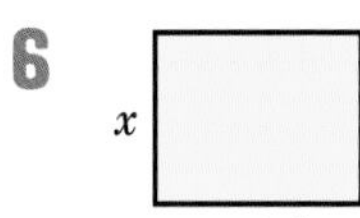

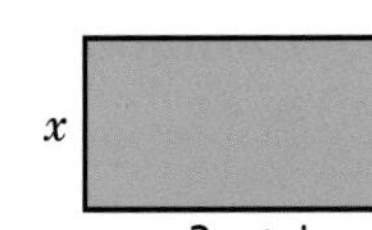

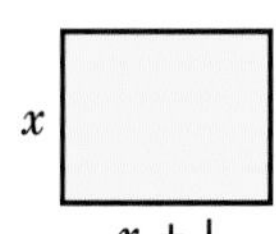

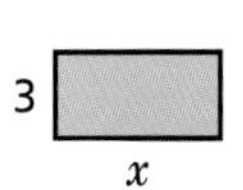

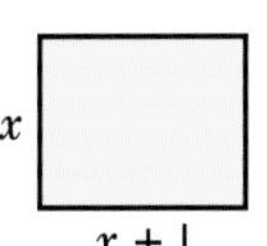

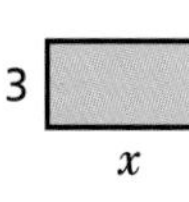

a Write an expression for the red area using brackets.

b Write an expression for the total yellow area using brackets.

c Write an expression for the total area of all the shapes, in its simplest form without brackets.

d A single rectangle has the same area as the total area you found in part c. Its width is $5x$. Use your answer to part c to help you find its length.

7 Find the value of each expression when $x = -2$ and $y = 4$.

a $3x(2x - 5y)$

b $3xy(2x^2 - 5y)$

c $x^2y^2(3x + 4y^3)$

8 Expand the brackets and simplify.

a $4(x + y) + 4(x - y)$ b $4(x + y) - 4(x - y)$

c $4x(x + y) - 4y(x - y)$

9 Expand the brackets and simplify where possible.

a $y(3x^2 + 5x + 6) + y^2(3x^2 + 5x + 6)$ b $y^2(3x^2 + 5x + 6) - y(3x^2 + 5x + 6)$

c $x(3x^2 - 5x - 6) + x^2(3x^2 + 5x + 6)$ d $x(3x^2 + 5x + 6) - x(3x^2 - 5x - 6)$

Level 7
7c I can substitute positive and negative numbers into expressions containing small powers and brackets
7b I can expand and simplify expressions with brackets and negative signs
7a I can expand and simplify expressions with indices, brackets and negative signs

Now try this!

A What's the area?

Write the total area of this shape in as many different ways as you can.

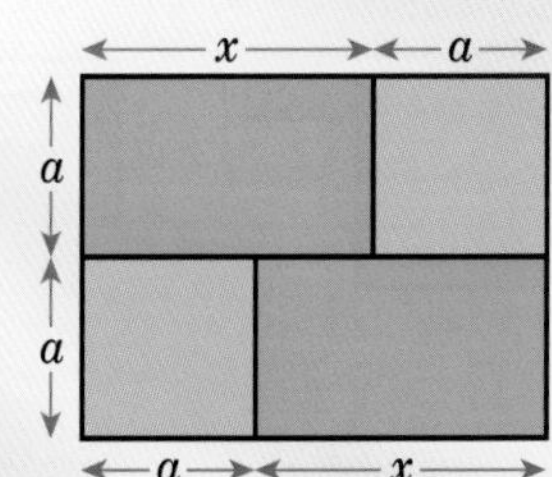

B Area variations

Work with a a partner. Draw a rectangle and split it into squares and rectangles to create a diagram similar to the one in Activity A. Now write its total area in as many different ways as you can. Extend to triangles if you are confident.

identity variable

2.2 Constructing and solving linear equations

➩ **Construct and solve equations with x on one or both sides in order to solve problems**

Why learn this?

Many objects are made up of different shapes. Lengths, areas and angles are linked. You can form and solve equations to find a missing length, area or angle from information you already know.

What's the BIG idea?

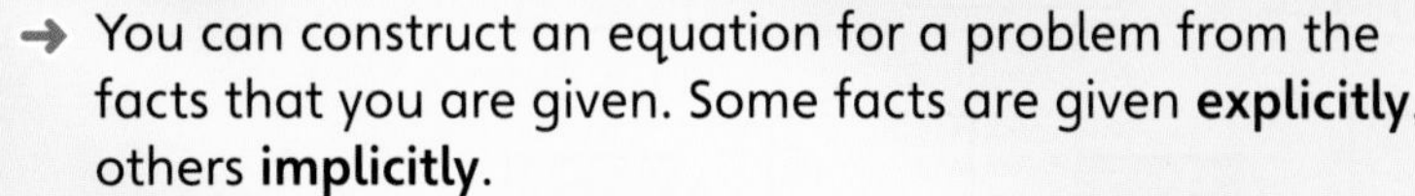

→ You can construct an equation for a problem from the facts that you are given. Some facts are given **explicitly**, others **implicitly**.

For example, find each angle of a triangle which has angles $(2x + 13)°$, $(3x + 4)°$ and $(4x + 19)°$.

The angles are given explicitly. The fact that the angles add up to 180° is given implicitly, by saying that they are the angles of a triangle.

You can use the facts to form an equation.

$$2x + 13 + 3x + 4 + 4x + 19 = 180$$

Now simplify the equation and solve it to find the unknown.

$$9x + 36 = 180$$

Subtract 36 from both sides $\quad 9x = 144$

Divide both sides by 9 $\quad x = 16$

Use the solution to the equation to solve the original problem.

When $x = 16$

$2x + 13 = 2 \times 16 + 13 = 45$

$3x + 4 = 3 \times 16 + 4 = 52$

$4x + 19 = 4 \times 16 + 19 = 83$

So the angles are 45°, 52° and 83°. **Level 6**

→ When you have an equation with x on both sides, you need to add or subtract the same number of xs from both sides in order to get x on one side only. **Level 6**

Watch out!

Once you have solved an equation, don't forget to link your answer back to the original problem, for example by substituting x back into each expression to find the angles in the triangle.

Practice, practice, practice!

Level 6

6C I can use diagrams to solve equations with x on both sides

1 The total areas of these two diagrams are the same.
The side lengths are in centimetres.

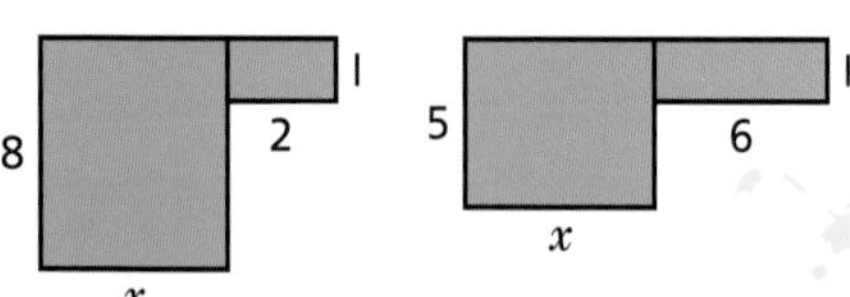

a Copy and complete this equation.

$8x + 2 = \square x + \square$

b An equal area is shaded green on both diagrams. What is this area?

c Copy and complete this equation based on the remaining purple area.

$\square x + 2 = \square$

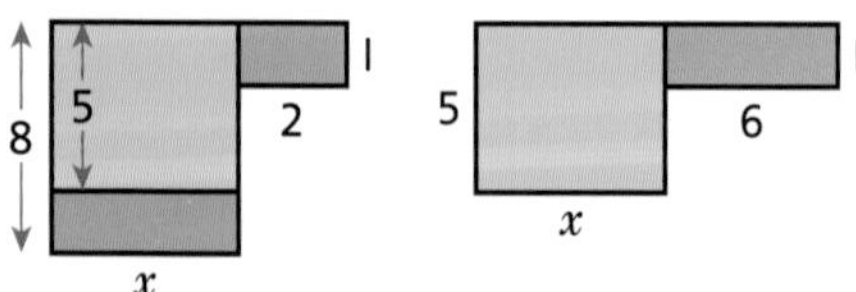

d Solve the equation to find the value of x.

explicitly implicitly interior angle

2 A triangle has angles of $x°$, $2x°$ and $(x + 40)°$.

- **a** Form an equation in terms of x and solve it to find x.
- **b** Work out the angles of the triangle.

Level 6

6b I can form and solve an equation with the variable on one side in order to solve a problem

3 A kite has sides, in centimetres, of length $y + 4$ and $3y - 1$. Its perimeter is 30 cm.

- **a** Form an equation and solve it to find y.
- **b** Work out the side lengths of the kite.

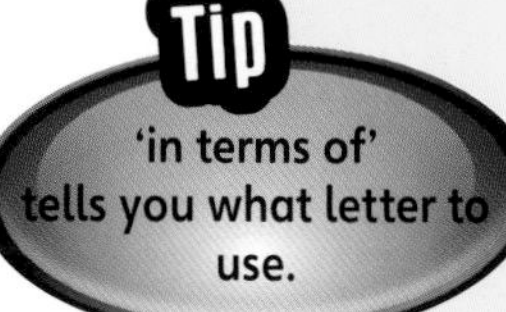

4 The sum of Paul, Jackie and Richard's ages is 97.
Paul and Jackie are the same age. Richard is 32 years younger.

- **a** Form an equation in terms of x. State clearly what x represents.
- **b** Solve your equation to find their ages.

Tip
Knowing the properties of triangles and quadrilaterals helps you to form equations.

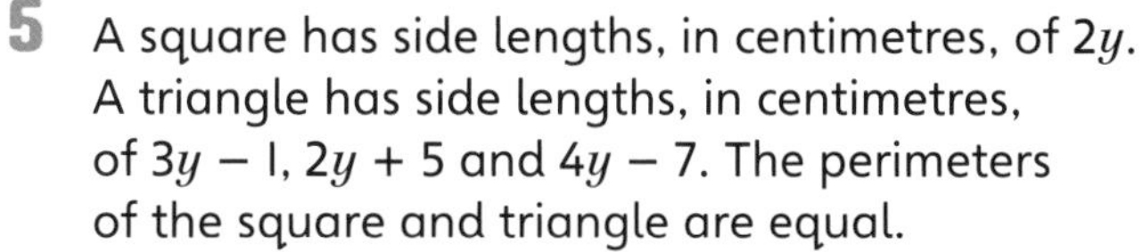

5 A square has side lengths, in centimetres, of $2y$.
A triangle has side lengths, in centimetres, of $3y - 1$, $2y + 5$ and $4y - 7$. The perimeters of the square and triangle are equal.

- **a** Form an equation in y and solve it.
- **b** Use your answer to find the side lengths of the triangle.

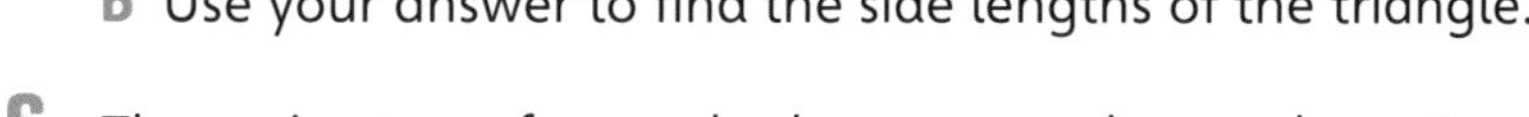

6a I can form and solve an equation with x on both sides in order to solve a problem

6 The perimeters of a regular hexagon and a regular octagon are equal.
The hexagon has side lengths, in centimetres, of $2x - 1$.
The octagon has side lengths, in centimetres, of $x + 5$.

- **a** Form an equation and solve it to find x.
- **b** Find the side lengths of the hexagon and the octagon.

7 I add 2 to a number and multiply the result by 4.
The answer is the same as when you add 8 to the number and multiply the result by 3. Find the number.

8 Jared starts with £56.50. Seth starts with £43.80.
Jared buys eight books and Seth buys six books.
The books are all the same price.
Jared and Seth have equal amounts of money left over.

- **a** Form an equation where x is the price of one book in pounds.
- **b** Solve your equation to find the price of one book.

Now try this!

A Area pairs

Draw a pair of diagrams like these.
The red lengths are equal and the two pink areas are equal.
Mark all dimensions on your diagrams.

B Equation generator

Think of a number for length x, for example 3.
Find numbers for the question marks so that the areas of the two pink shapes are equal.
Form an equation, for example $6x + 2 = 4x + 8$.
Ask a partner to solve your equation.

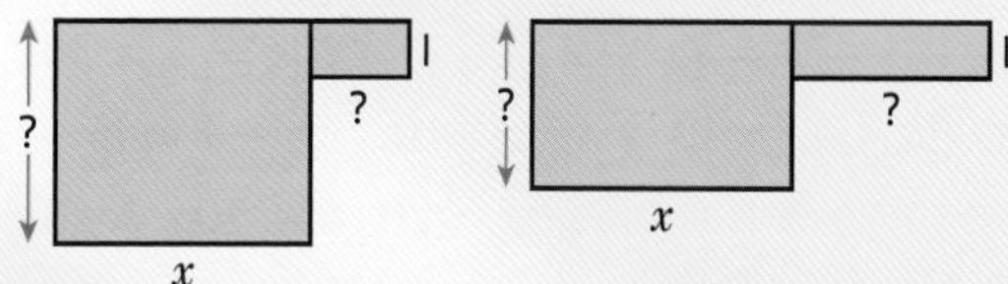

in terms of regular

2.3 Solving linear equations with x and brackets on both sides

- Identify equivalent equations
- Solve equations with x on both sides and which include negative signs and brackets on one or both sides

Why learn this?

Knowing how to solve linear equations helps you to solve practical problems, for example finding one length when given another.

What's the BIG idea?

- When you have an equation with x on both sides, you need to add or subtract the same number of xs from both sides in order to get x on one side only.
  For example, $x - 6 = 29 - 4x$
  The left-hand side has the higher number of xs. Add $4x$ to both sides to eliminate the xs on the right-hand side.
  $5x - 6 = 29$
  $5x = 35$
  $x = 7$ **Level 6**
- **Equivalent** equations have the same solution for x.
  For example, subtracting x from both sides of $3x + 10 = x + 4$ gives the equivalent equation $2x + 10 = 4$. The solution to both equations is $x = -3$. **Level 6**
- When an equation includes brackets, first expand the brackets and collect like terms. **Level 7**

Tip
Eliminate x from the side which has the lowest number of xs to begin with.

Watch out!
You cannot solve equations with x on both sides by simply reversing the operations which have been done to x. You need to operate on the unknown, for example by adding $4x$ to both sides.

Practice, practice, practice!

1 Match the equations into equivalent pairs.
Hint: You don't need to solve the equations to do this.

A $3x + 1 = 8x + 7$ B $11x = 8$ C $11x = 6$
D $3x + 1 = -8x + 7$ E $-3x + 1 = 2x + 7$ F $3x - 1 = -8x + 7$

2 Copy and complete these to make equivalent equations.

a $5x + 5 = 3x + 11$ $\square x + 5 = 11$
b $5x + 5 = 3x - 11$ $\square x = \square$
c $5(x - 1) = 3x - 11$ $\square x = \square$
d $5x + 5 = 11 - 3x$ $\square x + 5 = 11$
e $5 - 5x = 3x - 11$ $\square = \square x$
f $3(x - 5) = 2x - 6$ $3x = 2x + \square$

3 Solve these equations.

a $7x - 12 = 4x + 3$
b $8x - 14 = 3x + 1$
c $10x + 3 = 7x + 9$
d $20x - 4 = 30x - 6$
e $7x - 12 = 10 - 4x$
f $8x - 7 = 15 - 3x$
g $10x + 3 = 9 - 7x$
h $4 - 20x = 30x - 6$

Watch out!
To eliminate a term in x which has a negative number in front, you need to add. For example, to eliminate $-3x$, add $3x$.

Level 6

6b I can identify equivalent equations by operating on both sides

6b I can find an equivalent equation with x on one side only given an equation with x on both sides

6a I can solve equations with x on both sides

equivalent regular

4 Richard and Michael start with equal amounts of money.
Richard buys 11 cans of drink and has £3.50 left over. Michael borrows the extra £2.50 he needs in order to buy 19 cans. The cans are all the same price.

a Form an equation where x is the price of one can in pounds.

b Solve your equation to find the price of one can.

5 The diagram shows a kite. All lengths are in centimetres.

a Write down two equations using the facts given.

b Solve your equations to find the values of x and y.

c Find the perimeter of the kite.

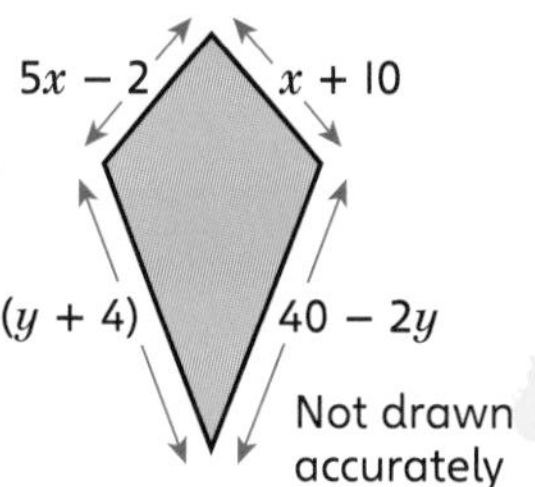

Level 6

6a I can form and solve an equation with x on both sides in order to solve a problem

6 Solve these equations.

a $5x + 3 = 2(x - 9)$ **b** $3(x - 2) = 4(x + 5)$

c $3(2x - 4) - x = 2(14 - 2x) + x$ **d** $5(3x - 2) = 2(x + 7)$

7 Solve these equations.

a $3(2x - 4) - 4 = x - 2(14 - 2x)$ **b** $2(x + 3) + 2x = 2x - 7(x + 3)$

c $30 - 2(x + 4) = 10 + 3(2x - 5)$ **d** $20 + 5(3x + 1) = 8 - 3(4x - 2)$

8 **a** Rose has a 50 cm length of red cord.
She makes a regular hexagon with sides of length y cm.
Write an expression, in terms of y, for the amount of cord that Rose has left over after she makes the hexagon.

b Rose takes a 75 cm length of blue cord.
She makes a regular pentagon with sides of length $(2y - 1)$ cm.
Write an expression, in terms of y, for the amount of blue cord left over.

c The lengths of blue and red cord left over are equal.
Form an equation and solve it to find y.

Level 7

7c I can solve equations with x on both sides and brackets on one or both sides

7b I can solve equations with x and brackets on both sides which involve multiplying a bracket by a negative number

7b I can form and solve equations with the unknown on both sides which involve multiplying a bracket by a negative number

Now try this!

A Equivalent equation extravaganza

Work with a partner. Begin with any linear equation with x on both sides. Take turns to create an equation that is equivalent to the previous one. Describe the operation you performed on both sides of the equation to generate your new one. If your partner disagrees with your result, ask another pupil or your teacher to check. A correct equivalent equation scores 1 point. For example, you could start with $4x - 6 = 5 - x$, and add $3x$ to both sides to generate $7x - 6 = 5 + 2x$ to score 1 point.
Keep a list of the operations you perform. Give another pair of pupils your starting equation and your list of operations. Do they end up with your final equation?

B Solutions in your head!

Play in a group of four, with one pair competing against the other. Pair 1 makes up a linear equation with a whole-number solution. Pair 2 solves it as follows:

- One person says any equation which is equivalent to the first.
- Then the other has to say the solution without writing anything down.

If the solution is correct Pair 2 scores 1 point. Swap roles and repeat.

maths!

Top profit

Simultaneous equations are used in manufacturing to work out the best way to maximise the profit of a factory, based on how many products they can produce. Here are two examples.

Callme Corporation

We are a company who manufacture two different models of phone: Xtra203 and You342i

Our problem
How to make the biggest profit from the order?

What we know

Fact 1 It takes us 4 hours to produce each Xtra203 which we sell to make a profit of £5.

Fact 2 It takes 5 hours to produce each You342i which we sell to make a profit of £11.

Fact 3 We can produce 300 of each phone at most, and we only have 1600 hours to complete the order.

Solving the problem
We need to write out four equations where x represents Xtra203 and y represents You342i

Equation 1	maximum number of hours	$1600 = 4x + 5y$
Equation 2	maximum number of Xtra203	$x = 300$
Equation 3	maximum number of You342i	$y = 300$
Equation 4	profit produced	$P =$ ______

This information can be shown on a graph. The unshaded area shows the possible numbers of phones that can be made within the given limits. The maximum profit will be made at one of the points marked A, B, C and D on the graph.

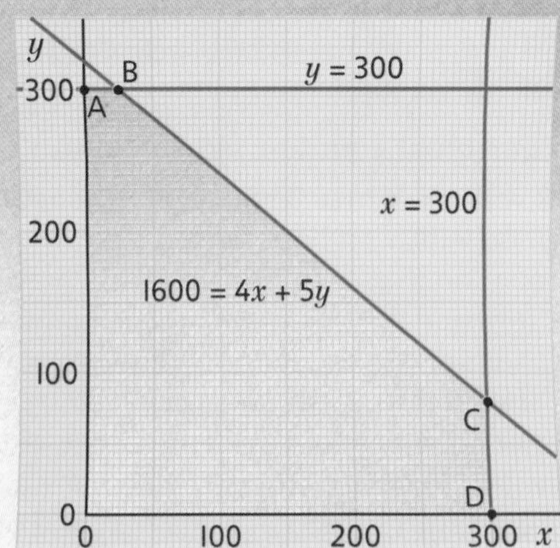

Problem solved

- Using the equation for *P*, which point gives the largest profit?

Gearup Corporation

We are a company who manufacture two different types of bicycle gear: GearX and GearY

Our problem
How to make the biggest profit from the order?

What we know

Fact 1 It takes us 4 hours to produce GearX which we sell to make a profit of £7.

Fact 2 It takes us 2 hours to produce GearY which we sell to make a profit of £3.

Fact 3 We can produce at most 400 GearX and 300 GearY, and we only have 2000 hours to complete the order.

Solving the problem
The four equations are:

Equation 1	maximum number of hours	2000 = ______
Equation 2	maximum number of GearX	x = ____
Equation 3	maximum number of GearY	y = ____
Equation 4	profit produced	P = ________

complete

- Plot the first three equations on a graph.
- Which point gives the maximum profit?

Problem solved
- What is the maximum profit?

2.4 Using trial and improvement to solve equations

→ Use ICT and trial and improvement to solve quadratic and cubic equations

Why learn this?

When mixing paint you try out different combinations, making small adjustments until you get the colour you want. Trial and improvement is useful for solving equations which you cannot easily solve using algebraic methods.

What's the BIG idea?

- → **Trial and improvement** involves making an initial 'guess' at a solution to an equation, trying it in the equation to find out whether the guess is too big or too small, and then improving the guess as required. Level 6, Level 7 & Level 8
- → You may need to form an equation, and then solve it using trial and improvement.
  For example, the length of a rectangle is 4 cm more than its width. Its area is 70 cm^2.
  Call the width w, so the length is $w + 4$. This gives the equation $w(w + 4) = 70$.
  To solve this, try different values of w on the left-hand side. Record in a table whether each area is too high or too low. Level 6, Level 7 & Level 8
- → When you have the solution sandwiched between two values of the required level of accuracy, you need to do the 'halfway' test.
  In the example $w(w + 4) = 70$, $w = 6.6$ is too low and $w = 6.7$ is too high.
  Next check $w = 6.65$. Here $w(w + 4) = 6.65 \times 10.65 = 70.8$.
  This is too high, so the correct value of w is below 6.65 and the answer to one decimal place is 6.6 cm. Level 6, Level 7 & Level 8

Super fact!

The artists' colour wheel, where the primary colours are arranged opposite their complementary colour, was first shown by Isaac Newton in 1666.

Practice, practice, practice!

1 The width of a rectangle is x cm.
Its length is 3 cm more than its width.
Its area is 90 cm^2.
Set up a spreadsheet like the one shown.
Use it to find the width to the nearest whole number.

	A	B
1	x	x(x+3)
2	7.0	70.00
3	7.5	
4	8.0	
5	8.5	
6	9.0	
7		

2 A rectangle is x cm wide.
Its length is 1 cm more than double the width.
Its area is 120 cm^2.
Use a spreadsheet to find its width.

Level 6

6c I can use ICT to find a whole number solution to a straightforward problem

cubic equation cubic expression

3 The perimeter of a rectangle is 20 units and its area is 18 square units.
Work out the width of the rectangle to one decimal place.

x	$10 - x$	$x(10 - x)$	Too big or too small?
4	6	$4 \times 6 = 24$	too big
3			
2			

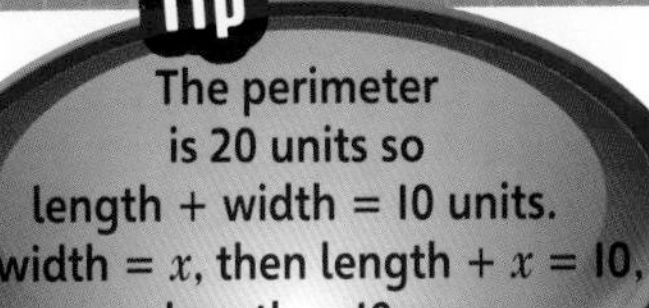

The perimeter is 20 units so length + width = 10 units. If width = x, then length + x = 10, so length = $10 - x$.

Watch out!

You must give your answers to the level of accuracy stated. Use the 'halfway' test to find out which side of the halfway mark the solution lies.

4 **a** A cube has a volume of $13\,\text{cm}^3$.
Use trial and improvement to find the length of a side to one decimal place.

b Solve the equation $x^3 = 200$ to one decimal place.

c Solve the equation $x^3 + 2 = 500$ to one decimal place.

Level 6

6b I can begin to solve more complex problems using trial and improvement

6a I can use trial and improvement to find a solution to one decimal place of an equation such as $x^3 = 30$

5 Use trial and improvement to find a solution to each equation to two decimal places.

a $x^3 + 3x = 80$ **b** $x^3 + 7x = 75$ **c** $x^3 - 7x = 75$

6 Use trial and improvement to find a solution to each equation to two decimal places.

a $x^3 + 2x^2 + 3x = 80$ **b** $x^3 + 2x^2 + 3x = 8$ **c** $2x^3 - x^2 - x = 11$

Level 7

7b I can use trial and improvement to find a solution to one decimal place of a more difficult equation involving x^3

7a I can use trial and improvement to find a solution to two decimal places of an equation involving x^3, x^2 and x

7 A cuboid-shaped carton has a face with an area of $(y^2 + 7y + 10)$ cm. The carton is y cm long. When full, the carton holds 1 litre of water. Form an equation and solve it, using trial and improvement, to find y to two decimal places.

Level 8

8c I can form and solve a cubic equation using trial and improvement

Now try this!

A Revealing rectangles

Sketch a rectangle. Label its width x and its length with an expression in terms of x, such as $x + 3$. Think of an area for the rectangle. Use trial and improvement to find x to one decimal place. Give your rectangle and its area to a partner and challenge them to find x. Compare your answers.

B Cubic solutions

Choose values for a, b and c to write a cubic expression of the form $ax^3 + bx^2 + cx$. For example you could choose $a = 2$, $b = 5$ and $c = 0$, giving $2x^3 + 5x^2$.
Now work out the value of your expression with a whole-number value for x. Substituting $x = 4$ in $2x^3 + 5x^2$ would give $128 + 80 = 208$.
Adjust the value slightly to give a cubic equation that does not have a whole-number solution, for example $2x^3 + 5x^2 = 222$. Swap equations with a partner and find solutions to one decimal place.

Super fact!

Think of a number between 1 and 1000. Your partner should be able to find it in no more than 10 questions by asking you each time if it is equal to, higher or lower than a given number. This method is called the 'binary chop'. Try it out!

in terms of trial and improvement

2.5 Solving simultaneous equations graphically

- Draw a straight-line graph
- Find the point of intersection of two straight-line graphs and link this to the solution of simultaneous equations
- Recognise that two parallel lines have no point of intersection

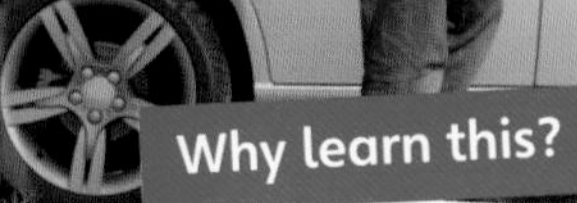

Why learn this?

You can use simultaneous equations to compare pricing policies for, say, mobile phones or car hire, to see which suits your needs.

What's the BIG idea?

- → You can construct a table of values to plot a graph. **Level 6 & Level 7**
- → **Simultaneous equations** are true 'at the same time', that is for the same values of x and y (or other letters). **Level 7**
- → You need two equations if there are two unknowns, three equations if there are three unknowns, and so on. **Level 7**
- → For two simultaneous equations, the point on a graph where the two lines cross is their solution. **Level 7**
- → There is no solution to the simultaneous equations of **parallel** lines because they do not cross. **Level 7**
- → If two simultaneous equations are equivalent equations, there is an infinite number of solutions to them because the lines coincide at every point.
  For example, $2y = 6x + 12$ and $y - 3x = 6$ are both equivalent to $y = 3x + 6$. **Level 7**

Practice, practice, practice!

Level 6

6a I can draw two straight-line graphs and find the point at which they cross

1 **a** Complete this table of values for the line $3y = 12 - 2x$.

x	0	1.5	3	4.5	6
y					

b Complete a table of values for the line $y = 10 - 2x$ for values of x from 0 to 6.

c Use your tables of values to draw the lines $3y = 12 - 2x$ and $y = 10 - 2x$.

d Write down the coordinates of the point where your lines cross.

2 **a** Plot the points (0, 10) and (10, 0) and join them with a straight line.
Use your graph to help you complete this table of values.

x	0	2	4	6	8	10
y						

b Look at the relationship between x and y in the table.
Use this to help you to write down the equation of the straight line.

c Draw the line $y = x + 4$ on the axes.

Hint: If you wish, draw up a table of values to help you.

d Write down the coordinates of the point where the lines cross.

Tip

The solution to a pair of simultaneous equations can be found graphically by finding where their lines cross. For example, if the lines cross at the point (3, −2) this means that the solution is $x = 3$, $y = -2$.

parallel

3 Use the graph to solve each pair of simultaneous equations.
Write 'no solution' if the lines do not cross and 'infinite number of solutions' if the lines are equivalent.

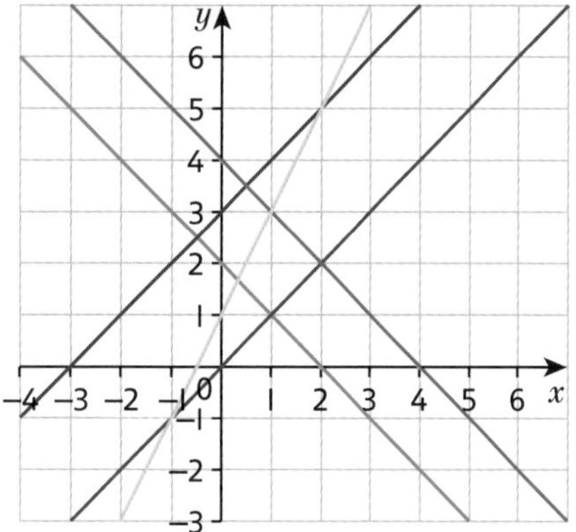

$y = 2x + 1$ and $y = x + 3$ $x = 2, y = 5$

- **a** $y = x$ and $x + y = 4$
- **b** $x + y = 2$ and $x + y = 4$
- **c** $y = 2x + 1$ and $y = x$
- **d** $y = x + 3$ and $y - x = 3$
- **e** $y + x = 2$ and $y - x = 3$

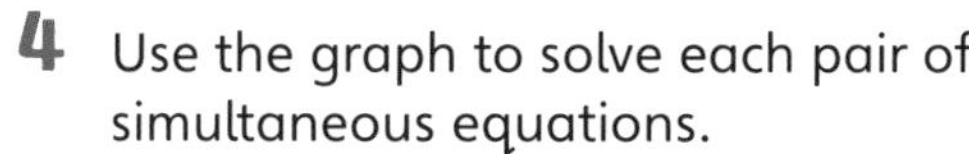

4 Use the graph to solve each pair of simultaneous equations.

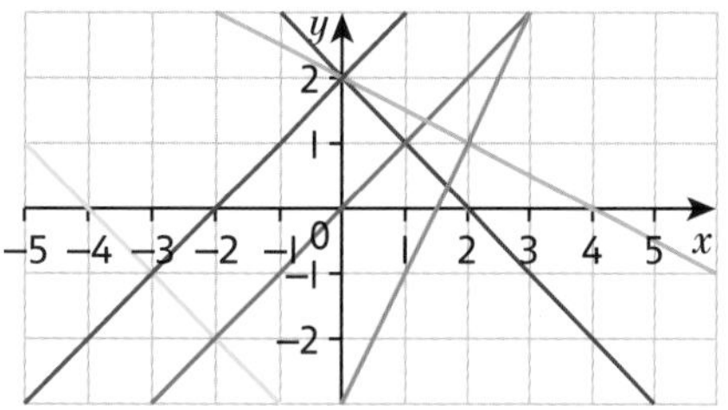

- **a** $y = x$ and $y = 2x - 3$
- **b** $x + y = 2$ and $x + 2y = 4$
- **c** $x + y = 2$ and $x + y = -4$
- **d** $x + 2y = 4$ and $2x - y = 3$
- **e** $x + y = -4$ and $x - y = -2$

5 Niall and Aiden have Saturday jobs gardening. Niall is paid £8.50 per hour. Aiden is paid £10 travelling costs and £6.28 per hour.

- **a** Construct a table of values for Niall and for Aiden, showing how much they would each earn for each whole number of hours up to 7 hours.
- **b** Use your tables of values to draw two straight-line graphs.
- **c** Write the equations of both lines.
- **d** Use your graph to find the number of hours' work for which Niall and Aiden are paid the same.

6 Solve each pair of simultaneous equations using a graphical method.

- **a** $y = 2x - 1$ and $y = x + 1$
- **b** $x + y = 4$ and $x + 2y = 3$
- **c** $x + 2y = 8$ and $3x + y = 9$
- **d** $y = 3x + 0.5$ and $y = 8 - 2x$
- **e** $x + 2y = 8$ and $3x - y = 13.5$

Watch out!
If straight lines are not parallel, they must cross somewhere. You may need to extend your lines into negative values.

Level 7

7b I can use graphs to find the solution to simultaneous equations

Did you know?
Parallel lines have the same gradient and never cross. Any other pair of different straight lines cross at exactly one point; this is the solution of their simultaneous equations.

7a I can form and solve simultaneous equations graphically to solve problems

7a I can solve a pair of simultaneous equations using a graphical method

Now try this!

A Crossed lines

On a graph draw two straight lines that cross. Write down the equations of the lines. Swap equations with a partner. Solve each other's simultaneous equations graphically by drawing the lines and finding where they cross.

B Crossroads

Roll two dice. Form a coordinate pair from the two numbers. Race your partner to write down the equations of two lines which cross at this point. Do not allow lines which are parallel to the axes.

Super fact!
Nearly 4000 years ago, the Babylonians could solve simultaneous equations. They used words for unknowns instead of letters because they did not yet have an alphabet.

2.6 Solving simultaneous equations algebraically

- Solve simultaneous equations with two variables algebraically

Why learn this?

One person buys two cod and three portions of chips for £9.40, and another buys three cod and four portions of chips for £13.70. Simultaneous equations can help work out the price of one portion of cod and chips.

What's the BIG idea?

- If you multiply every term on both sides of an equation by the same number you get an **equivalent equation**. Level 6
- You can add (or subtract) one equation to (or from) another, because what you are adding (or subtracting) is the same on both sides. Level 7
- **Simultaneous equations** are true 'at the same time'. Each equation has two unknowns, such as x and y. The solution is the value of x and y that satisfy both equations. Level 7 & Level 8
- To solve simultaneous equations:
  - If necessary, multiply one or both equations so that they have either the same number (ignoring the sign in front) of xs or the same number of ys.
  - Add or subtract the equations to eliminate one of the variables. Then solve the resulting equation to find the other variable.
  - Use **substitution** to find the remaining variable.

$3x + y = 29$ ①
$2x - y = 1$ ②
① + ② $5x = 30$
$x = 6$
Substitute in ① $3 \times 6 + y = 29$
$18 + y = 29$
$y = 11$

$4x - 3y = 22$ ①
$x - 3y = 1$ ②
① − ② $3x = 21$
$x = 7$
$7 - 3y = 1$
$3y = 6$
$y = 2$

$4x - 3y = 19$ ①
$x - 2y = -4$ ②
② × 4 $4x - 8y = -16$ ③
① − ③ $5y = 35$
$y = 7$
$4x - 21 = 19$
$4x = 40$
$x = 10$

Level 7 & Level 8

Did you know?

You can solve more than two simultaneous equations. You need the same number of equations as you have unknowns.

Watch out!

Remember to check by substituting your x and y values back into one of the original equations.

Practice, practice, practice!

Level 7

7b I can solve simultaneous equations by adding

7a I can solve simultaneous equations by subtracting

1 Solve each pair of simultaneous equations.

a $x + y = 15$ and $x - y = 11$
b $-x + y = 6$ and $x + 2y = 18$
c $3x + 2y = 15$ and $7x - 2y = 35$
d $x + y = 10$ and $3x - y = 6$
e $5x + 2y = -1$ and $-3x - 2y = -5$
f $-6x - 3y = -30$ and $-2x + 3y = -2$

2 Solve each pair of simultaneous equations.

a $x + y = 10$ and $x + 2y = 13$
b $x + 3y = 27$ and $x + 5y = 41$
c $3x + 2y = 23$ and $7x + 2y = 43$
d $5x - y = 29$ and $4x - y = 22$
e $6x - 5y = 45$ and $8x - 5y = 65$
f $-7x + y = -25$ and $-7x - 3y = -37$

equivalent equation simultaneous equations

3 **a** Copy and complete these to make equivalent equations.

$4x + 3y = 11$ $\quad$ $8x + \square y = 22$

b Use your answer to part **a** to help solve this pair of simultaneous equations.

$4x + 3y = 11$ and $7x + 6y = 20$

Level 7

7a I can begin to solve simultaneous equations where one of the equations needs to be multiplied

4 Solve each pair of simultaneous equations.

a $3x + y = 30$ and $5x - y = 34$ **b** $6x + 5y = 32$ and $6x + 3y = 24$

c $9x - 2y = 25$ and $7x - 2y = 19$ **d** $5x + 2y = 29$ and $7x + 4y = 49$

e $8x + y = 44$ and $10x + 2y = 58$ **f** $6x + y = 67$ and $5x - 2y = 36$

g $8x + 3y = 76$ and $11x - 6y = 145$ **h** $12x - 5y = -86$ and $17x - 15y = -201$

5 **a** Copy and complete these to make equivalent equations.

i $5x + 2y = 12$ $\quad$ $15x + \square y = 36$

ii $8x + 3y = 13$ $\quad$ $\square x + 6y = 26$

b Use your answers to parts **a** to help you solve this pair of simultaneous equations.

$5x + 2y = 12$ and $8x + 3y = 13$

6 Solve each pair of simultaneous equations.

a $2x + 3y = 19$ and $3x - 4y = -14$ **b** $3x - 2y = 9$ and $4x - 3y = 11$

c $4x - 3y = 21$ and $7x + 5y = 47$ **d** $2x + 3y = 49$ and $-5x + 4y = 4$

7 The diagram shows a trapezium.

a Use the fact that supplementary angles add up to 180° to write two equations.

b Solve your equations to find x and y.

c Use your values of x and y to work out the four angles in the trapezium.

$5x°$ $\quad$ $7x°$ $\quad$ $2y°$ $\quad$ $y°$

Not drawn accurately

8 **a** Write down a pair of simultaneous equations from this pie chart.

b Solve your equations to find x and y.

c Find all the angles of the pie chart.

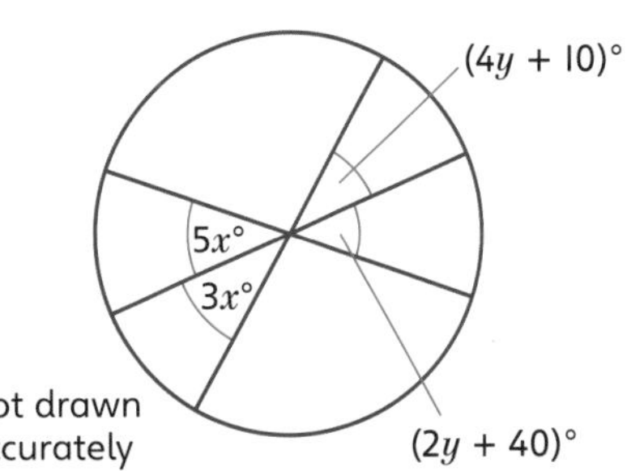

Level 8

8c I can solve simultaneous equations involving adding, subtracting and multiplying

8b I can begin to solve simultaneous equations where both equations need to be multiplied

Watch out!

When multiplying an equation, make sure that you multiply every term by the same number. Don't forget to multiply the constant term!

8b I can construct and solve simple simultaneous equations

Tip

8a I can construct and solve simultaneous equations

Now try this!

A Simultaneous equation swap

- Think of a number for x and a number for y. Work out the value of $2x + 3y$ and use this to write one equation. For example, if $x = 4$ and $y = -2$, $2x + 3y = 2$.
- Keep the same numbers for x and y. Write down a new equation, for example by working out $3x + 4y$.
- Swap equations with a partner. Solve each other's equations.

B Simultaneous solution swap

Think of a solution to a pair of simultaneous equations. Swap solutions with a partner. Now make up a pair of simultaneous equations with the solution you've been given.
Check your solution by plotting a graph.

substitute supplementary angles

Travelling equations

There are many everyday situations in which we solve equations informally without even realising it. For example, we do it when we compare two different ways of pricing something, such as mobile phone tariffs.

1 Paula plans to go to London by train. She wants her travel costs to be as low as possible, so she investigates whether it is cheaper to drive to the station and park there, or to take a taxi to the station.
These are the prices Paula has been charged in the past.

Taxi call-out charge	£1.86
Taxi mileage rate	£1.50 per mile
Station car park	£6.30
Paula's car costs	30p per mile

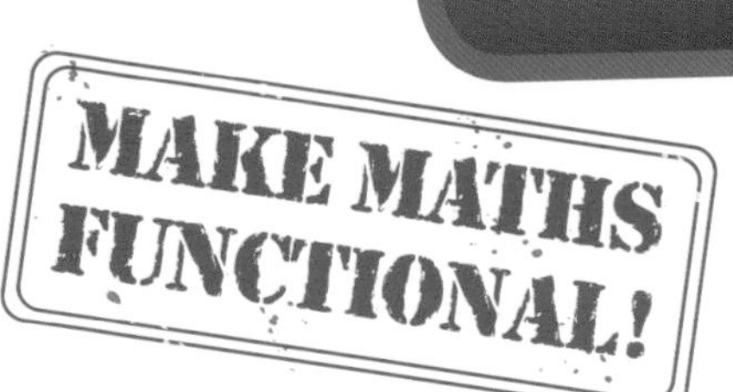

Paula lives 3.5 miles from the station.

Which option is cheaper, taxi or car, and by how much? **Level 6**

2 Form and solve an equation to find the distance for which the two options in Q1 would cost the same. **Level 7**

3 Quentin is travelling to London from a different station. The local minicab firm has a £1.50 call-out charge, but Quentin does not know how much they charge per mile.
He does know that calling out a minicab for a 3-mile journey costs £3.30 less than picking up a taxi from a taxi rank for a 7-mile journey.
There is no call-out charge when picking up a taxi from the rank, and the cost per mile is the same for a taxi as for a minicab.
Form and solve an equation to find the cost per mile. **Level 7**

4 Quentin's station car park costs £4.50 and it costs about 30p per mile to run his car.

- **a** Write an expression for the cost of calling out a minicab and travelling m miles.
- **b** Write an expression for the cost of travelling by car for m miles and using the station car park.
- **c** Draw two straight-line graphs to represent the cost of travelling by minicab and travelling by car. Plot miles travelled on the x-axis and cost on the y-axis.
- **d** Quentin lives 4 miles from the station. Use your graph to determine the cheaper option for him.
- **e** Use your graph to determine the distance for which both options would cost Quentin the same amount. **Level 7**

5 In London, Quentin buys an underground ticket from a machine. A single ticket costs £4 and a one-day Travelcard costs £6.80. In one day, the machine sells 100 tickets and takes £484.
How many tickets of each type were sold? **Level 8**

6 On a different day, Quentin used his own car to travel to a business meeting. He drove for 2 hours at one speed and then for 3 hours at another speed, and covered a distance of 252 km. If he had travelled 4 hours at the first speed and 1 hour at the second speed, he would have covered 244 km. Find the speeds at which Quentin drove. **Level 8**

The BIG ideas

- An **expression** has no equals sign. It is a way of expressing something general. Level 6
- **Equations** are true for particular values. For example, $2x + 2 = 8$ is only true for $x = 3$. Level 6
- You can construct an equation for a problem from the facts that you are given. Level 6
- When you have an equation with x on both sides, you need to add or subtract the same number of xs from both sides in order to get x on one side only. Level 6
- **Simultaneous equations** are true 'at the same time', that is for the same values of x and y (or other letters). Level 7
- For two simultaneous equations, the point on a graph where the two lines cross is their solution. Level 7
- To solve simultaneous equations:
  - If necessary, multiply one or both equations so that they have either the same number of xs or the same number of ys (ignoring the signs).
  - Add or subtract the equations to **eliminate** one of the variables. Then solve the resulting equation to find the other variable.
  - Use **substitution** to find the remaining variable. Level 7 & Level 8

Find your level

Level 6

Q1 Write, as simply as possible, an expression for the perimeter of this rectangle.

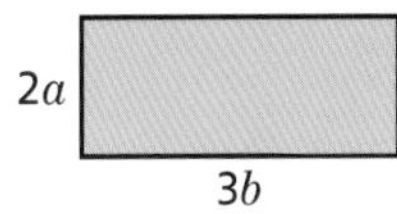

Q2 James multiplies out $4(3b + 4)$.
He writes: $12b + 8$
Explain why James's answer is incorrect.

Q3 An isosceles triangle's two equal sides are of length $2x + 1$ and $4x - 11$.

a Form an equation and solve it to find x.

b Write down the length of the equal sides.

Level 7

Q4 Look at these two expressions.
$3y - 5 \quad 4(y - 3)$

a What value of y makes these two expressions equal in value?

b What value of y makes the second expression twice as great as the first?

Q5 Multiply out the brackets.

a $a(a - 2)$

b $-b(2 - b)$

Level 8

Q6 a Solve the simultaneous equations.

$3x + y = 15$
$2x + 3y = 17$

b The solution to this pair of simultaneous equations is $x = 2$, $y = 5$.

$px + qy = 44$
$qx + py = 47$

Find the values of p and q.

Q7 Fiona has two juice cartons, each in the shape of a cube. Each carton is full of juice. Between them the cartons hold 80% of a litre.
The side length of the larger cube is 1 cm longer than that of the smaller cube.

a The side length of the smaller cube is l. What is the side length of the larger cube?

b What is the volume of

i the smaller cube

ii the larger cube?

c Form an equation in terms of l.

Hint: 1000 cm^3 holds 1 litre

d Using trial and improvement, solve your equation from part **c** to find the side length of the smaller cube.
Give your answer in centimetres to 1 decimal place.

3 Share and share alike

This unit is about some basic calculating skills. If you can do them in your head, you will be able to reach decisions and make plans quickly.

Tyrone is a personal trainer. His job involves a lot of calculations to work out fitness and nutritional assessments, and to prepare training plans for clients.

A few changes to what you eat and drink, or how you exercise, can make a big difference to your quality of life.

A careful plan will help you succeed in challenges like training for a half marathon or improving your fitness. To reach your target you may need to increase your intake of certain nutrients or the time you spend on a particular exercise.

Activities

A The table shows how much energy (in kilojoules) you use doing different levels of exercise.
I calorie is 4.184 kilojoules

- How many calories do you burn per minute of intensive exercise? Round your answer to the nearest whole number.
- Teenagers should aim to do at least 60 minutes of moderate exercise each day. How many calories would this burn?

B Do you meet the target of 60 minutes of moderate exercise each day? Use the table above to write your own exercise plan.

Level of exercise	Example activities	Average energy used
gentle	walking slowly	10–20 kilojoules/min
moderate	tennis gymnastics jogging cycling slowly walking quickly	21–30 kilojoules/min
intensive	football rugby swimming squash cycling fast disco dancing	31–40 kilojoules/min

1 Match each calculation in the first box with one in the second box that has roughly the same answer.

Level Up Maths 5-7 page 144

a 10% of 239	**d** 74% of 20
b $\frac{1}{2}$ of 61	**e** $\frac{1}{3}$ of 72
c $\frac{3}{5}$ of 25	**f** 20% of 150

2 Write each fraction in its simplest form.

Level Up Maths 5-7 page 54

a $\frac{35}{50}$ **b** $\frac{24}{60}$ **c** $\frac{12}{54}$ **d** $\frac{27}{81}$

3 Copy and complete these.

Level Up Maths 5-7 page 62

a $\frac{1}{2}$ of 180 = $\frac{1}{4}$ of ■

b $\frac{3}{5}$ of 100 = $\frac{2}{5}$ of ■

4 Which of these calculations gives the largest answer?

Level Up Maths 5-7 page 60

a 25% of 64 **b** 20% of 75

c 65% of 20 **d** 35% of 40

3.1 Decimals and fractions in order
3.2 Adding and subtracting fractions
maths! Joining the resistance
3.3 Multiplying and dividing fractions
3.4 Using percentages
3.5 Get things in proportion
3.6 Ratio and proportion
3.7 Brain power
3.8 More calculation strategies
3.9 Efficient calculation
Unit plenary: Towards a better lifestyle
Plus digital resources

Did you know?

About 70% of the energy from the food you eat is used to keep your body functioning. The other 30% gives you energy to move around.

3.1 Decimals and fractions in order

- Compare and order decimals and fractions
- Express one number as a fraction of another
- Understand and use inequality signs

What's the BIG idea?

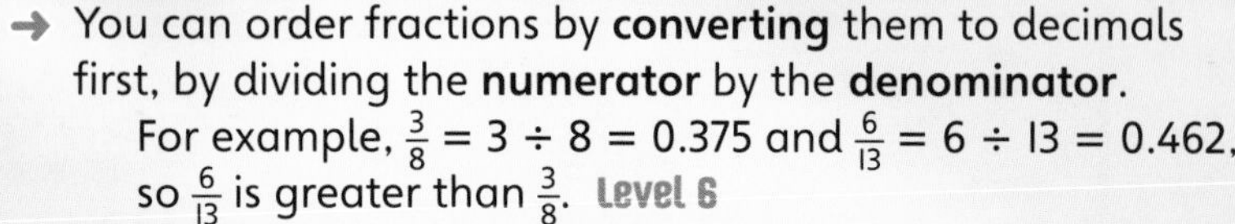

→ You can order fractions by **converting** them to decimals first, by dividing the **numerator** by the **denominator**.
For example, $\frac{3}{8} = 3 \div 8 = 0.375$ and $\frac{6}{13} = 6 \div 13 = 0.462$, so $\frac{6}{13}$ is greater than $\frac{3}{8}$. **Level 6**

→ You can also order fractions by converting them to **equivalent fractions** with a common denominator. **Level 6**

→ When you multiply or divide both sides of an inequality by a negative number, you need to reverse the inequality sign for the inequality to remain true.
For example, $2.5 > -3.9$ but $-2.5 < 3.9$ **Level 6**

→ You can use inequality signs to describe intervals.
For example, for temperature (T), the interval $-15 \leqslant T < -10$ includes all the temperatures between −10 and −15, and −15 but not −10. **Level 7**

Why learn this?
Comparing information, like biometric data, often means comparing decimal values.

Learn this
< means 'less than'
> means 'greater than'
⩽ means 'less than or equal to'
⩾ means 'greater than or equal to'

Practice, practice, practice!

Level 6

1 Use a fraction to represent each piece of data. Give your answers in their lowest terms.

Curlbridge High – Year 9 pupil data
- *20 out of the 25 pupils in 9B are taller than 1.6 m.*
- *Four of the six tutor groups have more boys in them than girls.*
- *120 of the 150 pupils live less than 10 miles from school.*
- *42 of the 70 girls are aged 14.*

(6c) I can express a smaller whole number as a fraction of a larger one

2 By converting the fractions to decimals, write these fractions in descending order.

$\frac{3}{8}$ $\frac{3}{4}$ $\frac{4}{5}$ $\frac{5}{8}$ $\frac{11}{50}$ $\frac{5}{6}$ $\frac{7}{20}$

Hint:

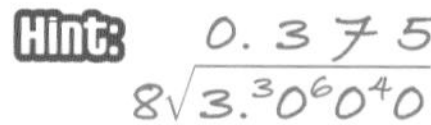

(6b) I can order fractions by converting them to decimals

3 Use equivalent fractions to put these fractions in order, starting with the smallest.

$\frac{7}{10}$ $\frac{5}{8}$ $\frac{3}{5}$ $\frac{17}{20}$

(6b) I can use equivalent fractions to order fractions

4 Pupils in class 9B measured the length of their longest finger. These are the lengths.

Ian $6\frac{1}{2}$ cm	Jan 6.92 cm	Zoë $6\frac{3}{4}$ cm	Jack 7.85 cm
Terry 6.84 cm	Carla $6\frac{1}{4}$ cm	Sharon $7\frac{3}{5}$ cm	Mayuri 6.8 cm

Put the data in order, from largest to smallest.

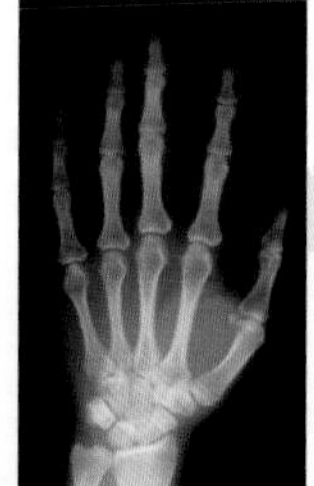

(6a) I can compare and order lists of fractions and decimals

compare convert denominator descending equivalent fraction

5 a Put these numbers in order, from largest to smallest.

−2.3 −3.24 −2.32 −3.2 −2.03 −2.38 −2.23 −2.34 −2.24

b How many of the numbers in a are greater than −2?

6 Put the correct inequality sign between each pair.

a 1.495 m ☐ 1.479 m b 1.645 m ☐ 1.655 m

c 1.075 m ☐ 1.57 m d 0.0354 m ☐ 0.034 97 m

7 a Which of these statements are *not* true?

A −5 > 3.4 B −4 < −5 C 2.467 < 2.5 D 3.56 > −3

b Look at the inequalities in a that are not true. Multiply both numbers by −1. Write in the correct sign. What do you notice?

c Repeat part b, but *divide* both numbers by −1. Write in the correct sign. What do you notice?

Level 6

6a I can order negative decimals (to 2 or 3 significant figures)

6a I can use inequality signs to compare positive decimals (to 4 or 5 significant figures)

6a I can multiply and divide both sides of an inequality by a negative number

8 Mayuri has collected temperature data for her statistics project.

a Put the correct inequality sign between the two temperature readings for each day.

b Explain, in words, the class interval $-15\,°C \geqslant T > -20\,°C$, where T is temperature in degrees Celsius.

c Write, mathematically, this class interval: the temperature, T, is less than or equal to −20°C and greater than −30°C.

	Temperature (°C)	
	9 am	4 pm
Monday	−11.1	−7.95
Tuesday	−0.79	−0.91
Wednesday	−2.97	−2.79

9 At 4 pm on Thursday, Mayuri recorded a temperature of T°C. The temperature was recorded to one decimal place. These two statements are both true. What are the possible values of T?

$\frac{1}{8} < T < \frac{5}{8}$

$\frac{2}{6} < T < \frac{4}{6}$

10 The numbers x, $\frac{1}{3}$, y, $\frac{1}{6}$ are in decreasing order of size. The difference between each number and the one next to it is the same in each case. What are the values of x and y?

Level 7

7c I can use inequality signs to compare negative decimals (to 2 or 3 significant figures)

7c I can order fractions to solve a problem

Super fact!

An amazing temperature drop was recorded in Montana, USA in 1916. The temperature dropped from 6.7°C to −48.9°C in 24 hours.

Now try this!

A Zoom in, in ten

A game for two players. Player 1 writes down a number between 5 and 10, with one or two decimal places. Player 2 has to guess the number, but they can only use the questions 'Is it less than...?' or 'Is it greater than...?'. Score 1 point if the number is guessed in 10 or fewer guesses. Swap roles and repeat.

B Prediction

A game for two players. Player 1 rolls a dice twice, divides the first number by the second and writes down the result. Player 2 has to predict whether repeating this with the next two numbers rolled will produce a result that is larger or smaller than this. Roll the dice to test the prediction. Every correct prediction scores 1 point. Swap rolls and repeat. First to 5 points wins.

lowest terms numerator order place value

3.2 Adding and subtracting fractions

- Add and subtract fractions with different denominators
- Use inverse operations to check fraction calculations
- Add and subtract mixed numbers
- Solve word problems involving fraction addition and/or subtraction
- Add and subtract simple algebraic fractions

Why learn this?

You need to be able to add together bits and pieces to find out how much you have altogether!

What's the BIG idea?

- To add or subtract fractions with different **denominators**, find **equivalent fractions** with a **common denominator**. Then add or subtract them. Level 6
- Answers that are **improper fractions** should be converted to **mixed numbers**. Level 6
- To add mixed numbers, add the whole-number parts and then add the fractional parts. You can use the same method for subtraction if the fractional part of the second fraction is smaller than the fractional part of the first. Level 6
- To subtract mixed numbers where the fractional part of the second fraction is larger than the fractional part of the first, convert them both into improper fractions before subtracting. Level 6
- To add or subtract more complex fractions, you may need to identify the **multiples** of each denominator to find the lowest common denominator. Level 7

Practice, practice, practice!

Level 6

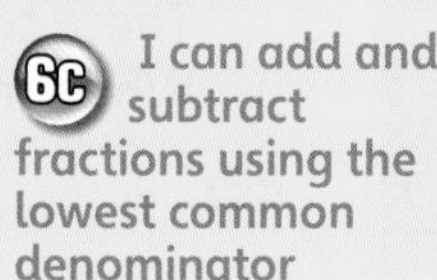

6C I can add and subtract fractions using the lowest common denominator

1 a Write the lowest common multiple of each set of numbers.

i 3 and 4 ii 8 and 4 iii 6 and 5 iv 9 and 12
v 8 and 12 vi 3, 4 and 5 vii 4, 5 and 6 viii 6, 9 and 10

b Use your answers from part a to work out these.

i $\frac{1}{3}+\frac{3}{4}$ ii $\frac{7}{8}-\frac{3}{4}$ iii $\frac{5}{6}-\frac{4}{5}$ iv $\frac{4}{9}-\frac{5}{12}$
v $\frac{1}{8}+\frac{5}{12}$ vi $\frac{2}{3}+\frac{1}{4}+\frac{2}{5}$ vii $\frac{3}{4}+\frac{2}{5}-\frac{1}{6}$ viii $\frac{5}{6}-\frac{2}{9}+\frac{3}{10}$

6C I can use inverse calculations to check fraction addition and subtraction

2 Use an inverse operation to check these pupils' calculations. Correct any that are wrong.

$\frac{2}{7}+\frac{1}{4}=\frac{13}{28}$ $\frac{13}{28}-\frac{1}{4}=\frac{13-7}{28}=\frac{6}{28}=\frac{3}{14}$, $\frac{2}{7}+\frac{1}{4}\neq\frac{13}{28}$

a $\frac{3}{5}-\frac{1}{4}=\frac{7}{20}$ b $\frac{3}{5}+\frac{1}{4}=\frac{19}{20}$ c $\frac{5}{6}-\frac{1}{5}=\frac{17}{30}$ d $\frac{2}{7}+\frac{3}{5}=\frac{32}{35}$

6C I can add and subtract fractions

3 Copy and complete.

a $\frac{1}{\square}+\frac{1}{\square}=\frac{3}{4}$ b $\frac{1}{\square}+\frac{1}{\square}=\frac{5}{6}$ c $\frac{1}{\square}+\frac{2}{\square}=\frac{11}{15}$

d $\frac{5}{\square}-\frac{2}{\square}=\frac{1}{6}$ e $\frac{2}{\square}+\frac{2}{\square}=1\frac{1}{15}$ f $\frac{7}{\square}-\frac{1}{\square}=\frac{3}{8}$

6C I can add mixed numbers

4 Work out these additions.

a $3\frac{2}{3}+2\frac{1}{6}$ b $1\frac{5}{6}+1\frac{5}{8}$ c $1\frac{7}{8}+2\frac{3}{4}$ d $5\frac{2}{7}+2\frac{4}{5}$

common denominator denominator equivalent fraction improper fraction fract

5 Work out these subtractions.

a $4\frac{3}{4} - 2\frac{1}{2}$ b $3\frac{7}{9} - 1\frac{2}{3}$ c $3\frac{3}{8} - 3\frac{1}{6}$ d $1\frac{5}{7} - 1\frac{3}{5}$

e $3\frac{4}{5} - 1\frac{2}{3}$ f $10\frac{4}{9} - 3\frac{1}{7}$ g $6\frac{7}{8} - 2\frac{4}{5}$ h $4\frac{5}{6} - 1\frac{1}{8}$

6 a Use improper fractions to show that $4\frac{1}{4} - 2\frac{3}{5} = 1\frac{13}{20}$.

b Copy and complete.

i $3\frac{1}{5} - \frac{1}{2} = \square$ ii $5\frac{1}{4} - \frac{2}{3} = \square$ iii $4\frac{2}{5} - 3\frac{3}{4} + 1\frac{1}{3} = \square$

iv $2\frac{1}{6} - 2\frac{3}{4} = \square$ v $3\frac{2}{3} + \square = 5\frac{8}{15}$ vi $\square + 2\frac{2}{3} = 4\frac{1}{6}$

Level 6

6b I can subtract mixed numbers by subtracting the whole numbers and fractional parts separately

6a I can subtract mixed numbers by converting them into improper fractions

7 Work out these.

a $\frac{3}{20} + \frac{4}{15}$ b $\frac{11}{12} - \frac{7}{30} + \frac{4}{15}$ c $4\frac{5}{18} + 3\frac{7}{24}$

d $3\frac{5}{18} - 1\frac{1}{15}$ e $4\frac{1}{15} - 1\frac{11}{24}$ f $2\frac{3}{40} - 1\frac{11}{15} - 1\frac{1}{3}$

Hint: List the multiples of each denominator to find the lowest common denominator.

8 A length of ribbon is $3\frac{5}{8}$ feet long.
A length of $1\frac{13}{16}$ feet is cut off.
What length of ribbon is left?

9 What is the perimeter of a rectangular rabbit run with length $11\frac{11}{18}$ m and width $6\frac{7}{12}$ m?

10 Rachel is working out her monthly expenses. Her mortgage accounts for $\frac{5}{12}$ of her monthly earnings, clothing $\frac{1}{6}$, and gas and electricity a total of $\frac{2}{9}$.
The remainder of her monthly earnings is spent on travel and entertainment.

a What fraction of her monthly earnings does she spend on travel and entertainment?

b Rachel spends the same amount on travel as on entertainment. What fraction of her monthly earnings does she spend on each?

Level 7

7c I can add and subtract more complex fractions

7b I can solve problems by adding and subtracting fractions

Now try this!

A Ancient Egyptian fractions

Ancient Egyptian fractions were written as unit fractions, that is with 1 as the numerator. Investigate how many different fractions you can make by adding any two of these unit fractions: $\frac{1}{2}, \frac{1}{3}, \frac{1}{4}, \frac{1}{5}$ and $\frac{1}{6}$. You may use the same fraction twice in a sum.

Did you know?

The Egyptians used fractions in calculations over 3000 years ago.

B Series investigation

Look at this series.

$$\frac{1}{1} + \frac{1}{2} + \frac{1}{3} + \frac{1}{4} + \frac{1}{5} + \frac{1}{6} + \dots + \frac{1}{n}$$

Try to predict what happens to the sum of the series s n gets bigger and bigger.

Will it ever reach a number greater than 4?

west common multiple (LCM) mixed number multiple numerator unit fraction

maths!

Joining the resistance

Resistors are electrical components that are used to restrict the flow of electric current. They are used in millions of devices, from brake lights to timing circuits.

Colour code

The value of a resistor is measured in ohms (Ω). Many resistors are colour coded using a four-band system. The colours of the first two bands represent a two-digit number, and the third band represents a power of 10 that acts as a multiplier. The fourth band gives the 'tolerance' – how close the resistance must be to this value.

For example, a resistor with the bands yellow, violet, yellow and brown would be $47 \times 10^4\,\Omega \pm 1\%$, or 470 kΩ ± 1%.

- What colour bands would you find on these resistors?

a 330 Ω ± 5%
b 22 Ω ± 1%
c 0.47 Ω ± 0.1%

Colour	Bands 1 and 2	Band 3	Band 4
Black	0	$\times10^0$	±20%
Brown	1	$\times10^1$	±1%
Red	2	$\times10^2$	±2%
Orange	3	$\times10^3$	±3%
Yellow	4	$\times10^4$	±4%
Green	5	$\times10^5$	±0.5%
Blue	6	$\times10^6$	±0.25%
Violet	7	$\times10^7$	±0.1%
Grey	8	$\times10^8$	±0.05%
White	9	$\times10^9$	
Gold		$\times10^{-1}$	±5%
Silver		$\times10^{-2}$	±10%

- Work out the maximum and minimum possible values of each of these resistors.

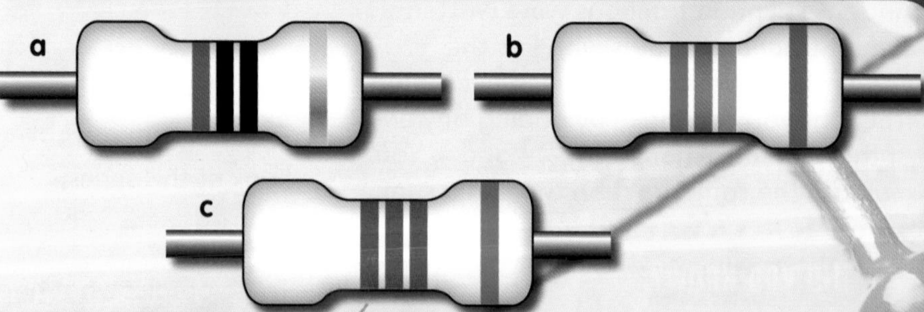

Series and parallel

Resistors R can be joined together in series or in parallel.

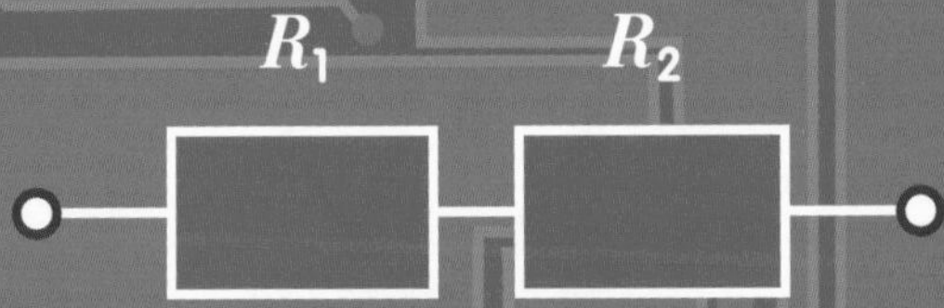

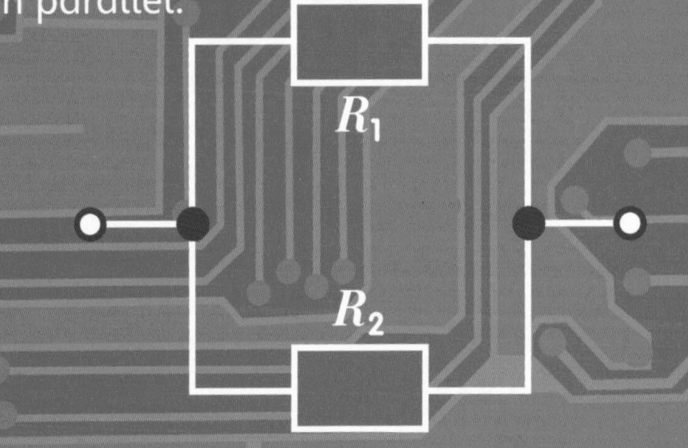

In series, $R_{\text{TOTAL}} = R_1 + R_2$

In parallel, $\dfrac{1}{R_{\text{TOTAL}}} = \dfrac{1}{R_1} + \dfrac{1}{R_2}$

Work out the combined resistance of each of these sets of resistors.

a

b

c

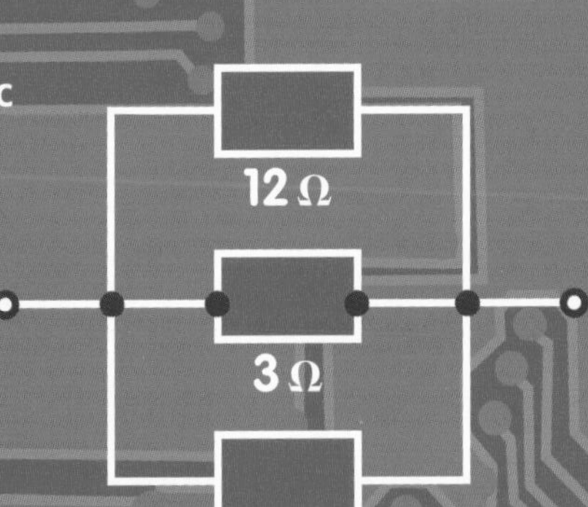

d

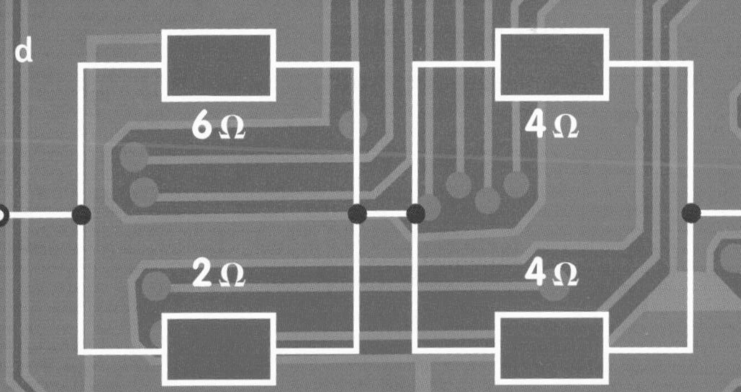

- Draw a diagram showing how a 2 Ω, a 3 Ω and a 4 Ω resistor might be connected together to produce a combined resistance of 5.2 Ω.
- Show using algebra that if two resistors with values x Ω and y Ω are connected in parallel, the resulting resistance is $\dfrac{xy}{x+y}$ Ω.

3.3 Multiplying and dividing fractions

- Multiply and divide integers and fractions by fractions
- Understand and use reciprocals
- Solve word problems involving fraction multiplication and/or division

What's the BIG idea?

Why learn this?

Being able to divide up cakes and pizzas between friends is useful, but you also need to understand fractions for many other areas of maths, such as probability.

- To multiply an **integer** by a fraction you can multiply by the **numerator** and then divide by the **denominator**.
  For example, $24 \times \frac{3}{4} = 24 \times 3 \div 4 = 72 \div 4 = 18$
  Or you can divide by the denominator and then multiply by the numerator.
  For example, $24 \times \frac{3}{4} = 24 \div 4 \times 3 = 6 \times 3 = 18$ **Level 6**
- To multiply a fraction by a fraction, multiply the numerators and multiply the denominators.
  $\frac{2}{3} \times \frac{3}{5} = \frac{2 \times 3}{3 \times 5} = \frac{6}{15}$ **Level 7**
- To divide fractions, invert the dividing fraction (turn it upside down) and change the division sign to multiplication.
  $\frac{2}{3} \div \frac{3}{5} = \frac{2}{3} \times \frac{5}{3} = \frac{2 \times 5}{3 \times 3} = \frac{10}{9}$ or $1\frac{1}{9}$ **Level 7**
- If you divide 1 by any number, you get the **reciprocal** of that number.
  For example, the reciprocal of 2 is $1 \div 2 = \frac{1}{2}$ or 0.5, and the reciprocal of $\frac{2}{5}$ is $1 \div \frac{2}{5} = \frac{5}{2}$ or 2.5. **Level 7**
- To multiply or divide **mixed numbers**, change them to **improper fractions** first. **Level 7**

Tip

$32 \div \frac{1}{4}$ means 'how many quarters are there in 32?' There are four quarters in 1, so there are 32×4 (= 128) quarters in 32.

Practice, practice, practice!

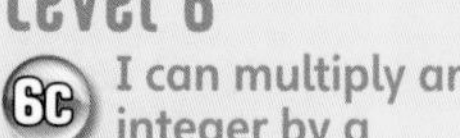

Level 6

6C I can multiply an integer by a fraction

1 **a** Work out

  i $24 \times \frac{1}{3}$ **ii** $24 \times \frac{1}{8}$ **iii** $24 \times \frac{1}{4}$ **iv** $24 \times \frac{1}{12}$

b Use your answers from part **a** to calculate

  i $24 \times \frac{2}{3}$ **ii** $24 \times \frac{5}{8}$ **iii** $24 \times \frac{3}{4}$ **iv** $24 \times \frac{7}{12}$

Tip

$30 \times \frac{2}{3} = \frac{2}{3}$ of 30

2 Here are some results of a survey of 6000 teenagers.

a $\frac{7}{25}$ want to own a large home.

b $\frac{27}{50}$ plan to have children.

c $\frac{21}{25}$ want to remain with one partner for life.

d $\frac{17}{20}$ say their parents approve of most of their friends.

e $\frac{14}{25}$ want to make a difference to the world.

How many teenagers are included in each result?

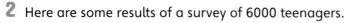

cancel denominator improper fraction integer inverse

3 Work out these. Simplify your calculations whenever possible.

a $\frac{5}{16} \times 64$ **b** $\frac{7}{12}$ of 60 kg **c** $240 \times \frac{13}{60}$ **d** $360 \times \frac{11}{90}$

4 **a** If you cut four pizzas into sixths, how many slices do you get?

b Calculate **i** $6 \div \frac{1}{3}$ **ii** $10 \div \frac{1}{5}$ **iii** $25 \div \frac{1}{4}$

c How many people can you give two thirds of a pizza to if you share six pizzas equally?

d Calculate **i** $3 \div \frac{3}{4}$ **ii** $12 \div \frac{2}{3}$ **iii** $6 \div \frac{3}{5}$ **iv** $3 \div \frac{2}{9}$

5 Steve has devised a method for dividing an integer by a fraction.

$6 \div \frac{1}{4} = 6 \times 4$

$6 \div \frac{3}{4} = 6 \times 4 \div 3$ or $\overset{2}{\cancel{6}} \times \frac{4}{\cancel{3}_1} = 8$

Use Steve's method to work out these.

a $25 \div \frac{5}{8}$ **b** $36 \div \frac{4}{5}$ **c** $60 \div \frac{12}{13}$ **d** $280 \div \frac{70}{110}$

Level 6

6b I can simplify before multiplying by a fraction

6b I can divide an integer by a fraction

Tip Write mixed numbers as improper fractions first.

6 Work out these, cancelling when possible.

a $\frac{3}{4} \times \frac{5}{6}$ **b** $\frac{2}{9} \times \frac{3}{4}$ **c** $\frac{7}{11} \times \frac{4}{14}$

d $\frac{2}{5} \times 2\frac{4}{13}$ **e** $\frac{3}{4} \times 1\frac{5}{12}$

7 Find the reciprocal of each number.
Give your answer as a mixed number if appropriate.

a $\frac{3}{4}$ **b** $\frac{1}{8}$ **c** $\frac{7}{8}$ **d** $4\frac{3}{5}$

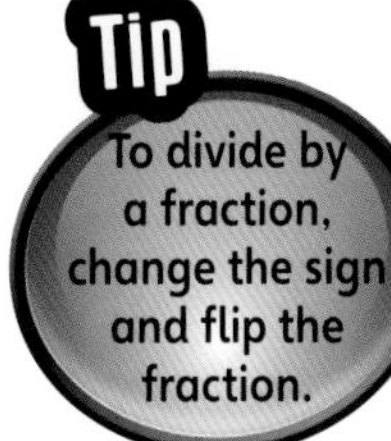

Tip
To divide by a fraction, change the sign and flip the fraction.

8 Work out

a $\frac{2}{5} \div \frac{7}{10}$ **b** $\frac{5}{8} \div \frac{25}{48}$ **c** $1\frac{2}{3} \div \frac{2}{3}$ **d** $1\frac{5}{6} \div \frac{22}{25}$

9 A group of friends visited Pizza Palace. They bought three pizzas to share. Maria took a quarter of the order, then Becky took a third of what was left, and Chris ate two thirds of what remained.

a What fraction of a single pizza did each person eat?

b They decided to cut up the remainder of the pizzas so each piece was exactly a quarter of a pizza. How many of them got an extra slice?

c If they are being fair, who should get an extra slice?

Level 7

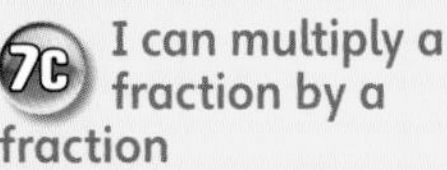
I can multiply a fraction by a fraction

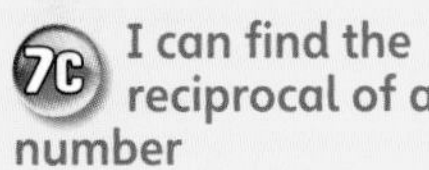
I can find the reciprocal of a number

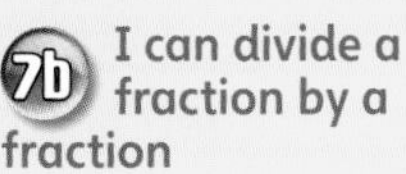
I can divide a fraction by a fraction

7a I can solve problems using fraction calculations

Now try this!

A Picture that

Draw a diagram to illustrate each of these calculations.
Make a poster, using your diagrams, to explain how you multiply and divide with fractions.

a $\frac{5}{6} \times 12 = 10$ **b** $\frac{2}{3} \times \frac{3}{4} = \frac{1}{2}$ **c** $3 \div \frac{1}{4} = 12$ **d** $\frac{3}{4} \div \frac{1}{8} = 6$

B Fraction fracas

A game for two people. Cut a sheet of A4 paper into 16 equally sized 'cards'. Using only the digits 3 to 9, write down a different fraction on each card (e.g. $\frac{4}{9}, \frac{3}{7}, \frac{5}{6}, \ldots$). Spread out the cards face down and take turns to turn over two cards. The first to correctly write down the product of the two fractions in its lowest terms scores 1 point. First to 10 points wins.

mixed number numerator reciprocal simplify

3.4 Using percentages

- Express one number as a percentage of another
- Find the outcome of a given percentage increase or decrease
- Solve percentage problems using the unitary method or an inverse operation
- Work with repeated proportional change
- Calculate compound interest

Why learn this?

You need to understand interest rates so that you can negotiate the best deal and so you don't pay more than you need to if you borrow money.

What's the BIG idea?

- To express one number as a **percentage** of another write the number as a fraction of the other, divide the numerator by the denominator and multiply by 100.
  For example, 12 out of 30 = $\frac{12}{30} \times 100 = 0.4 \times 100 = 40\%$ **Level 6**
- To compare fractions, decimals and percentages, change them to the same form. **Level 6**
- To find the original value after a percentage increase or decrease, find 1% from the new value and multiply that by 100. This is called the **unitary method.** **Level 6**
- You can calculate percentage change using a multiplier.
  Increase by 10%: multiplier is 110% = $\frac{110}{100}$ = 1.1
  new amount = 1.1 × old amount
  Decrease by 8%: multiplier is 92% = $\frac{92}{100}$ = 0.92
  new amount = 0.92 × old amount **Level 7**
- You can also find the original value after a percentage increase or decrease, by using a **multiplier** and the **inverse operation**. **Level 7**
- **Compound interest** is worked out on the **principal** (the original amount) plus any previous interest already earned. You can calculate compound interest by finding the multiplier and raising it to a power equal to the number of times the interest is added on. **Level 8**

Practice, practice, practice!

Level 6

6C I can use the equivalence of fractions, decimals and percentages to compare proportions

1 The table shows different estimates for the proportion of Burma's population affected by Cyclone Nargis in 2008.

Source	Statistic	Percentage (%) (to 1 d.p.)
radio	one in nine	
television	the probability of being affected is 0.09	
magazine	eight out of every hundred	
internet	4.2 million out of a population of 47 million	

a Copy and complete the table.

b Which source gave the highest estimate?

Super fact!

Relief and construction work in Burma after Cyclone Nargis cost over £500 million.

6C I can calculate a percentage increase

2 Mia is paid £6.20 an hour for her Saturday job. She has been promised a 5% increase on her next birthday. Joshua gets £6.00 an hour now, but he will get a 7.5% increase at the same time as Mia.

a How much extra will Mia get per hour after her birthday?

b What will Mia's hourly rate be after her birthday?

c After the pay rises, who will be paid more?

compound interest decrease increase inverse operation multiplies

3 The table shows average house prices in 2008 and the predicted percentage falls.

	Greater London	South East England	South West England	West Midlands	Wales	Scotland
Average house price in 2008	£350 000	£280 000	£230 000	£180 000	£170 000	£165 000
Predicted percentage fall	9%	7%	9%	11%	6%	4%
Predicted price in 2009	£318 500					

a Calculate the predicted average house prices after the given percentage falls.
b In which area are the average prices predicted to fall by more than £15 000?

4 Jay sold jeans and t-shirts for his Enterprise Project. He sold the t-shirts for £3.90 each. The selling price was 130% of the original cost.
a £3.90 represents 130%. What is 1%?
b Use your answer from part **a** to find the original cost of a t-shirt.
Jay made a loss of 10% on the jeans, which he sold for £10.80 a pair.
c What percentage of the original cost is £10.80?
d Use part **c** to find 1% and then the original price of a pair of jeans.

Level 6

6b I can calculate a percentage decrease

Learn this

To find the new value after a percentage increase or decrease, use the multiplier.

6a I can use the unitary method to solve percentage problems

5 **a** Match each percentage increase/decrease to the correct multiplier.
i 10% rise **ii** 25% increase **iii** 50% decrease
iv 15% rise **v** 15% fall **vi** 5% loss

0.85 1.1 0.95 0.5 1.25 1.15

b Calculate
i £45 plus a 15% increase **ii** £45 after a 15% decrease.
c Calculate the original price if the new price is
i £37.80 after a 5% increase **ii** £28.90 after a 15% decrease.

6 Elliot says, 'If you add 20% to £200 you get £240. So if you take 20% off £240 you will get £200'. Explain why he is wrong.

Level 7

7c I can use an inverse operation to solve percentage problems

Learn this

To find the original value from the new value, divide by the multiplier.

7 Mario has £1500 to invest.
a What calculation does $£1500 \times 1.05^4$ represent?
b How much would his investment be worth after 4 years at an interest rate of
i 3% **ii** 5% **iii** 10%?
c Explain why the increase in the investment after 4 years with a 10% rate, is not twice the increase in the investment with a 5% rate.

Level 8

8b I can calculate compound interest

Now try this!

A What a bargain!

Find out the interest rates that three different banks or building societies charge on loans. Find the price of a car you would like to buy and work out how much a loan would cost in total if you borrowed the money for 1, 3 or 5 years from each of the banks.

B How interesting!

Choose a percentage increase and guess how many times you would need to compound the interest before you double your money. Set up a spreadsheet to help you calculate the period for any percentage rate you choose.

percentage (%) **principal** **selling price** **unitary method**

3.5 Get things in proportion

- Identify situations where proportionality can be used to solve problems
- Recognise when sets of numbers are in proportion
- Understand and use proportionality

What's the BIG idea?

- Two **variables** are in **direct proportion** if they increase and decrease in the same ratio, i.e. with a constant multiplier and when one is zero so is the other. **Level 6**
- In a table of values for variables that are in direct proportion, the value of $\frac{y}{x}$ is always the same. **Level 7**

x	5	8	13	15
y	30	48	78	90
$\frac{y}{x}$	6	6	6	6

- The symbol for 'is **proportional to**' is $\propto$. $y \propto x$ means that y is directly proportional to x. For the table above, $y \propto x$ and $\frac{y}{x} = 6$, so $y = 6x$. **Level 7**
- If y is directly proportional to x
- When y is proportional to the **square** of x, then you write $y \propto x^2$. **Level 7**
- When y is **inversely proportional** to x, then you write $y \propto \frac{1}{x}$. **Level 8**

Why learn this?

Proportionality can be useful even when on holiday! You can understand exchange rates and convert kilometres to miles.

Watch out!

Not all conversion graphs use direct proportion (e.g. from °C to °F).

Practice, practice, practice!

Level 6

6a I can identify when proportion can be used to solve problems

1 a A teacher says skirts have to be lengthened by 10 cm. Give this increase as a fraction of each of these original lengths.
   i 50 cm   ii 55 cm   iii 60 cm

b The head teacher wants to increase all skirt lengths by a fifth. Work out the new lengths.

c Which of the proposals represents a proportional change? Explain your answer.

Level 7

7c I can recognise when data is in proportion

7b I can identify linear data that is directly proportional

2 Most school uniform prices have increased by 15% over the last three years. Which of these items have not increased by this proportion?

	Blouse	Shirt	Tie	Trousers	Skirt	Jumper	PE shirt
Old price	£4.80	£5.00	£2.60	£11.50	£12.00	£9.60	£6.00
Current price	£5.20	£5.75	£2.99	£13.00	£13.80	£11.50	£6.90

3 Which of these are not always in direct proportion? Give full reasons for your answers.

a the age of a person and their height
b the price in dollars and the price in pounds
c the distance run and the time taken
d the mass of a gold bar and its volume
e the time taken and distance travelled by the second hand on a clock face

Did you know? Most people's bodies have roughly the same proportions. A person's height is usually 7 to $7\frac{1}{2}$ times the height of their head.

constant   direct proportion   inverse proportion   proportionality

4 In which of these tables is y proportional to x^2?

a

x	1	2	3	4	5
y	2	8	18	32	50
x^2					
$\frac{y}{x^2}$					

b

x	1	2	3	4	5
y	2	5	10	17	26
x^2					
$\frac{y}{x^2}$					

c

x	1	2	3	4	5
y	10	40	90	160	250
x^2					
$\frac{y}{x^2}$					

d

x	1	2	3	4	5
y	2	4	6	8	10
x^2					
$\frac{y}{x^2}$					

Level 7

7a I can identify data that is proportional to the square of a variable

5 In each coordinate pair (a, b), b is proportional to the inverse of a, that is $b \propto \frac{1}{a}$.

a Use these coordinate pairs to work out a formula that expresses b in terms of a.
$(3, \frac{2}{3})$, $(4, \frac{1}{2})$, $(7, \frac{2}{7})$, $(5, \frac{2}{5})$

b Copy and complete these coordinate pairs using your formula from part **a**.
(11, ______), (6, ______), (______, $\frac{2}{23}$), (______, $\frac{1}{5}$), (30, ______)

c Which of these coordinate pairs do not fit your rule from part **a**?
A $(8, \frac{1}{4})$ B $(9, \frac{1}{3})$ C $(12, \frac{1}{6})$ D $(14, \frac{2}{7})$ E $(15, \frac{2}{15})$ F $(16, \frac{1}{8})$ G $(20, \frac{1}{10})$ H $(21, \frac{1}{7})$

Level 8

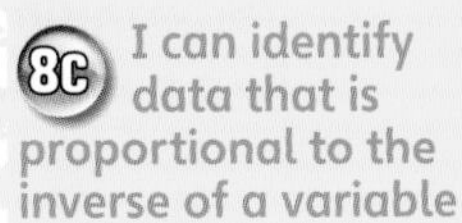

8c I can identify data that is proportional to the inverse of a variable

6 Choose one of the descriptions of proportionality to match each pair of variables in the table.

$y \propto x$ $y \propto x^2$ $y \propto \frac{1}{x}$ not proportional

	x	y
a	radius of a circle	area of a circle
b	length in centimetres	length in inches
c	diameter of a circle	circumference of a circle
d	height above sea level	distance from the top of the mountain
e	number of equal sectors in a pie chart	angle for each sector
f	width of a square patio	number of paving slabs needed

8b I can understand and use proportionality notation

Now try this!

A Reversal

Use two digits to make two different numbers, e.g. 18 and 81. Calculate one number as a proportion of the other, e.g. $81 \div 18 = 4\frac{1}{2}$. How many pairs of numbers can you find where one is $1\frac{3}{4}$ times bigger than the other?

B Conversion graphs

Work with a partner. Find out the exact exchange rate for the pound (£) against the euro (€). Use this to draw a conversion graph for any amount in euros up to €20. Choose an amount in euros and ask your partner to estimate the value in pounds and pence. Use your graph to check their conversion – any conversion within 10p of the exact amount scores 1 point.

proportional to ($\propto$) reciprocal square variable

3.6 Ratio and proportion

- Divide a quantity into two or more parts
- Simplify ratios
- Compare ratios by changing them to the form 1 : m or m : 1
- Find the area and volume of a shape after enlargement

What's the BIG idea?

- To divide a quantity into two or more parts in a given **ratio** (e.g. £300 in the ratio 2 : 3), divide the quantity by the total number of ratio parts (£300 ÷ (2 + 3) = £60), and multiply the result by each part of the ratio (£60 × 2 = £120 and £60 × 3 = £180). Level 6
- You can simplify a ratio expressed as fractions, or as decimals, by converting it to a whole-number ratio.

  For example, ×4 (0.75 : 2 → 3 : 8) ×4 Level 6
- You can **compare** ratios by changing them into the form 1 : m or m : 1.

  For example, ÷4 (28 : 4 → 4 : 1) ÷4 ÷3 (9 : 3 → 3 : 1) ÷3 Level 6
- You can find the **area** of a shape after an **enlargement** by multiplying the original area by the **square** of the **scale factor**. Level 8
- You can find the **volume** of a shape after an enlargement by multiplying the original volume by the **cube** of the scale factor. Level 8

Why learn this?

Using ratio is a good way to compare costs of similar items and identify the best value for money.

Practice, practice, practice!

1 Shortbread can be made with flour, sugar and butter in the ratio 6 : 2 : 4.
Susie wants to make 1.5 kg of shortbread for a party.
What quantities of each ingredient does she need?

Tip

When dividing a quantity into two or more parts, check that the sum of all the parts equals the original quantity.

2 Sherbet Dippy is sold in different size packets.

Size A: 20 g for £0.75 Size B: 80 g for £1.40 Size C: 60 g for £0.24

For each size, work out the ratio of weight (in grams) to price (in pounds).
Give each ratio in its simplest form. Which size is the best value for money?

3 Simplify these ratios.

a $\frac{1}{4} : 2$ **b** $5 : \frac{2}{3}$ **c** $\frac{5}{8} : 4$ **d** $6 : \frac{3}{7}$

Super fact!

In a recent survey carried out by RoSPA, 1 in every 40 car drivers was talking on a mobile phone whilst driving.

4 *Shop for a Mobile* has two price packages.

Package A: 500 text messages for £25
Package B: 4p per text message

a For each package, write a ratio to show the number of text messages to price (in pounds).
b Express each ratio in the form 1 : m. Which package gives the better value?
c Rewrite each ratio in the form m : 1. What does the m stand for in this ratio?

Level 6

6c I can divide a quantity into more than two parts in a given ratio

6b I can simplify a ratio expressed in decimal form

6b I can simplify a ratio expressed in fraction or decimal form

6a I can compare ratios by changing them into the form 1 : m or m : 1

area compare cube enlargement proportion ratio scale factor

5 Triangles *ABC* and *DEF* are mathematically similar. The ratio of *AB* to *DE* is 1 : 4.

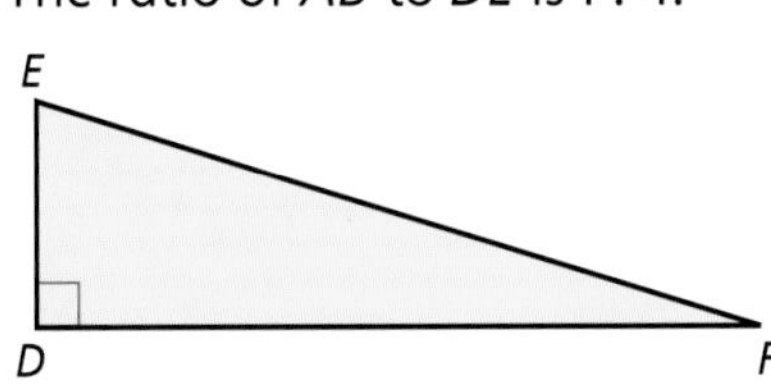

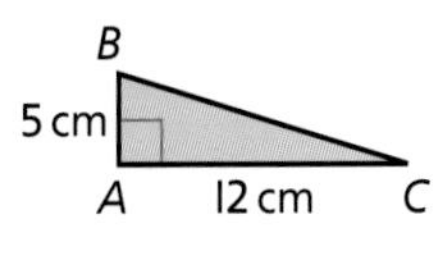

a How long is *DE*?

b What is the area of triangle *ABC*?

c Multiply the area of triangle *ABC* by 4^2.

d Find the length *DF* and use it to calculate the area of triangle *DEF*.

e What do you notice about your answers to parts c and d?

Learn this

Mathematically similar figures are exactly the same shape, but can be different in size. Corresponding sides are in the same ratio and corresponding angles are equal.

8b I can understand the effect on area of enlarging the lengths of a shape

6 The area of an enlarged triangle is 448 cm^2. Ollie has forgotten what the scale factor of enlargement was but he remembers that it was a whole number and the area of the original triangle was equal to a prime number.

a What was the area of the original triangle?

b Use the scale factor of enlargement to write the ratio:
base of the original triangle : base of enlarged triangle

8b I can use ratio to solve area problems and find the scale factor of enlargement

7 The volume of a cuboid is 60 cm^3. The cuboid is 4 cm long and 3 cm wide.

a How high is the cuboid?

b What is the surface area of the cuboid?

c The cube is enlarged using a scale factor of 3. What is the surface area of the enlarged cuboid?

8a I can use ratio to solve area problems

8 The ratio of the heights of two similar cones is 1 : 3.

a What is the ratio of the areas of the bases?

b What is the ratio of the volumes?

c The volume of the smaller cone is 45 cm^3. What is the volume of the larger cone?

8a I can understand the effect on volume of enlarging the lengths of a solid shape

Now try this!

A Scale models

Using a ratio of 1 : 8, write a specification for a scale model of a simple shape, like a cupboard, desk or filing cabinet. Measure the height, width and length of your chosen object. Write a description of the scale model, including information on lengths, area and volume.

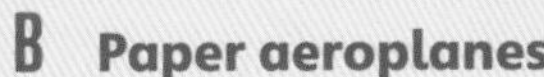

B Paper aeroplanes

The glide ratio is the ratio of the distance an aircraft moves forward to the loss in altitude. Make paper aeroplanes with a glide ratio greater than 5 : 1.

Super fact!

The longest recorded flight by a paper aeroplane indoors is 27.6 seconds.

similar simplest form simplify square unitary method volume

3.7 Brain power

- Do mental calculations with squares, square roots, cubes and cube roots
- Know and use the order of operations
- Simplify expressions containing powers and surds

Why learn this?

Mental calculations keep your brain active and build reserves of healthy brain cells.

What's the BIG idea?

- You can use the **square numbers** you know to calculate the squares of smaller and larger numbers.
  For example, $30^2 = 3^2 \times 10^2 = 900$ and $0.3^2 = 3^2 \div 10^2 = 0.09$ **Level 6**
- If a number has a pair of **factors** which are square numbers, you can use this fact to calculate the **square root** of that number.
  For example, $441 = 9 \times 49$, so $\sqrt{441} = \sqrt{9} \times \sqrt{49} = 3 \times 7 = 21$ **Level 6**
- You can use the **cube roots** you know to estimate other cube roots.
  For example, $\sqrt[3]{80}$ lies between 4 and 5 because $\sqrt[3]{64} = 4$ and $\sqrt[3]{125} = 5$. **Level 6**
- **Expressions** containing powers can sometimes be simplified to make calculations simpler.
  For example,
  $\frac{(8 \times 6)^2}{8 \times 4} = \frac{\cancel{8} \times 6 \times \cancel{8}^2 \times 6}{\cancel{8} \times \cancel{4}} = 6 \times 2 \times 6 = 72$ **Level 7**
- With negative numbers you may need to use brackets to make calculations clear. **Level 7**
  For example, $-6^2 + 74 = -36 + 74 = 38$, but $(-6)^2 + 74 = 36 + 74 = 110$ **Level 7**
- A **surd** is a square root that cannot be simplified further. $\sqrt{3}$ is a surd.
  $\sqrt{8}$ is not a surd because $\sqrt{8} = \sqrt{4 \times 2} = \sqrt{4} \times \sqrt{2} = 2\sqrt{2}$. **Level 8**
- An answer in surd form is exact. It is more accurate than an answer rounded to two decimal places. **Level 8**

Learn this

Order of operations
Brackets
Indices
Division and **M**ultiplication
Addition and **S**ubtraction

Practice, practice, practice!

Level 6

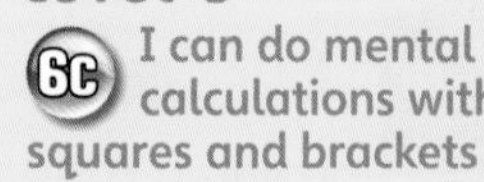

6C I can do mental calculations with squares and brackets

1 Use a mental method to evaluate each pair of expressions. Then insert the correct inequality sign between them.

$3^2 + 5^2 - 1$ ☐ 2×4^2 $3^2 + 5^2 - 1 = 33$ and $2 \times 4^2 = 32$ so $3^2 + 5^2 - 1 > 2 \times 4^2$

a $(2 + 5 \times 2)^2$ ☐ $10^2 + 6 \times 8$ **b** $2 + 2^2$ ☐ $(5 - 4 + 2)^2$

c $3^2 + 8 \times 5$ ☐ $2 + 3^2 \times 2^2$ **d** $65 - 15 \div 5$ ☐ $6 + 4^2 - 5$

6C I can use the square numbers I know to calculate the squares of decimals and large numbers

2 The areas of the different size square tiles in a matching set are

Size 1: 400 cm^2 Size 2: 0.09 m^2 Size 5: 640 000 mm^2

a What is the width of each tile, in millimetres?

b Size 3 tiles are 0.4 m wide. What is their area in square metres?

c Size 4 tiles have an area that lies between 3000 cm^2 and 4000 cm^2. What is the width of the tile? Give your answer in millimetres.

brackets   cube root   expression   factor   factorise   order of operations

3 a Use a written method to calculate

i 36×49 ii 25×81 iii 16×121 iv 64×9

b Use your answers from part a to find

i $\sqrt{1936}$ ii $\sqrt{2025}$ iii $\sqrt{1764}$ iv $\sqrt{576}$

Tip
Factorising means breaking up a whole number into two (or more) numbers which can be multiplied together to make the original number.

4 a Estimate the height of each of these cubes to 1 d.p.

b Explain how you can use the cube numbers you know to make an estimate.

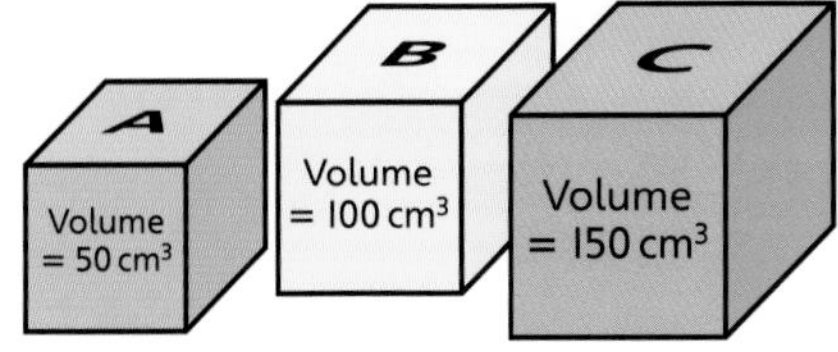

c Another cube has a volume of 1728 cm^3. Show how you can use the fact that $1728 = 27 \times 64$ to find the height of the cube.

5 a Four pupils have made mistakes in their calculations.

i $3 \times 4^2 + 8 - 2 = 70$ ii $\frac{5}{6} \times \sqrt{9} \times 4 + 6 = 15$

iii $6 + 2 \times (\frac{1}{2})^3 + 10 = 11$ iv $\sqrt[3]{64} + 13 \times 2 = 34$

Use a mental method to work out the correct answers.

b Add brackets to make each statement in part a correct.

6 Evaluate each expression when $a = 3$ and $b = -2$.

a $3a^2 - 2b$ b $3(a^2 - 2b)$ c $3(a - 2b)^2$ d $(3a - 2b)^2$

Level 6

6b I can find square roots by factorising

6a I can use the cubes and cube roots I know to solve problems

6a I can do mental calculations with brackets, cubes and square roots

6a I can work out which part of an expression to raise to a power

7 Work out these. Simplify expressions wherever possible.

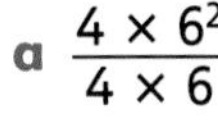

a $\dfrac{4 \times 6^2}{4 \times 6}$ b $\dfrac{2 + (-2)^2 - 2^2}{2^2}$ c $\dfrac{(5 \times 8)^2}{8 \times 5}$

d $\dfrac{6(-3)^3}{-2 \times 3}$ e $\left(\dfrac{6 - 4}{-2 \times 2}\right)^3$ f $\dfrac{(103 - 5^2 \times 4)^3}{(103 - 5^2 \times 4)^2}$

Watch out!
First simplify if you can. Then evaluate the numerator and denominator separately. Simplify again.

Level 7

7b I can work with expressions involving powers and brackets in the form of a fraction

8 Give an exact answer to each of these. Simplify your answer where possible.

a What is the length of the side of a square with an area of 7 cm^2?

b What is the perimeter of an equilateral triangle with a base of $\sqrt{2}$ cm?

c What is the area of a rectangle with length $\sqrt{3}$ cm and height $\sqrt{12}$ cm?

d What is the perimeter of a square with area 5 cm^2?

e What is the area of a triangle with height 8 cm and base $\sqrt{3}$ cm?

Level 8

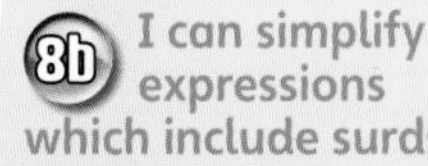

8b I can simplify expressions which include surds

Tip

Now try this!

A Bracket bother

Work with a partner. Each devise five calculations that include a set of brackets and work out the answers. Write out the calculations and answers but leave out all the brackets. Swap calculations and try to place the missing brackets in each other's calculations. If you disagree, check your answers with a calculator.

B Express yourself

Work with a partner. Each make four pairs of expressions that include brackets, powers and roots. Swap sets and try to pair up each other's expressions. First to finish wins.

power square number square root squared surd

3.8 More calculation strategies

- Use known facts to derive unknown facts
- Recognise equivalent fractions, decimals and percentages
- Do mental calculations with fractions, decimals and percentages
- Do mental calculations with powers and roots
- Use mental methods to solve word problems

Why learn this?

There are many situations, like shopping, when you need to be able to work out how good a deal is – without your calculator!

What's the BIG idea?

- You can use the **equivalence** of fractions, decimals and **percentages** to make calculations easier.
  For example, $\frac{1}{5} = 0.2$ and $\frac{1}{4} = 0.25$, so $\frac{1}{5} + \frac{1}{4} = 0.2 + 0.25 = 0.45$ Level 6
- To find a fraction of an amount, divide by the **denominator** and multiply by the **numerator**.
  For example, $\frac{4}{5}$ of $320 = 320 \div 5 \times 4 = 64 \times 4 = 256$ Level 6
- To find 10% of an amount, divide by 10. You can use this to find other percentages.
  For example, $40\% = 10\% \times 4$ and $5\% = 10\% \div 2$ Level 6
- You can calculate with percentages ending in 0.5 by **doubling** and **halving**.
  For example, 37.5% of 48 = 75% of 24 = 18 Level 6
- Identify the **strategy** and the **calculation** needed to solve a word problem.
  Problem: The square dance area in a club is covered with 32 wooden blocks, each measuring 0.75 m × 1.5 m. What is the length of brass needed to edge the dance area?
  Strategy: Find the total area, find the length of a side, and then calculate the perimeter. Level 6

Learn this

$\frac{1}{4} = 0.25 = 25\%$

$\frac{1}{2} = 0.5 = 50\%$

$\frac{3}{4} = 0.75 = 75\%$

$\frac{1}{5} = 0.2 = 20\%$

$\frac{1}{10} = 0.1 = 10\%$

$\frac{1}{100} = 0.01 = 1\%$

$\frac{1}{3} = 0.\dot{3} = 33\frac{1}{3}\%$

$\frac{1}{6} = 0.1\dot{6} = 16\frac{2}{3}\%$

Practice, practice, practice!

1 Work out the decimal equivalent of each fraction.

$\frac{5}{100}$ $\frac{1}{100} = 0.01$, so $\frac{5}{100} = 0.01 \times 5 = 0.05$

$\frac{1}{8}$ $\frac{1}{4} = 0.25$, so $\frac{1}{8} = 0.25 \div 2 = 0.125$

a $\frac{3}{20}$ **b** $\frac{3}{40}$ **c** $\frac{7}{20}$ **d** $\frac{3}{8}$ **e** $\frac{17}{200}$

f $\frac{4}{5}$ **g** $\frac{3}{25}$ **h** $\frac{11}{20}$ **i** $\frac{1}{50}$ **j** $\frac{7}{5}$

Watch out!

$\frac{1}{4} > \frac{1}{8}$

$\frac{1}{100} > \frac{1}{200}$

$\frac{1}{50} > \frac{1}{100}$

2 **a** What is 44% as a fraction? **b** Express 0.445 as a percentage.

3 Use an equivalent decimal to work out these.

$\frac{2}{5}$ of 12 $= 0.4 \times 12 = 4 \times 12 \div 10 = 4.8$

a $\frac{7}{10}$ of 12 **b** $\frac{4}{5}$ of 9 **c** $\frac{3}{20}$ of 4

d $\frac{3}{5}$ of £12 **e** $\frac{3}{10}$ of 24 m **f** $\frac{3}{25}$ of 4 kg

Super fact!

$\frac{17}{29}$ of $\frac{19}{23}$ is the same as $\frac{17}{23}$ of $\frac{19}{29}$. Can you see why?

4 Use an equivalent fraction to work out these.

$0.75 \times 680 = \frac{3}{4}$ of $680 = 680 \div 4 \times 3 = 170 \times 3 = 510$

a 0.6 × 25 **b** 0.15 × 60 **c** 0.05 × 840

d 0.4 × 350 **e** 0.7 × 210 **f** 0.125 × 88

Level 6

6c I can use equivalent fractions and decimals to calculate others mentally

6c I can use equivalent percentages and decimals to calculate others mentally

6b I can use equivalent decimals to find a fraction of an amount

6b I can use equivalent fractions to find a decimal of an amount

calculation convert denominator double equivalence

5 The table shows the annual interest rates on loans taken out in different years. Copy and complete it to show the cost of borrowing £800 for one year.

	1975	1980	1985	1990	1995	2000	2005
Rate	12.5%	17%	14.5%	16%	8.5%	8%	6.5%
Cost of borrowing £800	£100						

6 At an exchange rate of €1.12 = £1, you get €112 for every £100. How many euros do you get for these amounts and exchange rates?

a £50 at €1.5 = £1
b £30 at €1.2 = £1
c £70 at €1.4 = £1
d £200 at €1.35 = £1

7 a Which is the best deal? Why?
Deal A: $\frac{1}{3}$ off a price that has already been reduced by 10%.
Deal B: 20% off a price that already had 20% off
Deal C: 20% off a price that has already been reduced by a quarter.
Deal D: 35% reduction.

b Calculate the new price of a £2400 LCD television with each offer.

8 A train travelling at 125 mph travels about 55 metres in a second. How many metres would the train move in a second at each of these speeds?

a 100 mph
b 150 mph
c 75 mph
d 80 mph

9 A red, a green and a blue cube are used to build a tower. The red cube has a volume of $0.027\,m^3$, the green one $0.125\,m^3$ and the blue one $0.064\,m^3$.

a How tall is the tower?
b What is the surface area of each cube in square metres?
c The surface area of a yellow cube is a quarter of the surface area of the green cube. What is the height of the yellow cube?

Level 6

6b I can use equivalent fractions, decimals and percentages to find a percentage of an amount mentally

6a I can use the strategies of equivalence and conversion to solve problems mentally

6a I can do mental calculations with powers and roots

Now try this!

A Speed maths

Work with a partner. Your teacher will give you two piles of cards. Shuffle and deal a card from the 'percentage/fraction/decimal' pile and then a card from the 'amount' pile. The first person to correctly call out that percentage/fraction/decimal of the amount, scores 2 points. Incorrect calls lose 1 point. First to 5 points wins.

B Four parts make one

This 20 by 20 square has been divided into four sections to show that $\frac{1}{4} + \frac{3}{8} + \frac{1}{5} + \frac{7}{40} = 1$.
Choose four different fractions, percentages or decimals that add up to 1 and make a similar diagram to illustrate your calculation.

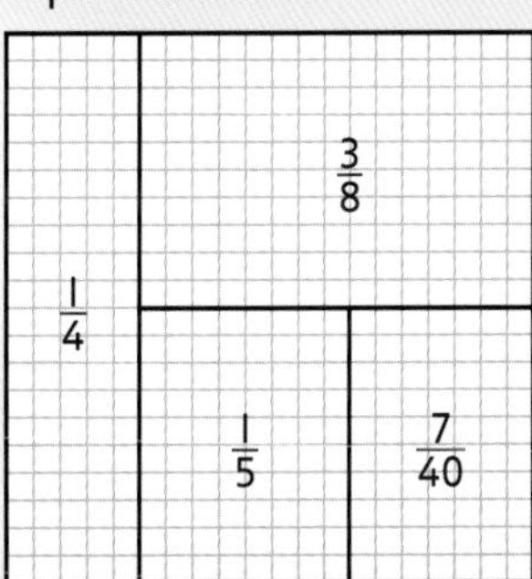

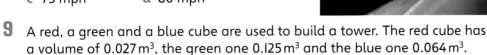

equivalent halve percentage (%) strategy

3.9 Efficient calculation

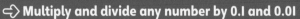

- Multiply and divide any number by 0.1 and 0.01
- Make estimates and approximations
- Multiply and divide a decimal by another decimal
- Carry out mental calculations with fractions and decimals
- Round numbers to one significant figure

What's the BIG idea?

Why learn this?

Being able to make quick calculations without a calculator is an essential skill when you need to make decisions instantly.

- Multiplying by 0.1 or 0.01 is the same as dividing by 10 or 100.
  Dividing by 0.1 or 0.01 is the same as multiplying by 10 or 100. **Level 6**
- You can **estimate** calculations involving decimals by **rounding** each value to the nearest whole number.
  For example, $250 \div 4.9^2 \approx 250 \div 5^2 = 250 \div 25 = 10$ **Level 6**
- You can use **equivalent fractions** to simplify decimal multiplications and divisions.
  For example, $1.6 \times 35 = \frac{8}{5} \times 35 = 8 \times 35 \div 5 = 8 \times 7 = 56$
  $0.042 \div 0.7 = (0.042 \times 1000) \div (0.7 \times 1000) = \frac{42}{700} = \frac{6}{100} = 0.06$ **Level 6**
- To multiply a decimal by a decimal, look for an equivalent calculation.
  For example, $0.35 \times 0.4 = (35 \div 100) \times (4 \div 10) = 35 \times 4 \div 1000 = 140 \div 1000 = 0.14$ **Level 6**
- The first **significant figure** of a number is the first non-zero digit in the number, counting from the left.
  For example, the first significant figure in 25 is 2
  the first significant figure in 0.042 is 4. **Level 7**
- To **round** to one significant figure (1 s.f.), look at the digit to the right of the first significant figure. If the digit is less than 5, round down. If the digit is 5 or more, round up.
  For example, 25 rounded to 1 s.f. is 30 and 0.042 rounded to 1 s.f. is 0.04. **Level 7**
- To calculate an **approximate** answer to a complex calculation, round all the values to 1 s.f.
  For example, $23.89 \times 0.048^2 \approx 20 \times 0.05^2 = 20 \times 0.0025 = 0.05$ **Level 7**

Practice, practice, practice!

Level 6

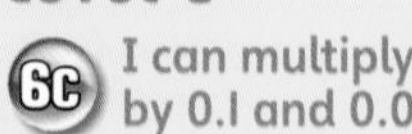

6C I can multiply by 0.1 and 0.01

1 **a** Work out.

i 3.4×0.1 ii 34×0.01 iii 340×0.1 iv 27×0.1

v 270×0.1 vi 27×0.01 vii 270×0.01

b Insert the correct sign, <, > or =, between each pair.

40×0.1 ☐ $45 \div 10$ *$40 \times 0.1 = 4$ and $45 \div 10 = 4.5$*
so $40 \times 0.1 < 45 \div 10$

i 64×0.1 ☐ $6.4 \div 10$ ii 0.71×0.1 ☐ $7.1 \div 10$

iii 53×0.01 ☐ $5.3 \div 10$ iv 0.201×0.01 ☐ 0.1×0.21

approximate approximately equal to (≈) equivalent fraction

2 Greg wants to fence off an area of about 50 m² for a vegetable patch.

a Estimate the area of each plot. Which plot is closest to 50 m²?

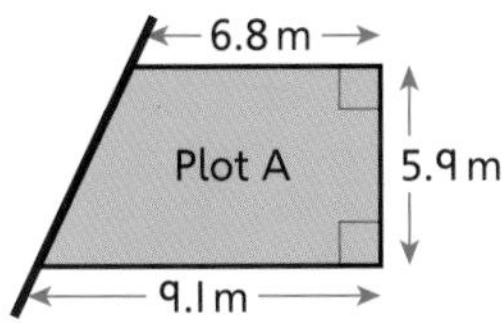

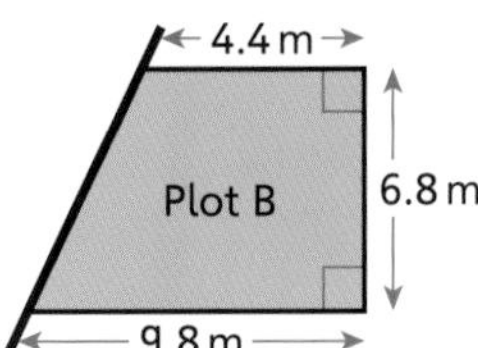

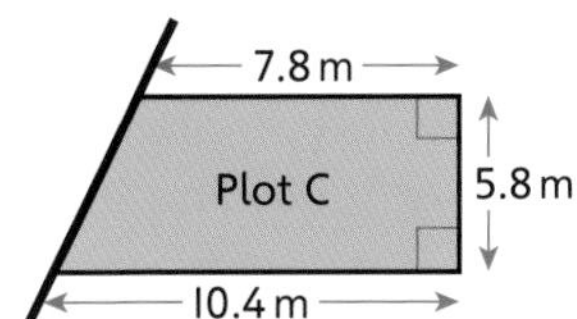

b Use a calculator to calculate the area of each plot, correct to 1 d.p.

c Change a measurement by 0.1 m to make a plot with area 49.0 m² (to 1 d.p.).

3 Copy and complete.

a $50 \div 0.1 =$ ____ **b** $500 \div$ ____ $= 5000$

c ____ $\div 0.01 = 500$ **d** $0.5 \div \frac{1}{10} =$ ____

e ____ $\div \frac{1}{100} = 5500 \div 0.1 =$ ____

Learn this

$\times 0.1 = \div 10$ and $\times 0.01 = \div 100$
$\div 0.1 = \times 10$ and $\div 0.01 = \times 100$

4 Work out these. Use a calculator to check your answers.

a 0.7×0.3 **b** 0.4×0.5 **c** 0.06×0.4

d 0.05×0.06 **e** 0.12×0.05 **f** 0.04×1.3

Tip
When you multiply with decimals, the number of decimal places in the answer is equal to the total number of decimal places in the question.

5 Copy and complete.

a $0.6 \div 0.03 =$ ____ **b** $6.4 \div$ ____ $= 16$

c $3.5 \div$ ____ $= 5$ **d** $0.35 \div 0.05 =$ ____

6 Use equivalent fractions to work out these.

$1.5 \times 2.5 = \frac{3}{2} \times \frac{5}{2} = \frac{15}{4} = 3\frac{3}{4}$

a $1.2 \div 0.16$ **b** 12.5×1.8 **c** 4×4.5^2 **d** $1.2 \div 1.5 \times 5$

Level 6

6c I can estimate answers to calculations involving several steps

Tip

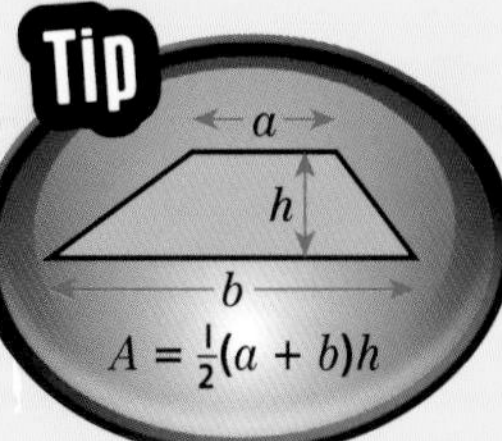

6b I can divide by 0.1 and 0.01

6b I can multiply two decimals mentally

6a I can divide decimals mentally

6a I can use equivalent fractions to simplify multiplication and division calculations

7 Write these numbers to 1 s.f.

a 5.2 **b** 7.8 **c** 0.42 **d** 87 000

e 0.056 **f** 350 **g** 42 768 **h** 0.98

8 **a** Estimate the answers to these calculations by rounding values to s.f..

i $13.4 + 37.71 \times 18.3$ ii 2.05×1.56^2

iii $\sqrt{389} \times 19.375$ iv $\frac{189.9 + 49.9}{46.3 + 8.2}$

b Use a calculator to check your answers.
Which of your estimations in part **a** overestimated the answer?

Level 7

7c I can round numbers to 1 s.f.

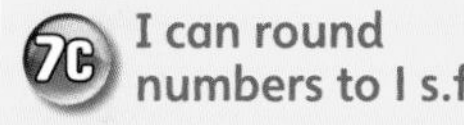

7c I can estimate using rounding to 1 s.f.

Now try this!

A Deal 5

Work with a partner. You will need digit cards 1–9. Shuffle the cards and deal five. Use the five cards to make two numbers like this:

Write down an estimate for the product of the numbers. Use a calculator to check the answer. The closest estimate scores 1 point.

B Three makes six

Work with a partner. You will need digit cards 1–9. Shuffle the cards and deal three. Arrange them to make six different calculations like this:

Calculate the answers. The first person to get all six calculations correct scores 1 point.

estimate product round significant figures

Towards a better lifestyle

Since starting work, Maria has had plenty of money to spend on a great social life, but she is beginning to realise that she smokes and drinks too much. She decides to make an appointment with Tracey, a lifestyle adviser, to discuss her concerns.

In the first interview, Tracey discovers that Maria's eating habits are chaotic and she hasn't done any regular physical exercise for more than two years.

Tracey needs to set up a programme to help Maria get fit.

MAKE MATHS FUNCTIONAL!

1 Currently Maria smokes one pack of 20 cigarettes a day and wants to stop. Which of Tracey's suggestions would help Maria reduce the amount she smokes the most? Explain your choice.

A Only buy six packs of cigarettes a week
B Buy 200 cigarettes and make them last for two weeks
C Smoke 20% fewer cigarettes over a two-week period **Level 6**

2 Maria was advised to reduce the amount of money she spends on junk food by 20% and increase the amount she spends on fruit and vegetables by 250%. Look at Maria's last shopping bill.

a Which food item is roughly 20% of the total bill?

b How much extra would Maria have to spend on fruit and vegetables to meet her target? **Level 6**

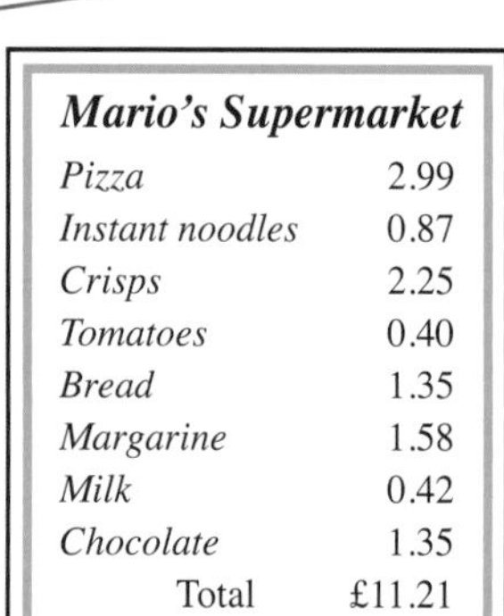

Mario's Supermarket

Pizza	2.99
Instant noodles	0.87
Crisps	2.25
Tomatoes	0.40
Bread	1.35
Margarine	1.58
Milk	0.42
Chocolate	1.35
Total	£11.21

3 Maria was advised to divide up the money she would save by not smoking between eating healthily, getting fit and treating herself, in the ratio 3 : 2 : 1.

a Cigarettes cost £3.99 a pack. Approximately how much money would Maria save in a month if she stopped smoking?

b From the money saved, how much would she be able to spend on each activity? **Level 6**

4 Most prices at Maria's local leisure centre have increased by 5%. Which items have not increased by this proportion?

	Activity				
	Swimming	Aerobics	Pilates	Circuits	Spinning
Old price	£2.80	£3.20	£3.40	£3.70	£3.60
Current price	£2.94	£3.36	£3.74	£3.80	£3.78

5 As a result of the healthy-living programme, Maria's monthly spending on fitness activities has increased by 12%. If she now spends £28 per month, how much did she originally spend? **Level 7**

6 A friend of Maria's tells her that if she reduces the number of cigarettes she smokes by 10% each month, she will have practically stopped smoking after two years.

a Is Maria's friend correct?

b How many times would she have to cut her smoking by 10% to reduce the number of cigarettes she smokes by a half?

c Would smoking one fewer cigarette per day each month for 20 months be a better option? **Level 8**

The BIG ideas

- To express one number as a **percentage** of another, write the number as a fraction of the other, divide the numerator by the denominator and then multiply by 100. Level 6
- To divide a quantity into two or more parts in a given **ratio**, divide the quantity by the total number of the ratio parts, and multiply the result by each part of the ratio. Level 6
- To find the original value after a percentage increase or decrease, find 1% from the new value and multiply that by 100. This is called the **unitary method**. Level 6
- You can also find the original value after a percentage increase or decrease by using a **multiplier** and the **inverse operation**. Level 7
- You can calculate percentage change using a multiplier.
  Increase by 10%: multiplier is 110% $= \frac{110}{100} = 1.1$
  new amount = 1.1 × old amount Level 7
- To calculate an **approximate** answer to a complex calculation, round all the values to one **significant figure**. Level 7
- In a table of values for variables that are in direct proportion, the value of $\frac{y}{x}$ is always the same. Level 7
- **Compound interest** is worked out on the **principal** (the original amount) plus any previous interest already earned. You can calculate compound interest by finding the multiplier and raising it to a power equal to the number of times the interest is added on. Level 8

Find your level

Level 6

Q1 **a** Which of these fractions are larger than $\frac{3}{11}$?

$\frac{3}{10}$ $\frac{5}{11}$ $\frac{3}{14}$ $\frac{1}{4}$ $\frac{7}{15}$

b Which decimal is closest to $\frac{3}{11}$?

0.3 0.25 0.11 0.311

c Write a fraction to complete this statement.

$\frac{3}{11}$ is half of ____

Q2 Add $\frac{4}{10}$ and $\frac{4}{5}$.

Q3 How many sixths are there in $1\frac{1}{3}$?

Level 7

Q4 Work out $5\frac{1}{4} \div \frac{7}{8}$. Show your working.

Q5 Stuart is on a four-day cycling holiday.

Day	Distance cycled (km)
1	45
2	38
3	42

On day 4, he has $\frac{1}{6}$ of his journey left. What is the total length of his cycling journey?

Q6 A hospital monitors the number of patients who miss their appointments.

In 2007, 5702 patients missed their appointment. This was a 40% reduction on the previous year.
How many patients missed their appointment the previous year?

Level 8

Q7 **a** The price of a coat is reduced by 6% in the sales. The original price was £60. Which calculation gives the sale price of the coat?

A: £60 × 1.6 B: £60 × 0.06
C: £60 × 0.94 D: £60 × 1.06

b The coat is further reduced by 2.5%. What is the sale price of the coat after the second reduction?

Q8 Look at this statement.

> A 10% reduction followed by a further 10% reduction is *not* the same as a 20% reduction.

What is the total percentage reduction?

4 Be constructive

This unit is about angles, triangles and constructions.

Stonehenge is the most famous stone circle in the British Isles. It was constructed over a period of more than 1000 years.

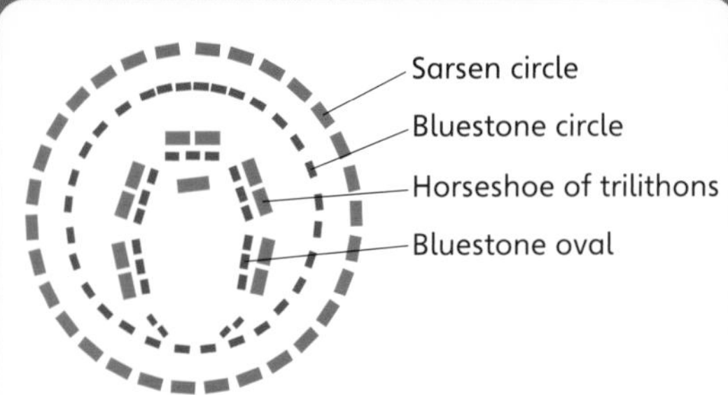

Diagram of Stonehenge 4000 years ago

Approximate date	Constructions
3100 BC	Circular ditch and bank.
2600 BC	Bluestones arranged in double circle at centre of ditch.
2400 BC	• Sarsen circle built – a circle of vertical stones supporting horizontal stones in a circle round the top. • Horseshoe of trilithons built – a trilithon consists of two large vertical stones with a horizontal stone across the top. • Bluestones arranged into positions shown in the diagram.

This structure remained intact for around 500 years before it began to fall into disrepair.

Activities

A How could the first builders at Stonehenge have dug a ditch in a circle? Work out a way of drawing a circle using a length of string, two people and some chalk. Use your method to draw a circle on a suitable surface outside.

B Use compasses to construct a circle with radius 5 cm. Use a ruler to draw a radius. Use a protractor to measure an angle of 36° from this radius. Draw another radius at 36° to the first.

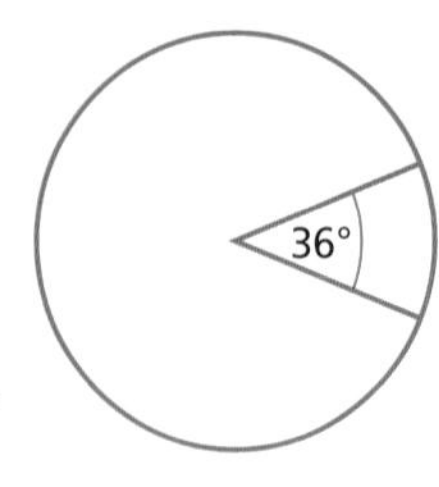

Repeat until you get back to your starting position. Join the points where the radii cut the circle. What shape have you constructed?

Write instructions for constructing a regular nonagon (9 sides). Give them to a partner to follow.

No one knows what Stonehenge was used for. Some theories suggest it may have been built as an astronomical observatory or for rituals linked to the Sun or crop fertility.

Before you start this unit...

1 Work out the missing angles.

Level Up Maths 5-7 page 22

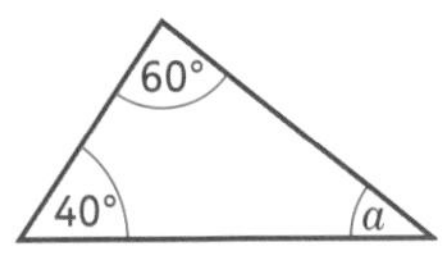

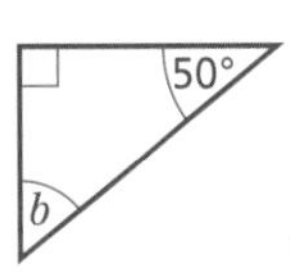

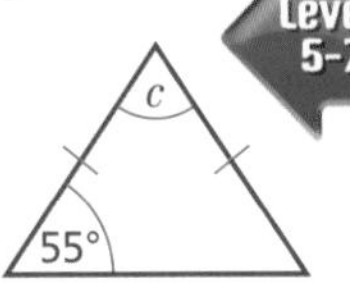

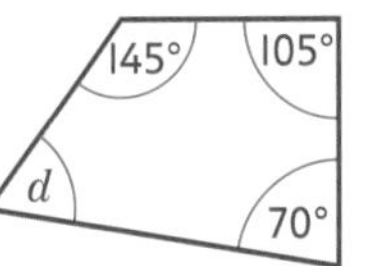

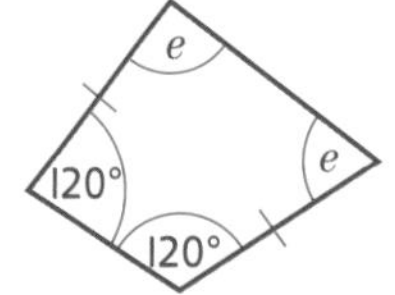

2 **a** Which of these angles are equal to 50°? Explain your reasoning.

Level Up Maths 5-7 page 33

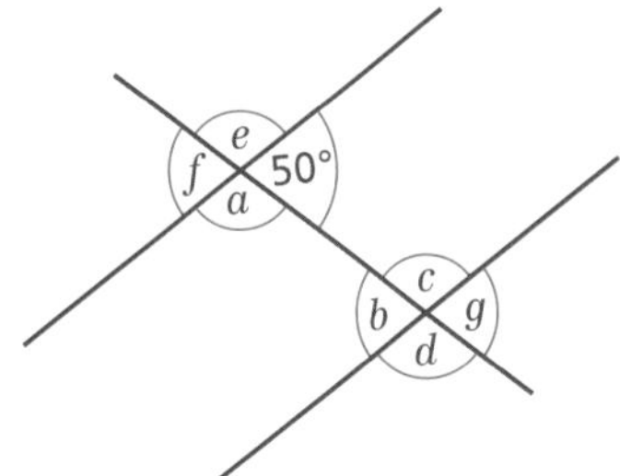

b Work out the missing angles.

- 4.1 Quadrilaterals, angles and proof
- 4.2 Interior and exterior angles in polygons
- 4.3 Angles in triangles and quadrilaterals
- 4.4 Pythagoras' theorem
- 4.5 Using Pythagoras' theorem to solve problems
- 4.6 Constructions
- 4.7 Construction problems
- 4.8 Circles and tangents
- 4.9 Perilous paper round (World's Greatest Maths)
- Unit plenary: Engineering calculations

Plus digital resources

Did you know?

The bluestones were brought all the way from the Preseli Mountains – a distance of about 385 km.

4.1 Quadrilaterals, angles and proof

- Identify alternate and corresponding angles
- Understand proofs that the sum of the angles of a triangle is 180° and of a quadrilateral is 360°
- Understand a proof that the exterior angle of a triangle is equal to the sum of the two interior opposite angles
- Classify quadrilaterals by their geometric properties
- Know the difference between a practical demonstration and proof
- Recognise the difference between conventions, definitions and derived properties

Why learn this?

The shapes of clothes are made up of lines, curves and angles. Pattern makers in the fashion industry need to make sure that the different pieces fit together and the angles match.

What's the BIG idea?

- **Corresponding angles** are equal. Level 6
- **Alternate angles** are equal. Level 6
- The **interior angles** in a triangle sum to 180°. Level 6

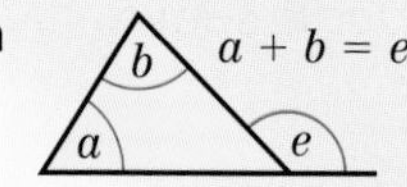

- The interior angles in a **quadrilateral** sum to 360°. Level 6
- The **exterior angle** of a triangle is equal to the sum of the two interior opposite angles. Level 6

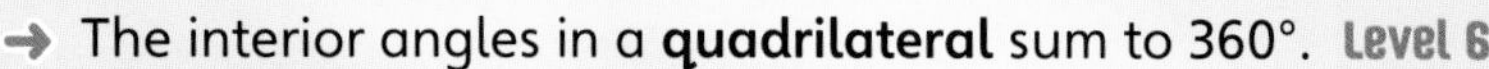

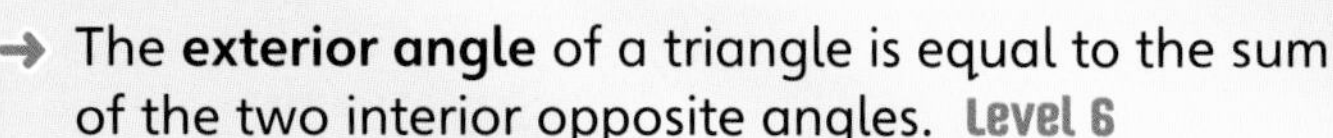

- Quadrilateral properties:

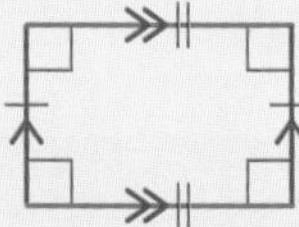
rectangle

square

parallelogram

rhombus

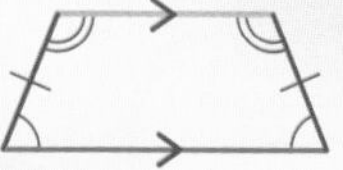
isosceles trapezium

kite

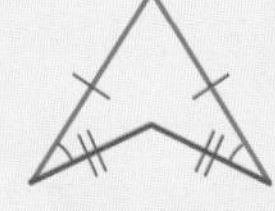
arrowhead

The black lines show diagonals that intersect at 90°. Level 6

- When solving problems using the properties of shapes you must explain your reasoning. Level 7
- A convention is an accepted mathematical way to show some information. A definition is a precise description of something. A derived property is information that can be worked out from a description. Level 7

Did you know?

The word 'isosceles' derives from the Greek word 'isos' (same) and 'skelos' (leg). Isoscles triangles and isosceles trapeziums have two equal sides and one line of symmetry.

Practice, practice, practice!

1 **a** Sketch this isosceles triangle. Reflect the triangle along the side PR. Label the new corner S.

b $\angle QPR = \angle RPS$. Write down the other pairs of angles that are equal.

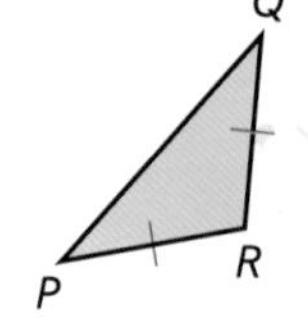

2 In the diagram, which pairs of angles are corresponding angles.

angles a and e.

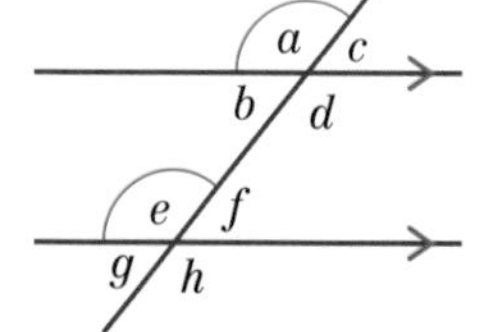

Level 6

6c I can use naming and labelling conventions for lines, angles and shapes

6c I can identify corresponding angles

alternate angles arrowhead corresponding angles exterior angles

3 Calculate the size of the lettered angles. Give reasons for your answers.

a

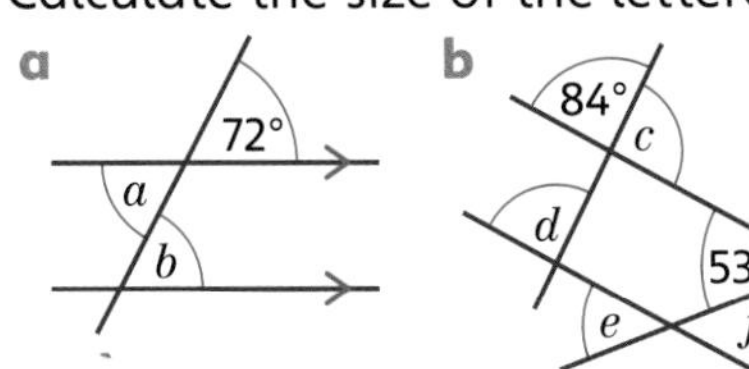

b

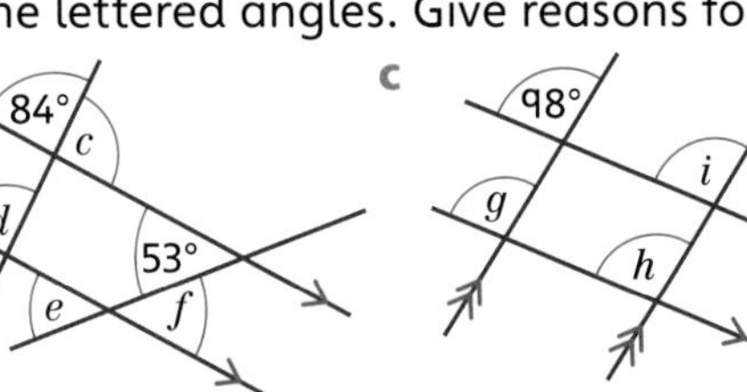

c

Tip Alternate angles form a Z-shape on a pair of parallel lines.

Level 6

6c I can identify alternate and corresponding angles on the same diagram

4 Sketch this diagram. Copy and complete this proof.

a and d and c and e are ____________ angles.
$d + b + e =$ ____________ because they lie on a ____________ . $a =$ ____________ and $c =$ ____________
so $a + b + c =$ ____________ $+ b +$ ____________ .
Therefore angles in a triangle sum to ____________ .

6c I can follow a proof that the sum of the angles of a triangle is 180°

5 Use this diagram. Copy and complete this proof to show that the interior angles of a quadrilateral sum to 360°.
The quadrilateral has been split into two ____________ .
$a + b +$ _____ = 180° because the angles in a triangle sum to _____ . $d + e + f =$ _____ . Therefore $(a + b + c) + (d + e + f) =$ _____ .

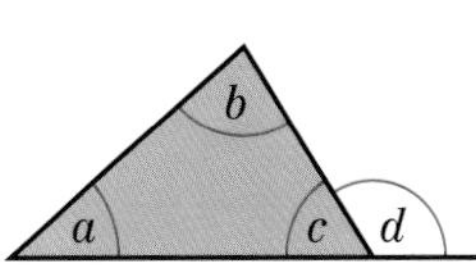

6b I can follow a proof that the sum of the angles of a quadrilateral is 360°

6 Describe three different geometrical properties of each of these quadrilaterals.

a a square
b a rhombus
c an isosceles trapezium
d a kite

6b I can identify properties of quadrilaterals

7 The interior angles of a triangle sum to 180°
so _____ $+ b +$ _____ = _____
Angles on a straight line sum to 180° so _____ $+ d =$ _____ .
Continue the proof to show that $a + b = d$.

6a I can follow a proof that the exterior angle of a triangle is equal to the sum of the two interior opposite angles

Level 7

8 Decide whether each statement is a definition, a convention or a derived property.

Angles on a straight line sum to 180°. *Derived property*
An interior angle is an angle inside a shape. *Definition*

a A quadrilateral has four sides and four interior angles.
b The interior angles in a rhombus sum to 360°.
c Angles are marked with arcs.
d Unknown angles are represented by letters.

7c I can recognise the difference between conventions, definitions and derived properties

9 Decide whether each of these is a practical demonstration or a proof.

a Alison cuts out a triangle. She then tears the corners off and arranges them on a straight line to show that the angles sum to 180°.
b There are 360° around a point. The turn of a straight line is half that around a point. Therefore the angle on a straight line is 180°.

7b I can recognise the difference between a practical demonstration and a proof

Now try this!

A Investigating reflection

Investigate what shapes are made when you reflect different triangles (isosceles, scalene and equilateral) along one of their sides. Investigate the angle and symmetry properties of the shapes that are formed.

B Reflection poster

Make a poster showing and explaining what you found out in Activity A.

interior angles isosceles parallel lines proof quadrilateral

4.2 Interior and exterior angles in polygons

- Calculate interior and exterior angles of triangles, quadrilaterals and regular polygons
- Find the sum of the interior and exterior angles of polygons
- Use the interior and exterior angles of regular and irregular polygons to solve problems

Why learn this?

You can spot polygons in nature. The wax honeycombs that bees make are made up of tessellating hexagons.

What's the BIG idea?

→ The sum of the interior angles in an n-sided polygon is $(n - 2) \times 180°$. **Level 6**

→ The sum of the exterior angles in any polygon is always 360°. **Level 6**

→ The interior angle of a **regular polygon** $= \frac{\text{sum of interior angles}}{\text{number of sides}}$.
In a regular polygon all sides are equal and all angles are equal. **Level 6**

→ You can use interior and exterior angles in polygons to solve problems. **Level 7**

Practice, practice, practice!

Level 6

6C I can calculate interior and exterior angles of triangles and quadrilaterals

1 Calculate the missing angles in these shapes. State any angle facts that you use.

a

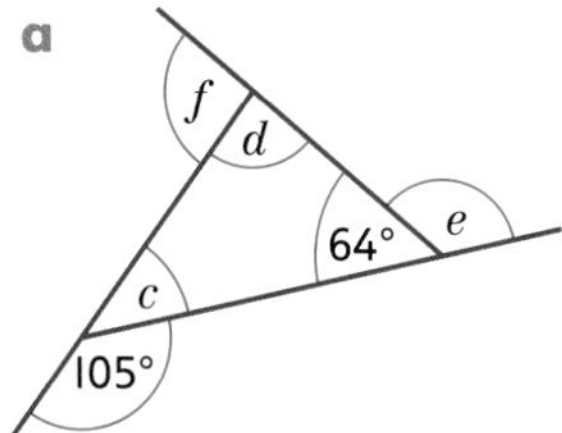

b

c

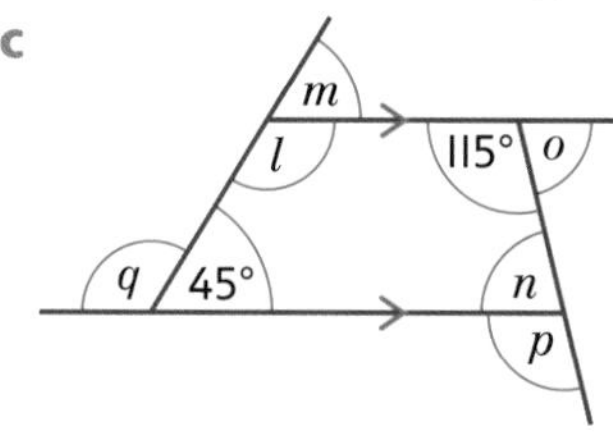

6C I can explain how to find the sums of the interior and exterior angles of triangles, quadrilaterals and pentagons

2 Find the sum of the exterior angles of each of the shapes in Q1. Write down anything that you notice.

3 **a** Draw a pentagon and mark the exterior angles.
b Use a protractor to measure each exterior angle.
c Find the sum of the exterior angles.

4 Explain how to find the sum of the interior angles of a pentagon.

exterior angle hexagon interior angle irregular polygon octagon

5 Copy and complete this table. **Hint:** Imagine drawing diagonals to split each polygon into triangles.

Polygon	Number of sides	Number of triangles	Sum of interior angles	Sum of exterior angles
triangle	3	1	$1 \times 180° = 180°$	
quadrilateral	4	2	$2 \times 180° = 360°$	
pentagon				
hexagon				
heptagon				

6 Explain how to find the sum of the interior angles in an n-sided polygon.

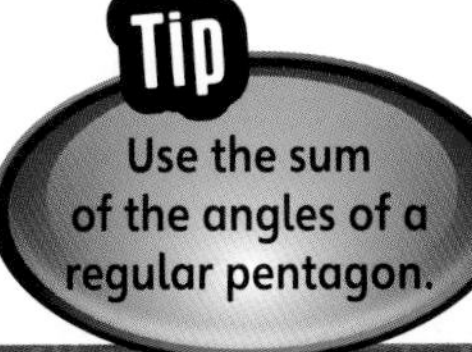

Tip
Look for a relationship between the number of sides and the number of triangles.

7 Explain how to find, in a regular hexagon, the size of

a an interior angle **b** an exterior angle.

8 Calculate the size of an interior and an exterior angle of each of these polygons.

a regular pentagon **b** regular octagon

c regular heptagon **d** regular 14-sided polygon

9 Each interior angle of a regular pentagon is $(2x - 3)°$. Form an equation and solve it to find x.

Tip Use the sum of the angles of a regular pentagon.

Level 6

6b I can calculate the sums of the interior and exterior angles of irregular polygons

6b I can explain how to find the interior and exterior angles of regular polygons

6a I can calculate the interior and exterior angles of regular polygons

10 Each exterior angle of a regular polygon is 30°.

a How many sides does the polygon have?

b Calculate the size of one of the interior angles.

11 Each interior angle of a regular polygon is 160°. How many sides does this polygon have?

12 Is it possible to draw a polygon which has interior angles that sum to 1300°? Explain your reasoning.

13 The diagram shows a regular hexagon with centre B.

a Calculate the sizes of angles x, y and z.

b What type of triangle is ABC?

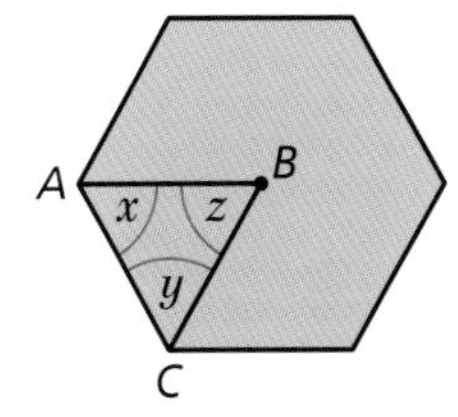

Level 7

7c I can use the interior and exterior angles of regular polygons to solve problems

Now try this!

A Tessellating

A tessellation (or tiling) is made by fitting together shapes in a repeating pattern without any gaps or overlaps.
Equilateral triangles tessellate.
Investigate other regular polygons that tessellate.
Try using more than one type of regular polygon.

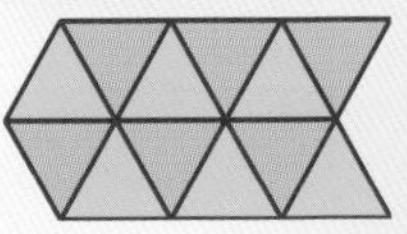

B Drawing polygons

Work with a partner. Think of a regular polygon. Give your partner instructions to draw your polygon using angles and lengths only.
Have they drawn the correct polygon?
Swap roles and repeat.

pentagon quadrilateral regular polygon triangle

4.3 Angles in triangles and quadrilaterals

➪ Solve geometrical problems involving different polygons
➪ Solve simple geometrical problems, showing and explaining reasoning

Why learn this?

Imagine a world without polygons. Polygons can be used to make buildings look more interesting.

What's the BIG idea?

→ **Equilateral triangles** have three equal sides and three equal angles. **Level 6**
→ **Isosceles triangles** have two equal sides and two equal angles. **Level 6**
→ You can use the **properties** of **quadrilaterals** to solve problems. **Level 6 & Level 7**
→ When solving problems involving different shapes, look at the different shapes individually first. **Level 6 & Level 7**
→ It is important to explain your **reasoning** when solving shape problems. **Level 6 & Level 7**

Practice, practice, practice!

Level 6

6c I can solve problems using the properties of special triangles

1 Imagine an equilateral triangle reflected along one of its sides.
- **a** Write down the name of the new shape that is formed.
- **b** What size are the interior angles of the new shape formed?

2 Find the size of the missing angles in these triangles.

a
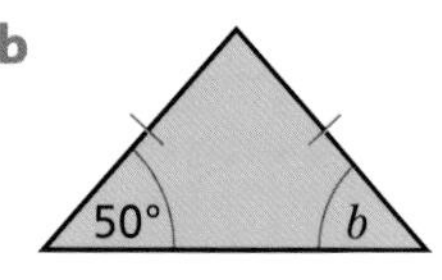

b
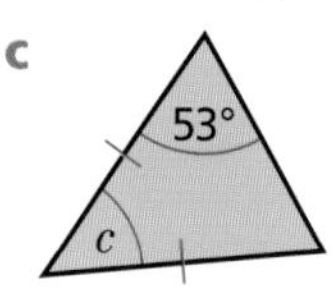

c

Tip For each angle you find, write a reason.

6c I can solve simple problems using properties of quadrilaterals

3 *ABCD* is a rhombus.
$\angle ABC = 48°$.
Calculate $\angle BAD$.

A D 48° B C

6b I can solve problems using the properties of triangles and quadrilaterals

4 Look at this arrowhead.
$\angle CBD = 26°$, $\angle BDC = 104°$
Work out the size of

a $\angle ADB$ **b** $\angle BCD$ **c** $\angle BAD$

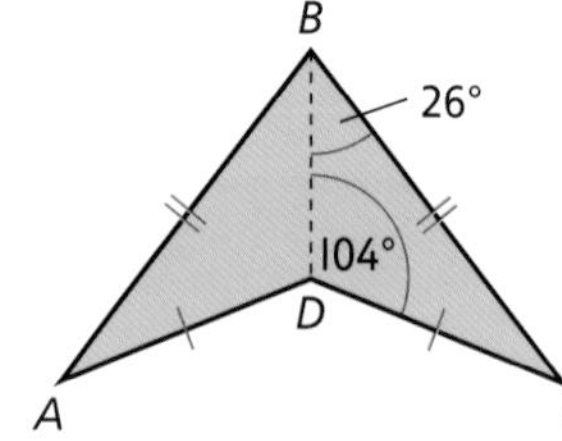

6a I can use reasoning to solve geometrical problems

5 Here is a parallelogram.
- **a** Calculate
  - **i** $\angle BAC$ **ii** $\angle ABC$
- **b** For each calculation in **a**, explain your answer.

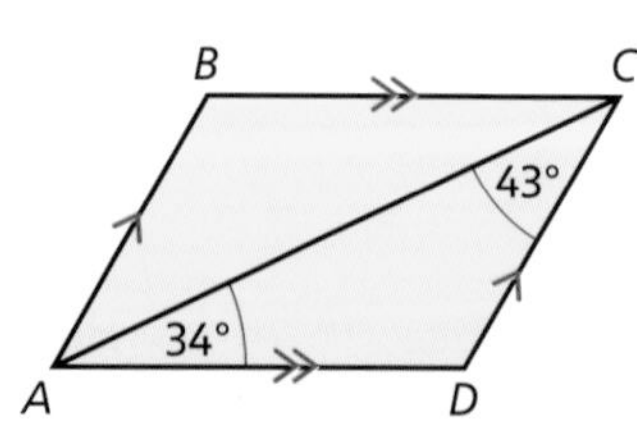

arrowhead equilateral triangle isosceles triangle parallelogram

6 Calculate ∠USW in this rectangle. Show your steps for solving this problem and explain your reasoning for each step.

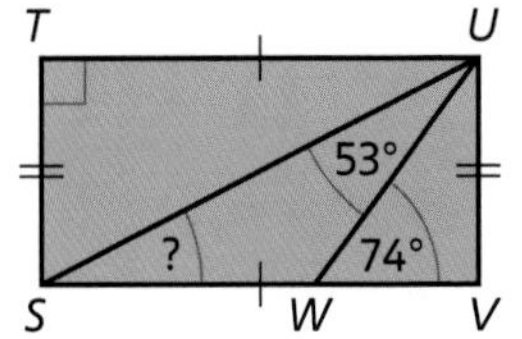

Level 6

6a I can solve simple geometrical problems

7 Work out the sizes of these angles. Explain your reasoning.

a ∠BAF **b** ∠ABE

c ∠CBE **d** ∠BCD

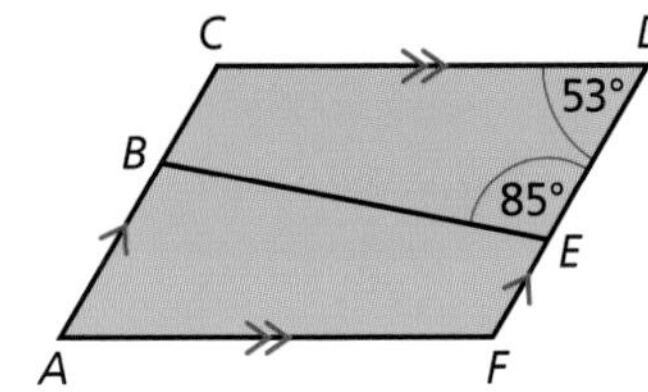

Level 7

7c I can solve more complex geometrical problems

8 In this shape, ∠TSV = 42° and ∠TUV = 36°.

Calculate

a ∠STV **b** ∠VTU.

Explain your reasoning.

9 This equilateral triangle and isosceles triangle fit together to make another triangle.

Work out the size of the angles in the new triangle.

Explain your reasoning

7b I can justify and explain my reasoning with diagrams and text

10 Calculate the value of x in this diagram. Explain your reasoning.

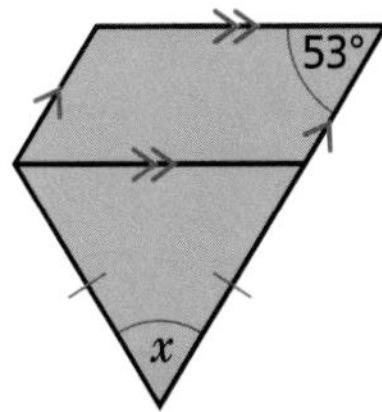

Now try this!

A Missing angles 1

Work with a partner.

1 Each draw a right-angled triangle or an isosceles triangle.
2 Label one of the angles with its size.
3 Challenge your partner to work out the size of the missing angle or angles.
4 Check that they are correct by using a calculator to work out the sum of all the interior angles in the triangle.

B Missing angles 2

Work with a partner.

1 Each draw a parallelogram.
2 Draw in one of the diagonals.
3 Label two of the angles with their sizes.
4 Challenge your partner to work out the sizes of the missing angles.
5 Check that they are correct by using a calculator to work out the sum of all the interior angles in the parallelogram.

Learn this

Angles in a triangle sum to 180°, and angles in a quadrilateral sum to 360°.

properties quadrilateral reasoning rhombus right-angled triangle

4.4 Pythagoras' theorem

- Be able to identify the hypotenuse
- Know the formula for Pythagoras' theorem and how to use it
- Calculate the length of a line segment *AB* given the coordinates of points *A* and *B*

Why learn this?

You can use Pythagoras' theorem to check that two walls are at right angles to each other.

What's the BIG idea?

→ The **hypotenuse** is the longest side of a **right-angled triangle**. It is the side opposite the right angle. **Level 6**

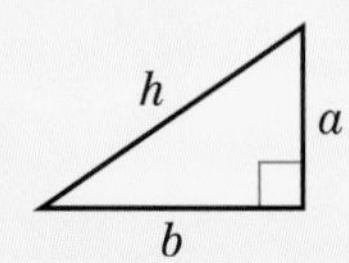

→ **Pythagoras' theorem** is $h^2 = a^2 + b^2$, where h is the length of the hypotenuse and a and b are the lengths of the shorter sides. **Level 7**

→ Pythagoras' theorem only works for right-angled triangles. **Level 7**

→ You can find the distance between two points using Pythagoras' theorem. **Level 7**

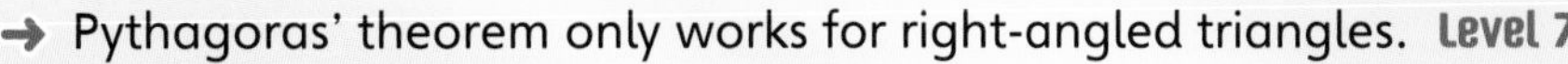

Practice, practice, practice!

Level 6

6C I can identify the hypotenuse in a right-angled triangle

1 These triangles are all right-angled triangles.

a 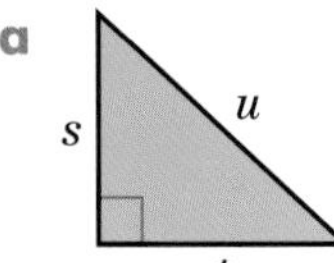

b 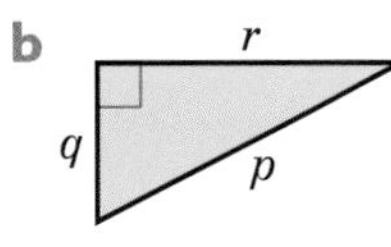

c 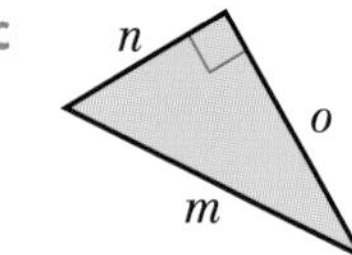

d 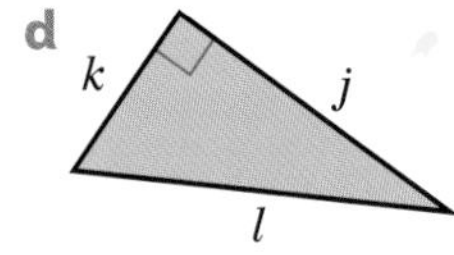

Write down the letter that marks the hypotenuse of each triangle.

Level 7

7C I can investigate Pythagoras' theorem in different ways

2 **a** Using a protractor and a ruler draw each of these right-angled triangles. The lengths are for the two shorter sides.

i 3 cm, 4 cm **ii** 5 cm, 12 cm **iii** 6 cm, 8 cm **iv** 2.5 cm, 6 cm

b Label the shorter sides a and b, and measure the hypotenuse, h.

c Using a calculator, work out a^2, b^2 and h^2.

d Work out $a^2 + b^2$ for each of the triangles. Write down any patterns that you notice.

7C I can use Pythagoras' theorem to find the length of the hypotenuse

3 Calculate the missing lengths in these right-angled triangles. Give your answers to 1 d.p.

a 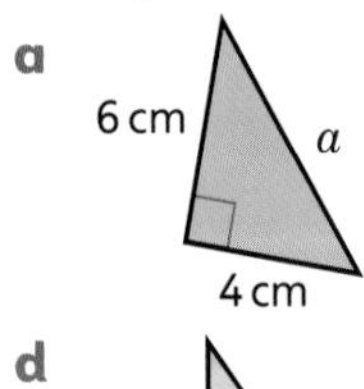

b 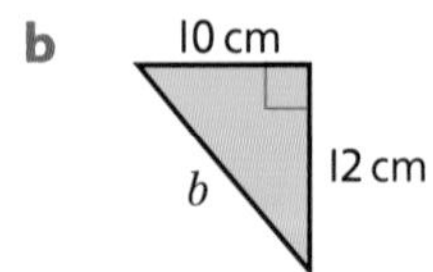

c 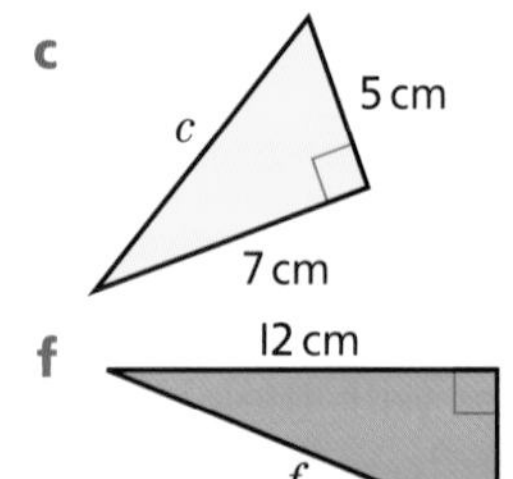

d

e 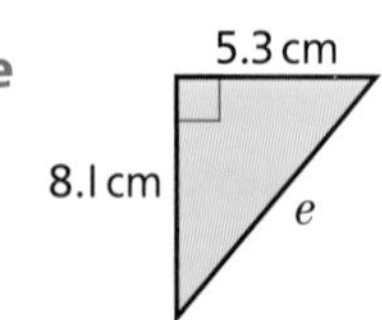

f

Watch out!

Don't forget to find the square root when finding the length of the side.

coordinates hypotenuse Pythagoras' theorem

4 Calculate the length labelled x in this isosceles triangle.

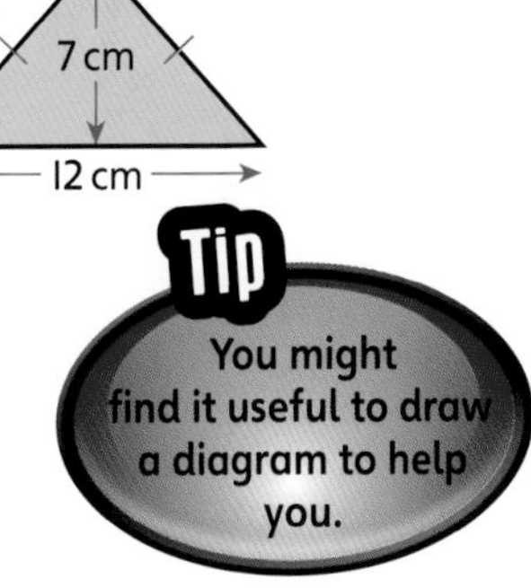

Did you know?

Pythagoras was a Greek mathematician and philosopher who was born on the island of Samos about 580 BC.

Tip

You might find it useful to draw a diagram to help you.

5 A football pitch is 85 m long and 65 m wide. A player runs along one of the diagonals. How far does he run?

6 When Jamie travels from home to school, he has to walk 0.8 km east and 1.3 km north.

a Draw a diagram to show the route Jamie takes to school.

b Find the shortest distance from Jamie's house to his school.

7 Billy is making a kite out of wooden rods. He has made a cross-shape inner frame. Calculate the total length of rod he will need for the perimeter of the kite.

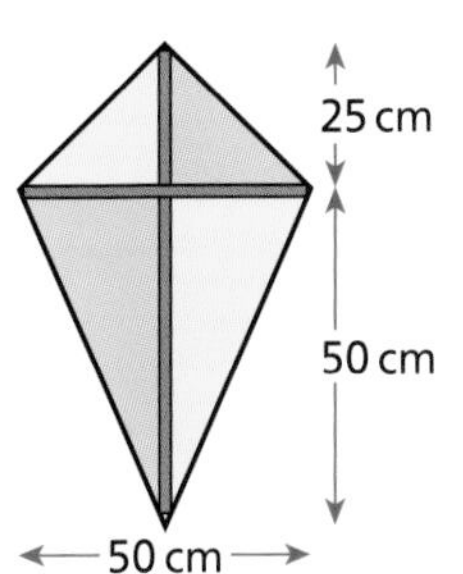

8 The coordinates of A are (2, 1). The coordinates of B are (5, 3).

a Write down the horizontal distance, a.

b Write down the vertical distance, b.

c Use Pythagoras' theorem to find the straight-line distance between A and B.

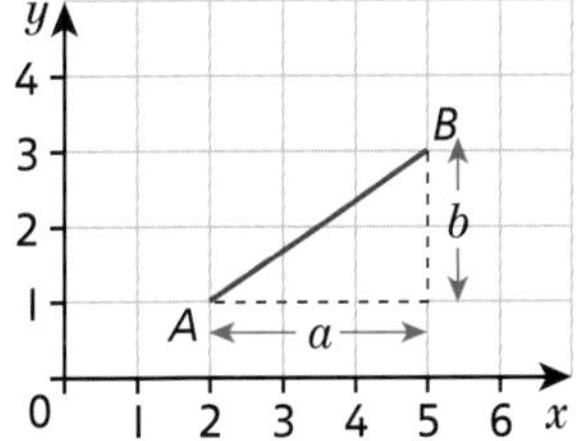

9 Find the shortest distance between each of these pairs of points.

a (1, 2) and (4, 7) **b** (2, 4) and (6, 8) **c** (2, 3) and (5, 9)

d (4, 2) and (8, 9) **e** (−1, 3) and (4, 9) **f** (3, −4) and (−5, −1)

Level 7

7c I can use Pythagoras' theorem to find the length of the hypotenuse

7b I can apply Pythagoras' theorem when solving problems in two dimensions

7a I can calculate the length of a line segment *AB* given the coordinates of points *A* and *B*

Now try this!

A Squares on triangles

You will need Resource sheet 4.4. Colour one of the smaller squares green and the other blue. Cut out the two smaller squares and then cut them up into individual squares.
Arrange the small squares on top of the largest square.
What does this tell you about the triangle?

B Distances between points

Work with a partner. Each draw a right-angled triangle accurately.
Measure and write down the length of two of its sides. Challenge your partner to calculate the missing length.
Check that they are correct by measuring the length of the third side.

right-angled triangle square square root

4.5 Using Pythagoras' theorem to solve problems

- Know and find Pythagorean triples
- Use Pythagoras' theorem to find the length of a shorter side of a right-angled triangle
- Use Pythagoras' theorem to solve problems

Why learn this?

SatNav systems use Pythagoras' theorem to calculate the distance between two places.

What's the BIG idea?

→ A **Pythagorean triple** is a group of three whole numbers that satisfy **Pythagoras' theorem**. For example, 3, 4 and 5 are a Pythagorean triple because $3^2 + 4^2 = 5^2$. **Level 7**

→ To find the length of a shorter side of a right-angled triangle, rearrange Pythagoras' theorem to give $a^2 = h^2 - b^2$ or $b^2 = h^2 - a^2$. **Level 7**

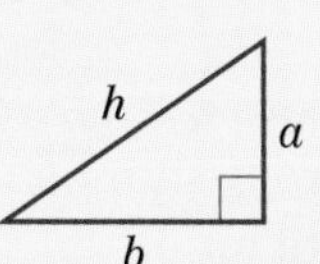

Practice, practice, practice!

Level 7

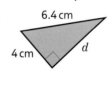

7c I can recognise and use Pythagorean triples

1 Which of these groups of numbers are Pythagorean triples?

A 7, 8, 10 B 6, 8, 10
C 16, 20, 24 D 5, 12, 13
E 10, 15, 21 F 12, 16, 20

7b I can use Pythagoras' theorem to find the length of a shorter side

2 Calculate the missing length in each of these triangles. Give your answers to 1 d.p.

a

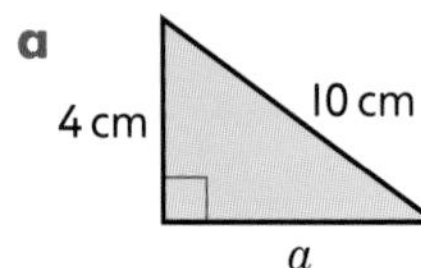

b 5.2 cm, 12.4 cm, b

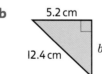

c

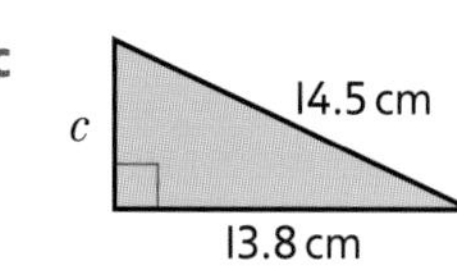

d 8.4 cm, 25.3 cm, d

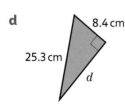

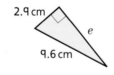

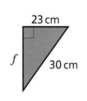

e 2.9 cm, 9.6 cm, e

f 23 cm, 30 cm, f

3 Calculate the missing lengths in these triangles. Give your answers to 1 d.p.

a

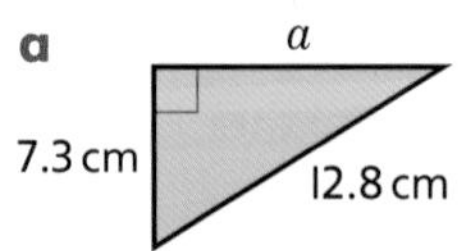

b 8 cm, 5 cm, b

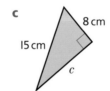

c 8 cm, 15 cm, c

d 6.4 cm, 4 cm, d

hypotenuse Pythagoras' theorem Pythagorean triple

4 A ladder is placed so that it reaches a window which is 4 m above the ground.
The ladder is 4.1 m long.
Calculate the distance between the bottom of the ladder and the wall.

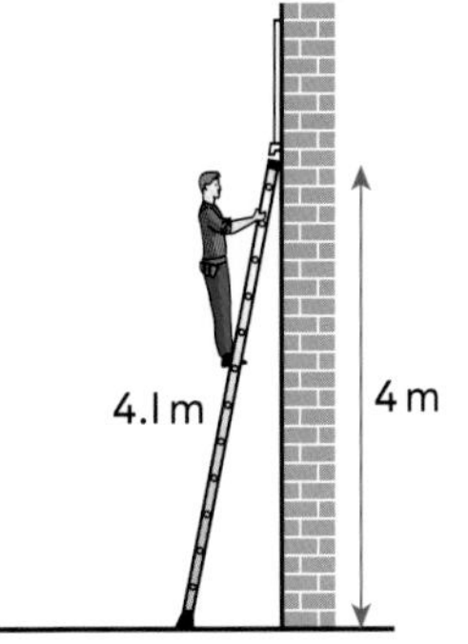

5 Calculate the width of a rectangle with diagonal of 8 cm and length 6 cm.
Give your answer to 1 d.p.

6 A taxi leaves an art museum.
It drives 7 miles due south and then turns due east and drives to the airport.
A crow flies 11 miles in a straight line from the art museum to the airport.
How many miles east does the taxi travel?

7 Calculate the missing length in this isosceles triangle.

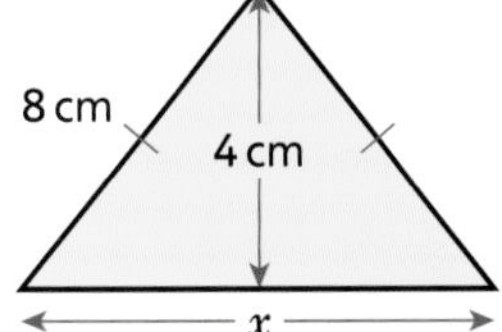

8 Look at this rectangle.
Work out

a the height of the rectangle

b the area of the rectangle.

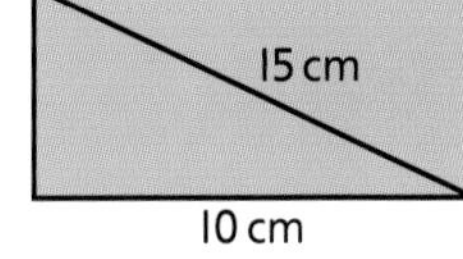

9 Find the area of this triangle.

18 cm

10 cm

10 Calculate the values of x and y in this shape.

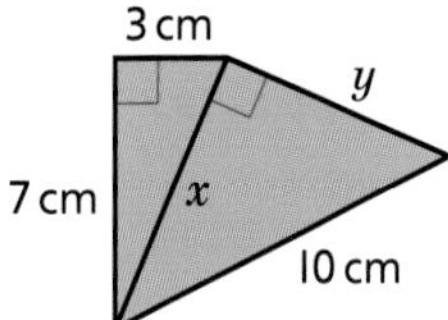

Level 7

7a I can apply Pythagoras' theorem when solving problems

Tip

Be careful – do you need to find the hypotenuse or a shorter side?

Super fact!

Pythagoras wrote nothing down, and there are no contemporary accounts of his work and life. The first detailed account of Pythagoras was written 150 years after his death, but only fragments survive. From that time onward, many books were written claiming to be the original Pythagorean texts.

Now try this!

A Pythagorean triples

A game for any number of players.
Each find as many Pythagorean triples as you can in 5 minutes.
Each number must be less than 100. Compare your Pythagorean triples with each other's.
The winner is the player who finds the most Pythagorean triples.

B Right-angled triangles

Draw three different triangles: one with a right angle, one with three acute angles and one with an obtuse angle. Label the shorter sides a and b, and the longest one h in each triangle.
Measure and write down all three lengths in each triangle.
Use a calculator to work out $a^2 + b^2$ and h^2 for each triangle. What do you notice?

right-angled triangle   square   square root

4.6 Constructions

- Use a straight edge and compasses to construct the perpendicular bisector of a line and an angle, the perpendicular to a line, and a triangle
- Use a straight edge and compasses to investigate the properties of overlapping circles

Why learn this?

Architects and engineers use constructions when drawing accurate scale drawings.

What's the BIG idea?

→ Lines that meet at **right angles** are **perpendicular** to one another. Perpendicular lines can be constructed using **compasses** and a ruler. **Level 6**

→ The angle **bisector** cuts an angle in half and can be constructed using compasses and a ruler. **Level 6**

→ The perpendicular from a point to a line segment is the shortest distance to the line. **Level 6**

→ You can construct a triangle using a ruler and compasses if you know the lengths of the three sides. **Level 6**

→ You can construct a right-angled triangle using a ruler and compasses if you know the lengths of the hypotenuse and one of the shorter sides. **Level 7**

Practice, practice, practice!

1 Using only a ruler and compasses, draw the perpendicular bisector of each of these lines. Mark the mid-point of each line *M*.

a a straight line *AB* of length 7 cm

b a straight line *CD* of length 9 cm

2 Use a ruler and compasses to draw the bisector of

a an acute angle **b** an obtuse angle.

3 Make an accurate drawing of this diagram. Use a ruler and compasses to construct the perpendicular at the point *X* on the line.

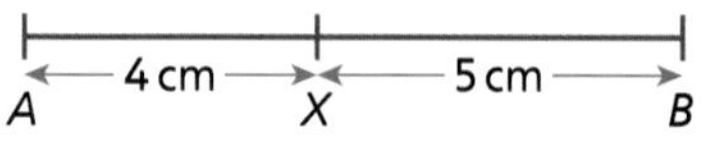

4 **a** Make an accurate drawing of this triangle using only a ruler and compasses.

b Use a ruler to measure the length *AB*.

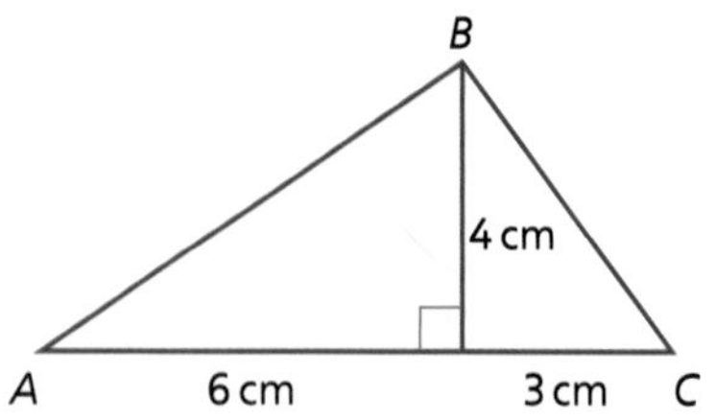

5 Make a copy of this diagram. Construct the perpendicular from the point *X* to the line.

Tip

'Construct' means use a ruler and compasses.

Level 6

6c I can construct the mid-point and perpendicular bisector of a line segment

6b I can construct the bisector of an angle

6a I can construct the perpendicular from a point on a line segment

6a I can construct the perpendicular from a point to a line segment

arc bisect chord circle compasses

6 Use a ruler and compasses to construct a triangle with side lengths of 5 cm, 4 cm and 7 cm.

7 Make an accurate drawing of an equilateral triangle with sides of length 6 cm using a ruler and compasses.

8 Use a ruler and compasses to draw the net of this regular tetrahedron.

5 cm

Level 6

6a I can construct a triangle given three sides

9 Paul needs to draw a scale diagram of the cross-section of a roof.

5 m

4 m

a Make an accurate scale drawing of the cross-section of the roof.

Hint: Use a scale of 1 cm : 1 m.

b Use a ruler to measure the height of the roof.

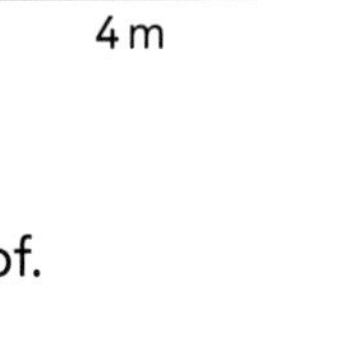

10 Laura sails 6 km North and then 5 km East then returns to her starting point along the shortest route.

Draw her route to scale and measure how far she travelled on her return journey.

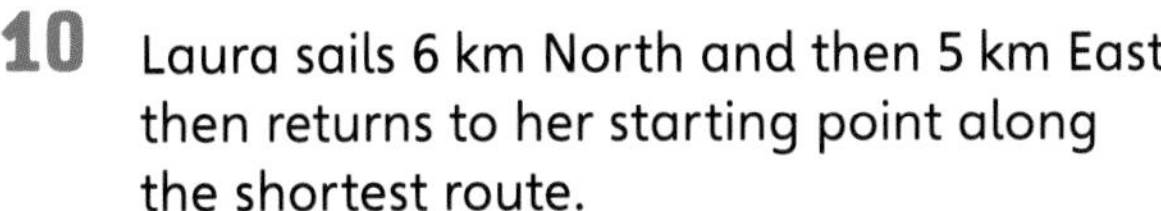

Level 7

7c I can construct a right-angled triangle when I know the lengths of the hypotenuse and one of the sides

Super fact!

In 2005, Ellen Macarthur broke the record for the fastest non-stop solo sail around the world. She sailed the 27 354 nautical miles in 71 days, 14 hours, 18 minutes and 33 seconds. Her average speed was 15.9 knots.

11 a Using compasses, draw two overlapping circles, both with radius 5 cm.

b Draw the radii from the centre of each circle to the points where the circles intersect.

c What is the name of the special quadrilateral that is formed? Explain your reasoning.

7a I can explain how standard constructions using a ruler and compasses relate to the properties of two intersecting circles with equal radii

Did you know?

Geometrical constructions were used in ancient Greek mathematics. Plato (c.427–c.347 BC) was a Greek philosopher whose school had a sign on the entrance that said 'Let none ignorant of geometry enter here'.

Now try this!

A Drawing triangles

Work with a partner.
Each secretly draw any triangle using a ruler.
Measure the sides of your triangle using a ruler.
Describe your triangle to your partner by telling them the lengths of its three sides.
Challenge your partner to construct the triangle you have described using only a ruler and compasses. Check to see if your partner's triangle matches yours.

B Triangle tessellation

Draw this tessellation of equilateral triangles using only a ruler and compasses.

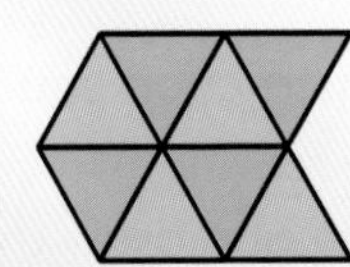

diameter perpendicular radius right angle

4.7 Construction problems

- Accurately draw triangles using a protractor and a ruler, or construct using compasses and a ruler
- Recognise which triangles are unique from their angles and sides

What's the BIG idea?

- You can **construct** a **triangle** using a ruler and **compasses** when
  - three **sides** are given (SSS) or
  - two **angles** and the side between them (ASA) or
  - two sides and the angle between them (SAS). **Level 6**
- You can construct a right-angled triangle using a ruler and compasses if you know the lengths of the hypotenuse and one of the shorter sides (RHS). **Level 7**
- Triangles drawn using the information SSS, SAS, ASA or RHS are unique. **Level 8**
- Triangles drawn using the information SSA or AAA are not unique.
  If you are only given three angles (AAA) you can draw lots of different triangles so three angles does not define a unique triangle.
  Similarly for SSA you can draw two different triangles given two side lengths and one angle.
  Level 8

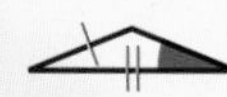

Why learn this?

Designers of fairground rides use triangles to add strength.

Practice, practice, practice!

Level 6

6a I can construct triangles

1 Draw these triangles accurately, using a ruler and a protractor.
Determine the values of a, b, c and d.

a 5 cm, 7 cm, 50°, a

b 10 cm, 56°, 37°, b

c 3 cm, 127°, 30°, c

d 8 cm, 8 cm, 2 cm, d

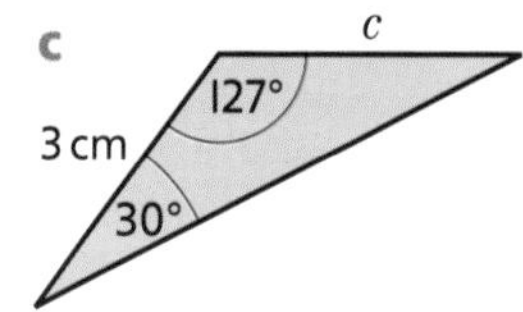

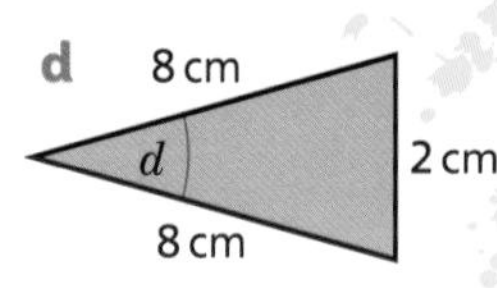

2 Using a ruler and compasses, construct these quadrilaterals.
Use a protractor to measure the size of the lettered angles.

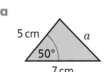

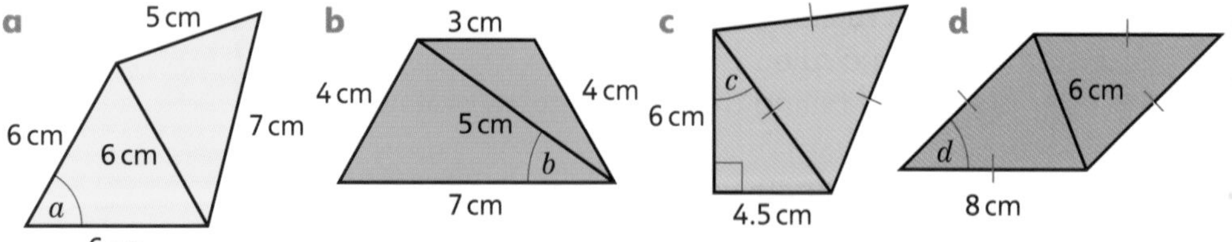

Watch out!
Don't rub out your construction arcs as they show that you have used the compasses correctly.

3 **a** Using a ruler and compasses, construct an equilateral triangle with sides of length 5 cm.

b Reflect the triangle along one side and draw the resulting shape.

angle compasses construct

4 Using a ruler and compasses, construct the net of this square-based pyramid.

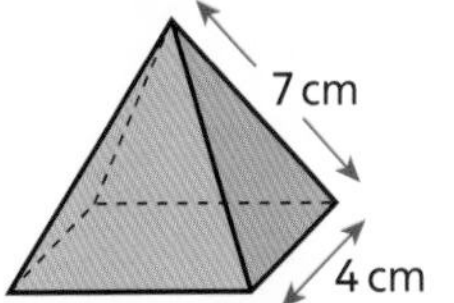

Level 6

6a I can construct nets of solid shapes

5 Claire has designed this tile pattern. Using only a ruler and compasses, make an accurate drawing of the tile.

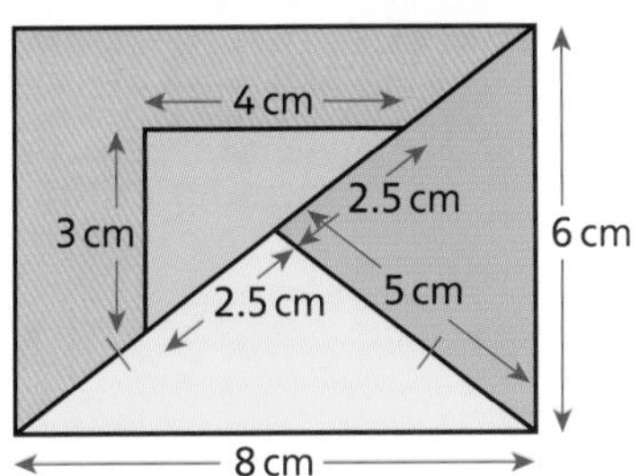

6 Which of these triangles can you construct from the information given?

- **a** Triangle ABC with $\angle A = 80°$, $\angle B = 60°$, $\angle C = 30°$
- **b** Triangle ABC with $\angle A = 70°$, $\angle B = 35°$, $\angle C = 75°$
- **c** Triangle ABC with $AC = 8$ cm, $AB = 4$ cm, $BC = 3$ cm
- **d** Triangle ABC with $AB = 10$ cm, $AC = 12$ cm, $\angle A = 50°$
- **e** Triangle ABC with $AC = 6$ cm, $\angle A = 54°$, $\angle C = 63°$
- **f** Triangle ABC with $AB = 4$ cm, $\angle B = 63°$, $BC = 5$ cm

Level 7

7a I can construct right-angled triangles

7a I can say whether a triangle can be constructed from the information given

7 **a** Using a protractor, draw three different triangles with angles of 70°, 60° and 50°.

b Measure the sides of each of the triangles. What do you notice about the lengths of the sides?

Hint: Compare the ratios of the sides of the triangles.

8 Which of the possible triangles in Q6 are unique triangles? Explain your answers.

Tip

'Unique' means that there is only one possible triangle that can be drawn.

9 Some sets of information are sufficient to describe a unique triangle ABC. Others are not. For each of these sets of information, construct either the unique triangle described, or two different triangles that both fit the description.

- **a** $AB = 5$ cm, $AC = 7$ cm, $BC = 8$ cm
- **b** $\angle ABC = 30°$, $\angle BCA = \angle CAB = 75°$
- **c** $AB = 6$ cm, $\angle ABC = 37°$, $BC = 8$ cm
- **d** $\angle ABC = 54°$, $BC = 8$ cm, $\angle BCA = 67°$
- **e** $AB = 7$ cm, $BC = 10$ cm, $\angle BCA = 32°$
- **f** $AB = 4$ cm, $BC = 5$ cm, $\angle BAC = 90°$

Level 8

8b I can understand which properties make a triangle unique

8b I can say if the angle and side properties of a triangle define a triangle uniquely

Now try this!

A Constructing hexagons

Draw a regular hexagon with side lengths of 4 cm by constructing six regular triangles using a ruler and compasses.

B Tile patterns

Make up your own tile pattern similar to the one in Q5 using a ruler and compasses.

net protractor side triangle

4.8 Circles and tangents

- Know the names of parts of a circle
- Explain why inscribed regular polygons can be constructed by equal divisions of a circle
- Know that the tangent at any point on a circle is perpendicular to the radius at that point
- Know that the perpendicular from the centre to a chord bisects the chord

Why learn this?

Tangents to circles are all around us. The surface of a road is a tangent to a car's wheels.

What's the BIG idea?

→ An **inscribed regular polygon** is a polygon that can fit exactly inside a **circle** with its vertices touching the **circumference** of the circle. Level 6

→ Lines that meet at **right angles** are **perpendicular**. Perpendicular lines can be constructed using compasses. Level 6 & Level 8

→ A **tangent** to a circle is a straight line at right angles to the **radius** of the circle at just one point. Level 8

→ A **chord** is a line drawn across a circle, not passing through the centre. Level 8

→ When you draw the perpendicular from the centre of the circle to a **chord** in the circle, it will **bisect** the chord. Level 8

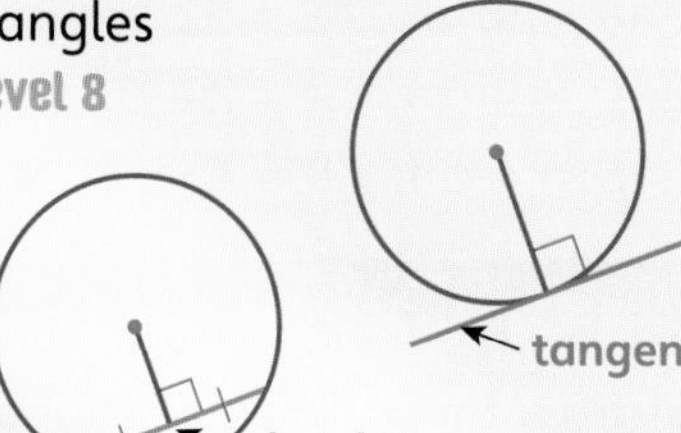

Did you know?

The Maracanã football stadium in Rio de Janeiro is circular. In any given row all the spectators are the same distance from the centre of the pitch. In the 1950 FIFA World Cup it hosted 199 500 people, and it is still one of the largest football stadiums in the world.

Practice, practice, practice!

Level 6

6b I can define a circle and name its parts

1 a Draw a circle with radius 5 cm.
b Draw and label on it
i a radius ii a chord iii an arc iv a diameter.

2 Write a definition of the circumference of a circle.

6a I can explain why inscribed regular polygons can be constructed by equal divisions of a circle

3 This diagram shows a regular pentagon inscribed in a circle.
a Calculate the size of $\angle BOC$. Explain your reasoning.
b What type of triangle is OCB?
Explain how you know.
c Calculate the size of $\angle BCO$.

B, A, O, C, E, D

4 Draw a circle of radius 4 cm.
Make an accurate drawing of an inscribed regular hexagon.
a Write down the size of $\angle DOE$.
b Measure each side of the hexagon.
What do you notice?
c Calculate the size of $\angle DEF$.
d What shape is $CDEF$?

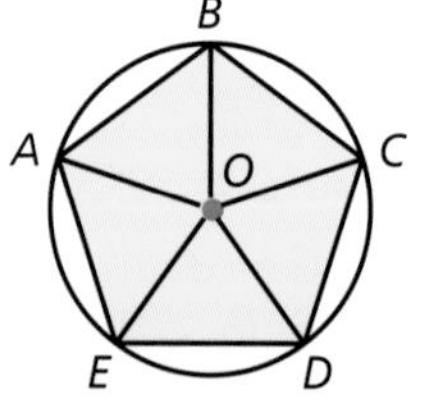

Learn this

An inscribed polygon in a circle has vertices that touch the circumference of the circle.

bisect chord circle circumference inscribed

5 Claire says that it is possible to draw an inscribed regular polygon in a circle by dividing the circle into equal parts on the circumference.
Is Claire correct? Explain your answer.

6 **a** Draw a circle of radius 6 cm.

b Mark any four points on the circumference of the circle and join them up to make a quadrilateral.

c Construct the perpendicular bisector of each side of your quadrilateral.

d Write down what you notice.

Level 6

6a I can explain why inscribed regular polygons can be constructed by equal divisions of a circle

6a I can construct perpendicular lines

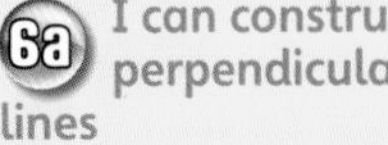

7 **a** Draw a circle of radius 4 cm. Draw a radius.

b Using a straight edge and compasses, draw the perpendicular to the radius at the point where the radius meets the circumference.

c Label this line as a tangent to the circle.

d At what angle do the tangent and radius meet?

8 The line *AC* is a tangent of the circle with centre *O*.

a What is ∠*OBA*? How do you know?

b Calculate the size of ∠*OAB*.

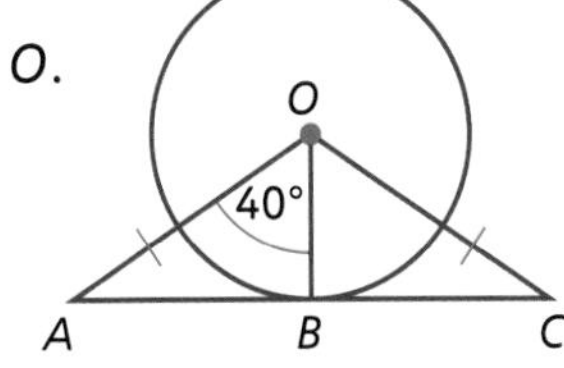

9 *TS* and *TU* are tangents to the circle.

a Calculate the size of obtuse ∠*SOU* when ∠*STU* is 50°.

b Write an expression for the size of ∠*SOU* when ∠*STU* is x°.

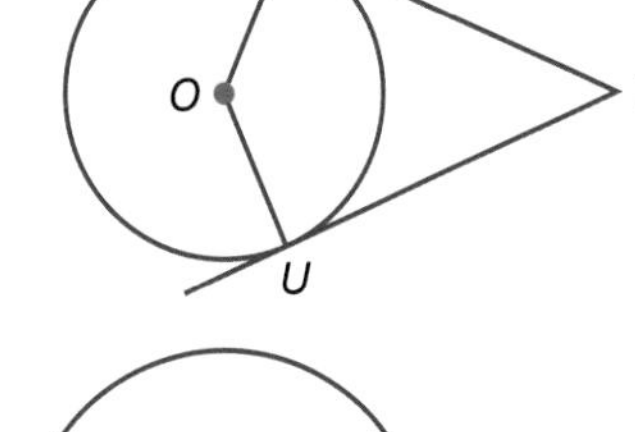

10 **a** Calculate the size of ∠*BOC*.

b What can you say about the lengths *AB* and *BC*? Explain your reasoning.

c Calculate the size of ∠*OAB*.

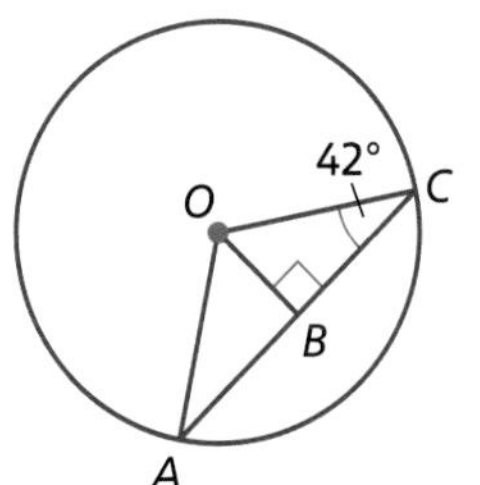

Level 8

8b I can explain the link between the tangent at any point on a circle and the radius at that point

8a I can explain that the perpendicular from the centre to the chord bisects the chord

Now try this!

A Inscribed polygons

A game for two players.
Draw a circle with an inscribed regular octagon.
Label the centre of the circle *O* and label the vertices of the octagon *A* to *H*.
Take turns to ask questions about the angles in the octagon. For example, 'What is ∠*AOB*?... ∠*AOC*?'
Check whether the answers are correct by measuring using a protractor.
Score 1 point for the correct angle.
The winner is the person with the most points after several turns each.

B Cyclic quadrilaterals

Draw a circle.
Mark any four points on the circumference of the circle and join them up to form a quadrilateral.
This is called a cyclic quadrilateral.
Measure each of the interior angles in order around the quadrilateral using a protractor.
Repeat for other cyclic quadrilaterals.
Write down anything that you notice about the angles in a cyclic quadrilateral.

perpendicular radius regular polygon right angle tangent

World's Greatest Maths! 4.9

Perilous paper round

You've taken a part-time job delivering newspapers. Unfortunately, every house has at least one dog in the front garden!

Barkingside Terrace

Use your knowledge of loci to find a safe route through each garden in Barkingside Terrace.

Copy each garden onto a coordinate grid, or use Resource sheet 4.9a.

For each one:

a construct and shade the locus of points the dogs can reach

b draw a safe route from *G* to *L* using straight lines

c use a list of coordinates to describe your safe route.

This dog is tied to a post at (1, 2) by a lead that is 3 m long. The shaded area shows the locus of points the dog can reach. A safe route from the gate (*G*) to the letterbox (*L*) is marked as a dotted line.

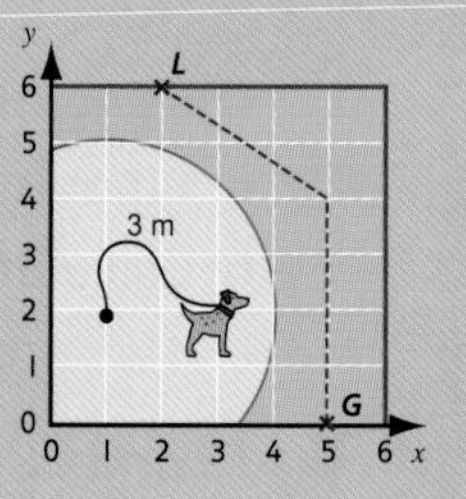

You could describe the safe route in the example as (5, 0), (5, 4), (2, 6).

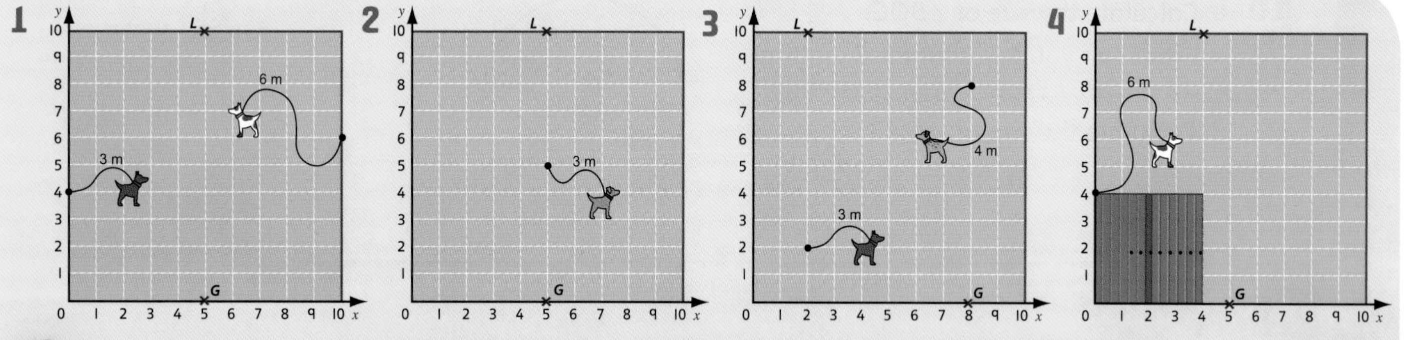

The dog-owners in Q5 to Q8 have attached their dogs to sliding rails! These are marked with a blue line. The end of the dog lead can move to any position on the rail.

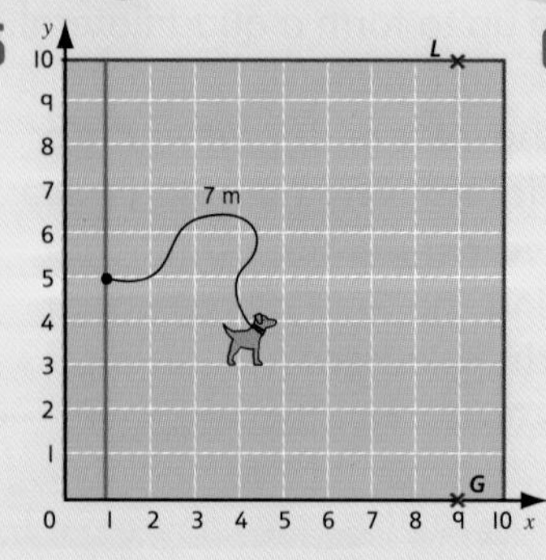

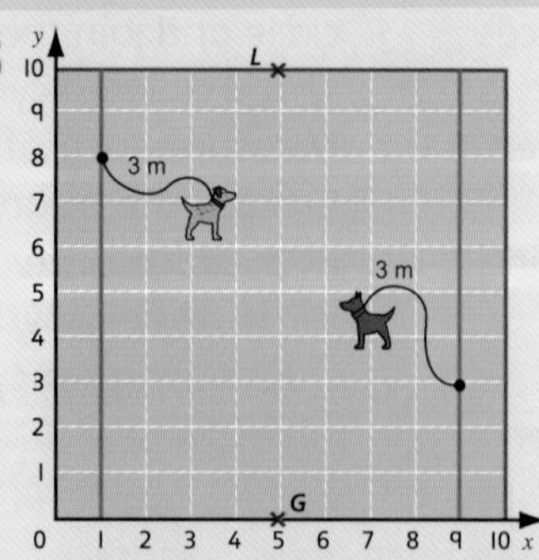

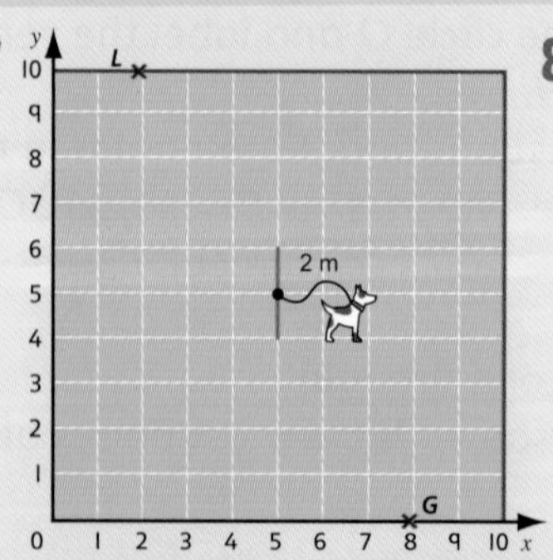

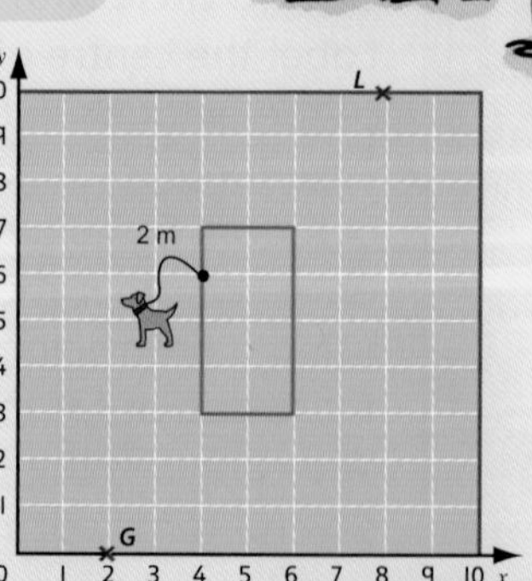

Woofbury Avenue

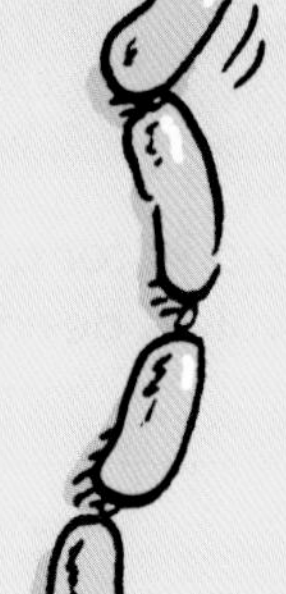

Uh oh! The dog-owners in Woofbury Avenue have managed to completely block the routes through their front gardens. Luckily, you've got a string of sausages which you can use to distract the dogs.

You have to put the sausages at a point where all the dogs can reach them.

Copy each garden onto a coordinate grid, or use Resource sheet 4.9b.

For each one:

a shade the locus of points each dog can reach

b mark a point where you could put the sausages, and write down its coordinates.

If you put the sausages at (4, 5), only one dog would be distracted. You could distract both the dogs by putting the sausages at (3, 4).

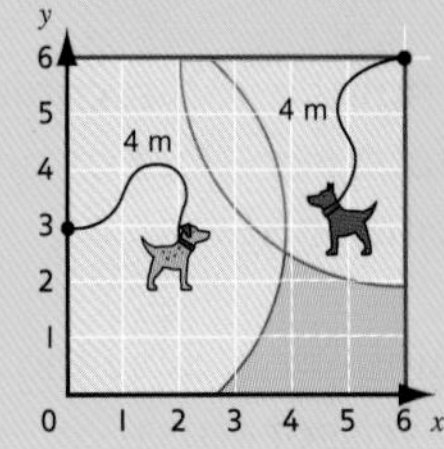

1

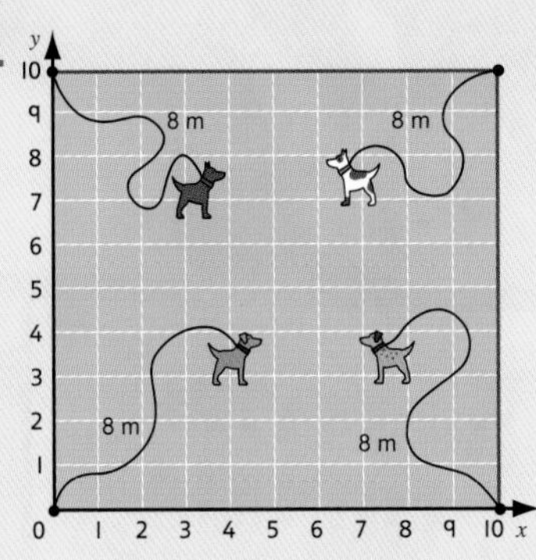

2

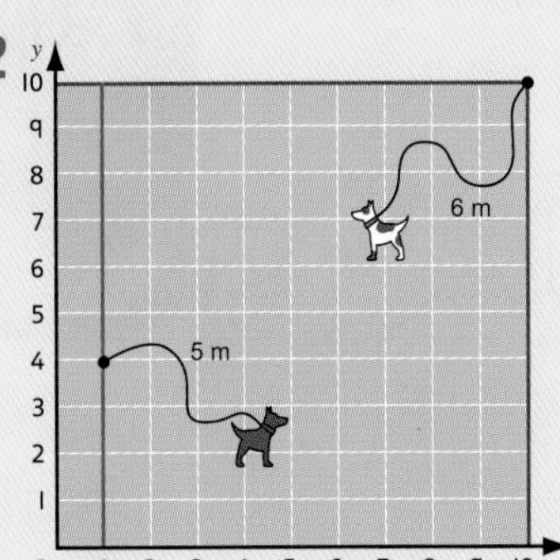

3

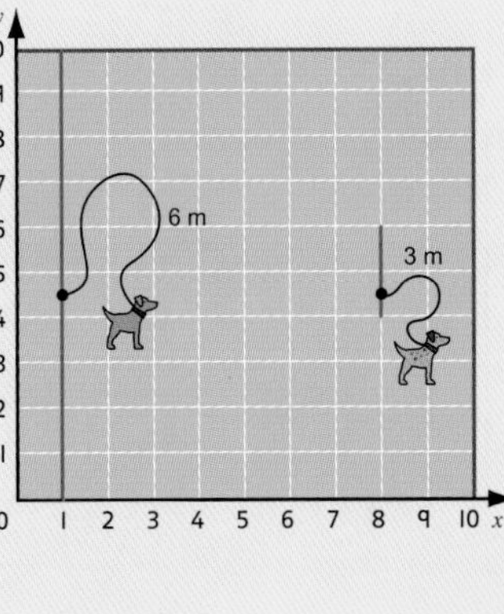

4

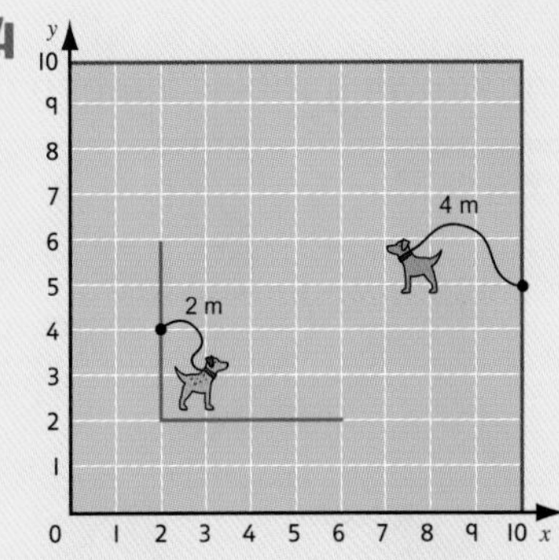

Wacky wheels

You're trying out two new 'wheels' for your delivery bike.

For each 'wheel', copy the diagram using the dimensions shown, or use Resource sheet 4.9b.

Construct the locus that the point P will trace as the 'wheel' is rolled along a straight line.

1

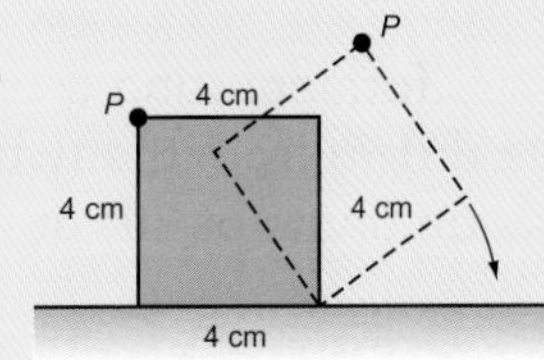

2

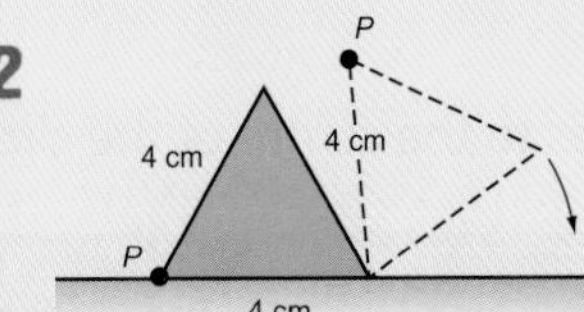

Engineering calculations

Erica is an engineering technician who works for a roof truss manufacturer. Part of her work involves designing project-specific roof trusses.

1 This engineering drawing shows a scissor truss. The scissor truss is made up of two congruent isosceles triangles.

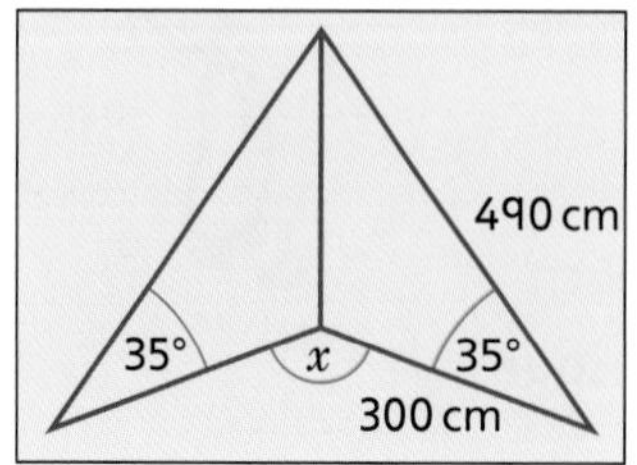

a What is the size of angle x?

b Using an appropriate scale, construct an accurate scale drawing of the scissor truss. **Level 6**

2 This engineering drawing shows an attic truss.

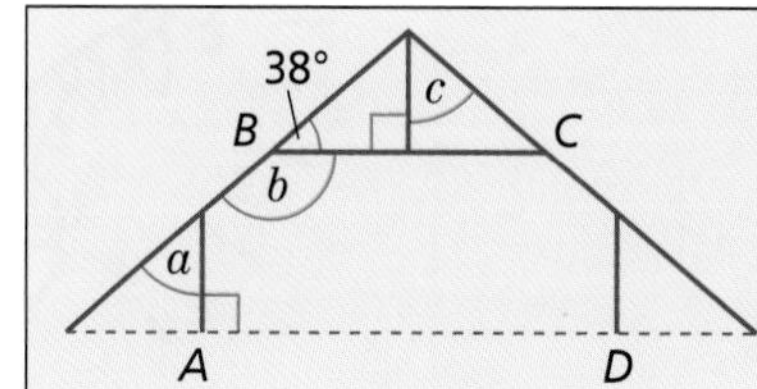
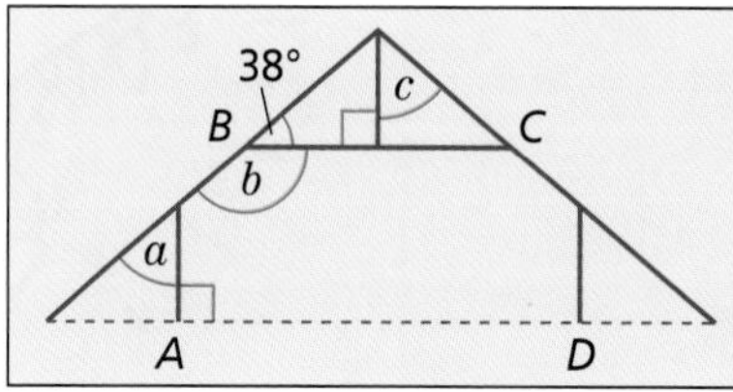

a Calculate angles a, b and c.

Hint: The attic truss has a line of symmetry.

b What shape is $ABCD$?

c What is the sum of the interior angles of $ABCD$? **Level 6**

3 This engineering drawing shows a mono roof truss.

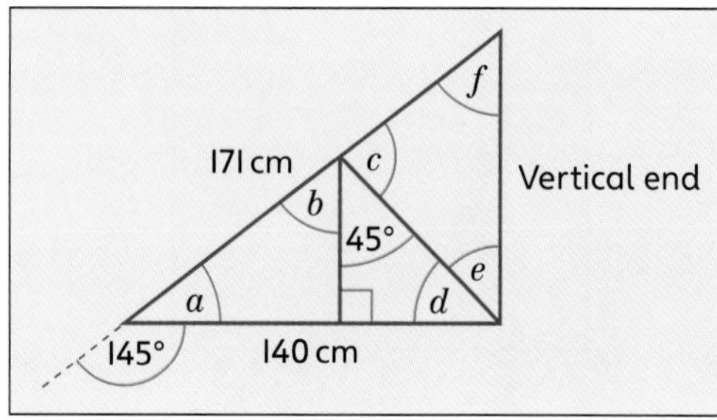

a Work out the sizes of angles a, b, c, d, e and f.

b Using an appropriate scale, construct an accurate scale drawing of the mono roof truss.

c Use your drawing to work out the vertical end height of the truss.

d Use a mathematical calculation to check your answer to part **c**. **Level 7**

4 The entrance to a building is formed using a circular construction. Erica is told that a triangle drawn from the two ends of a diameter will always make an angle of 90° at the edge of the circle.

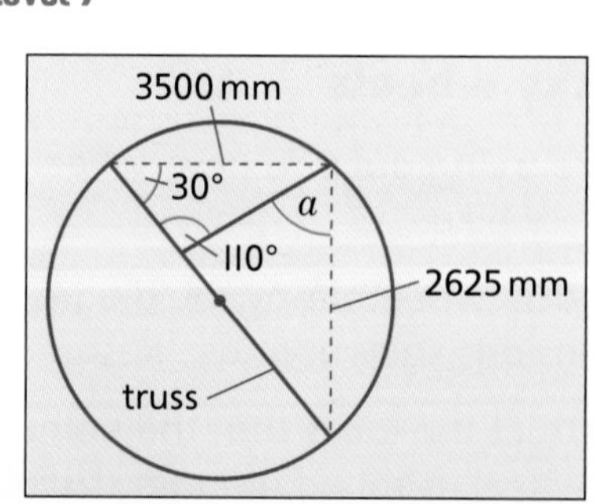

Use the information in the diagram to find

a the size of angle a

b the length of each truss to the centre of the circular hub. **Level 8**

The BIG ideas

- The **interior angles** in a triangle sum to 180°. **Level 6**
- The interior angles in a **quadrilateral** sum to 360°. **Level 6**
- The sum of the interior angles in an n-sided polygon is $(n - 2) \times 180°$. **Level 6**
- **Isosceles triangles** have two equal sides and two equal angles. **Level 6**
- You can **construct** a triangle using a ruler and **compasses** when you are given
  - three **sides** (SSS) or
  - two **angles** and the side between them (ASA) or
  - two sides and the angle between them (SAS). **Level 6**
- The **hypotenuse** is the longest side of a **right-angled triangle**. It is the side opposite the right angle. **Level 6**
- **Pythagoras' theorem** is $h^2 = a^2 + b^2$, where h is the length of the hypotenuse and a and b are the lengths of the shorter sides. **Level 7**

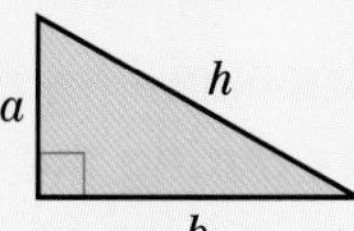

- A **tangent** to a **circle** is a straight line at right angles to the **radius** of the circle at just one point.

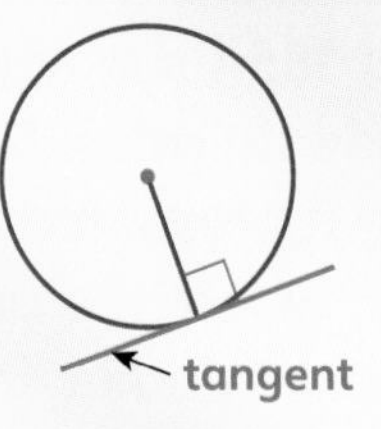

- A **chord** is a line drawn across a circle, not passing through the centre. **Level 8**
- When you draw the **perpendicular** from the centre of the circle to a chord in the circle, it will **bisect** the chord. **Level 8**

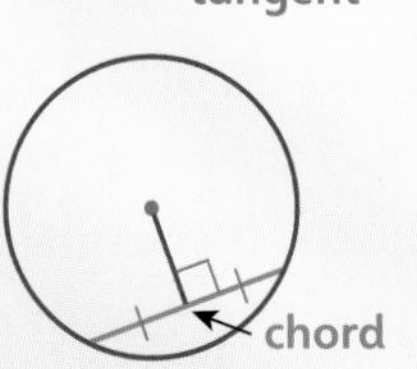

Find your level

Level 6

Q1 Look at the diagram. What is the size of angle x?

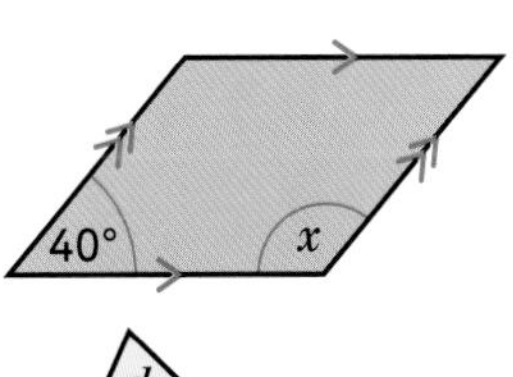

Q2 Look at the diagram. Work out the size of angles a, b, c and d.

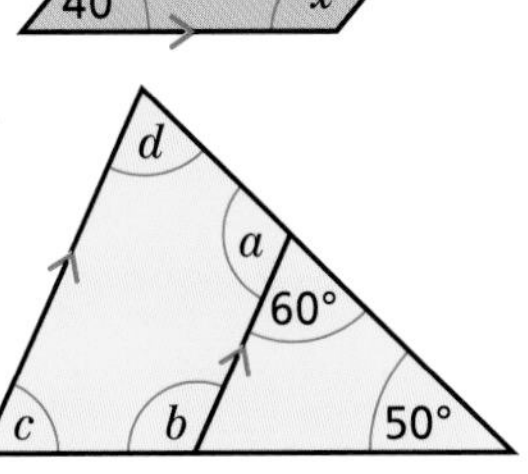

Level 7

Q3 **a** Calculate the missing length in the right-angled triangle.

b Work out the area of the triangle.

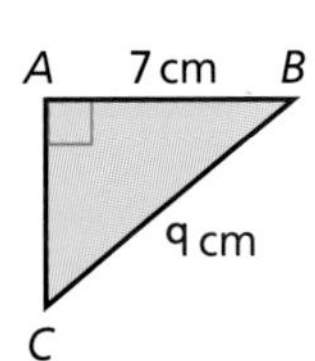

Q4 *ABCD* is a parallelogram. Work out the size of angles s and t. Give reasons for your answers.

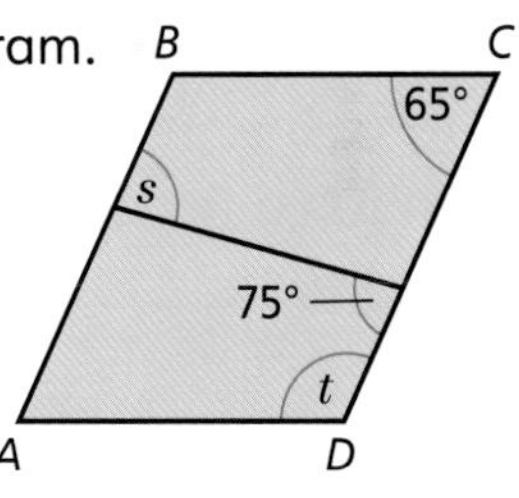

Q5 On a map of Spain, Burgos is 247 km north of Madrid and Teruel is 220.6 km east of Madrid. Calculate the direct distance between Burgos and Teruel.

Level 8

Q6 **a** Show that the perimeter of triangle *ABC* is 39 cm.

b Show that triangle *ABC* is right-angled.

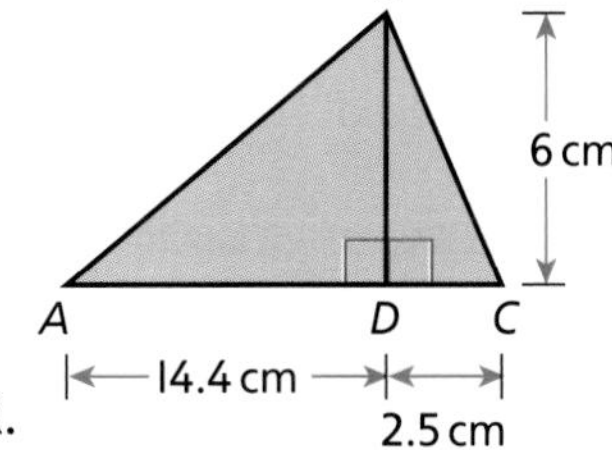

5 Stat's entertainment

This unit is about collecting and representing data in charts and graphs.

Florence Nightingale is famous for nursing injured British soldiers during the Crimean War (1853–56). When she first arrived at the military hospital in Scutari in November 1854 she was horrified by the dirty conditions and high death rates. She collected data on causes of death and represented it in charts called 'polar area diagrams', where the area of each section represented the number of deaths from that cause. Her diagrams showed very clearly that most deaths were due to preventable disease rather than wounds. Convinced by her evidence, the British Army made improvements to hygiene and the death rate fell.

This was the first time that statistical data had been used in this way – to show where changes were needed. After the Crimean War ended, Florence Nightingale continued to collect data from hospitals in Britain, and her findings led to further improvements in conditions and to falling death rates. In 1859 she became the first woman to be elected to the Statistical Society.

Activities

A Newspapers and television news reports often use graphics to represent information. Generally, people tend to understand the information more easily when it is presented as a diagram.
Find as many examples of this type of graphic as you can. See if you can find out the name of each type of chart or diagram.

B Collect some data from your class – about favourite types of TV show, hair colour, or anything you like. Design your own chart to represent this data.

Did you know?
Many British soldiers died of cholera in the Crimean War. Jamaican Mary Seacole, who had successfully nursed and treated cholera victims in an epidemic in Kingston, offered her services to the British Army but was rejected.

Did you know?
The pie chart was first invented by William Playfair, an engineer and economist, in 1801.

He also invented the line graph and bar chart.

Before you start this unit...

1 Decide whether these types of data are discrete or continuous

Level Up Maths 5-7 page 294

a weights of apples on a tree

b numbers of books in school bags.

2 Calculate the mean, median, mode and range of this set of numbers.

Level Up Maths 5-7 page 190

30, 29, 31, 29, 30, 32, 33, 30, 31, 36

3 A quiz question has four possible answers, A, B, C or D. Here are the answers given by 12 contestants.

Level Up Maths 5-7 pages 42,192

A, C, C, B, A, C, D, C, B, D, C, A

a Draw a frequency table for this data.

b Draw a bar chart for the data.

5.1 Planning an investigation
5.2 Organising data
5.3 Calculating and using ststistics
5.4 Pie charts
maths! Power for the future
5.5 Line graphs
5.6 Frequency diagrams and frequency polygons
Unit plenary: Data on disease

Plus digital resources

5.1 Planning an investigation

- Find sources of secondary data
- Identify possible sources of bias in samples and questionnaires

What's the BIG idea?

- If you carry out a **survey**, the data you collect is called **primary data**. Alternatively, you can use data that is already available in a newspaper or on a website. This is called **secondary data**. Level 6
- A survey is carried out on a **sample** of people. A sample should **represent** a whole population. A sample might be **biased** if it only represents part of the population, for example, if you only ask one age group or gender. Level 7
- Badly written questions can lead to bias. For example, questions that start 'Do you agree...' might push the respondent to say 'Yes'. Level 7

Why learn this?

Biased surveys can give misleading results.

Practice, practice, practice!

Level 6

6C I can identify possible sources of secondary data

1 Look at websites to find the best source of secondary data to investigate these.

- **a** how house prices have changed over time
- **b** bank savings rates
- **c** changes in pollution levels in UK rivers

2 Look at a local or national newspaper.

- **a** What secondary data can you find?
- **b** What could you investigate by using it?

3 Rio is buying a new car.
He would like the car to be as environmentally friendly as possible.
Also he doesn't want the car's value to depreciate too quickly.
Suggest possible sources of secondary data that could help Rio choose which car to buy.

Level 7

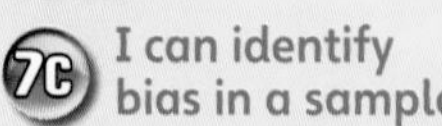
7C I can identify bias in a sample

4 Sasha is doing a survey to find out how often people go ten-pin bowling and how much they spend when they do.
She waits outside the bowling alley and asks people as they come out.
Give a reason why this method of sampling could be biased.

5 Nathan is doing a survey to find out people's views on global warming.

- **a** He asks the other members of staff at the charity shop where he works on Saturdays. Give a reason why this sample could be biased.
- **b** He picks a page at random from the telephone directory and phones everyone on that page. Why might this sample be biased?

bias primary data questionnaire random

6 Lisa wants to know whether most people eat at least five portions of fruit and vegetables every day. She asks the members of her netball club.
Give three reasons why her sample could be biased.

7 Karim is doing a survey on road congestion in his town. He stands by the main road and counts the cars that pass him between 2 pm and 3 pm one Wednesday
Give a reason why his sample could be biased.

8 A local councillor wants to know whether people will support her plan to provide free bus travel to under-18s.
What is wrong with each of these samples?

a members of a youth club **b** people at a bus station

9 Peter is doing a survey on nutrition.
He has written this question.

> What is your favourite fruit?
> Apple ☐ Pear ☐
> Banana ☐ Strawberry ☐

What is wrong with Peter's question?

Watch out!
Make sure you've covered all the possible answers when writing a question for a questionnaire.

10 Molly is investigating local facilities for young people.
She has written this question for a questionnaire.

> Do you think the council should build a skate park?

a What is wrong with Molly's question?

b Write a better question to find out people's views on local facilities.

11 The table shows the numbers of pupils in Years 7 to 9 at Parkway School.
The school council want to interview 10% of the pupils for a survey on after-school activities.

Year 7	400
Year 8	300
Year 9	200

We can interview 30 pupils from each year.
Suzy
There are twice as many Year 7s as Year 9s, so our 10% sample should have twice as many Year 7s as Year 9s.
Tom

a Who is correct? Explain why.

b How many pupils from each year should be surveyed?

Level 7

7c I can identify bias in a sample

7b I can identify bias in a questionnaire

7a I can plan to minimise bias in a questionnaire

7a I can plan to minimise bias in a sample

Did you know?
Phone-in polls are biased because people select themselves to take part and are likely to have strong opinions. People who don't have a strong opinion are unlikely to respond.

Now try this!

A Secondary data
Think of something you would like to investigate – it could be relevant to your family, hobbies or local area.
What secondary data could you collect to help you with your investigation?

B Logo design
Sanjay is designing a logo for his skateboard company.
What questions could Sanjay ask in a questionnaire to help him with his design? Who should he ask?

representative sample secondary data survey

5.2 Organising data

- Construct tables for discrete and continuous data
- Choose suitable class intervals for continuous data
- Design and use two-way tables

Why learn this?

Research produces large amounts of data. This needs to be organised clearly.

What's the BIG idea?

- You can organise **raw data** by sorting it into a table. Level 6
- A **frequency table** can be used to show a number of observations and their frequency. Level 6
- The **class interval** 150 cm ≤ height < 155 cm includes all the heights from 150 cm up to, but not including, 155 cm. Level 6
- A **two-way table** shows data sorted according to two sets of categories. Level 6

Practice, practice, practice!

Level 6

6b I can design tables to record discrete data

1 Gavin has counted the seeds in 60 seed packets.

33	41	28	32	37	49	23	32	43	26
37	24	21	42	46	37	34	39	30	47
31	38	43	33	27	39	40	28	45	36
35	26	42	22	35	40	31	43	46	24
33	29	39	37	27	31	40	37	44	23
50	39	32	36	24	30	32	32	36	38

Construct a frequency table for Gavin's data.

Watch out! Discrete data sometimes needs to be grouped in a frequency table to reduce the number of categories.

6b I can design and use two-way tables

2 Petra has recorded the football shirts sold at the club shop where she works. Football shirts are available in two designs: home (H) and away (A). They are available in four sizes: small (S), medium (M), large (L) and extra large (X).

HL	AM	HL	AL	AL	HS	HM	HL	HX	HL
AL	HL	HM	HM	HM	HL	AL	HM	HX	HL
AM	AL	HL	HX	HM	HM	HS	AX	AM	HM
HL	HX	AX	HL	AL	AM	HM	AL	HL	HL

Organise the information in a two-way table.

Watch out! In a two-way table, the row totals must add up to the same as the column totals.

3 The table shows the numbers of trophies won by four English football clubs. Copy and complete the table.

	League champions (Premier League and predecessors)	FA cup	League cup	Total
Nottingham Forest	1	2		
Sheffield Wednesday			1	8
West Bromwich Albion	1	5		7
Wolverhampton Wanderers	3		2	9
Total	9		8	

class interval frequency table raw data

4 A pharmaceutical company has developed a new treatment for migraines. They carried out a trial on 310 people. During the trial, 150 people were given the new drug. The rest were given a placebo. Of those who were given the placebo, 40 people said their symptoms improved. Altogether, 130 people said their symptoms improved. Construct a two-way table to show this information.

Hint: A placebo is a tablet that does not contain the drug that is being trialled.

Level 6

6b I can design and use two-way tables

5 The table shows the results of 50 people's driving tests.

	Men	Women
Pass	11	10
Fail	15	14

a What percentage of the drivers passed their driving test?
b What fraction of the men failed their driving test?
c What is the pass : fail ratio for all the drivers?

6b I can make statements using data from two-way tables

6 A refuse company uses a datalogger to record the weights of wheelie bins before they are emptied. The weights are recorded in kilograms to 1 d.p.

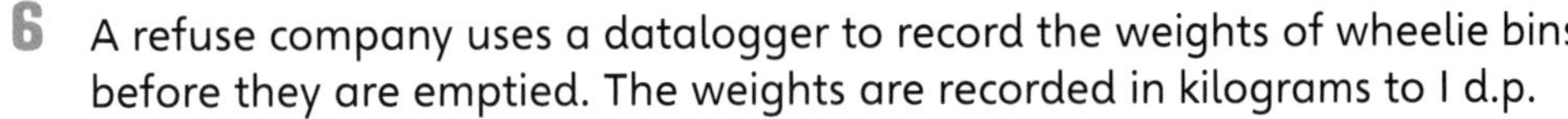

18.5	16.1	20.0	19.7	21.6	14.3	18.9	16.0	19.1	21.2
17.8	16.4	18.6	19.9	23.8	15.6	19.0	16.7	18.4	22.3
20.9	17.8	18.5	19.2	15.1	21.6	19.3	16.2	17.3	19.9

An empty wheelie bin weighs 14 kg. Construct a frequency table to show the weight of rubbish inside each wheelie bin.

Tip
Use tallies to help you complete a frequency table.

6a I can design tables to record continuous data

7 The tables show the populations of Asia and Europe, in millions, by age and gender.

Asia

Age, a	Male	Female
$0 \leqslant a < 20$	757	700
$20 \leqslant a < 40$	655	615
$40 \leqslant a < 60$	415	400
$60 \leqslant a < 80$	155	180
$80 \leqslant a < 100$	17	33

Europe

Age, a	Male	Female
$0 \leqslant a < 20$	84	80
$20 \leqslant a < 40$	105	103
$40 \leqslant a < 60$	100	104
$60 \leqslant a < 80$	54	71
$80 \leqslant a < 100$	8	18

Write a short report comparing the populations of Asia and Europe, based on the information in these tables.

6a I can make more complex statements using data from two-way tables

Now try this!

A Gender and hair colour

Design a two-way table to show data about gender and hair colour. Fill in your table for the pupils in your class.

B Missing numbers

Design a two-way table on any topic. Include a 'Total' row and column. Choose numbers to fill your table. Re-draw the table but miss out one number from each row, and one other number. Challenge a partner to complete your table.

two-way table

5.3 Calculating and using statistics

- Calculate statistics, knowing when to use the mean, median, mode and range
- Calculate the mean using an assumed mean
- Compare two sets of data using an average and the range
- Construct and use stem-and-leaf diagrams
- Find the modal class of a set of data

Why learn this?

If you don't know what's average, it can be hard to judge whether something is big or small.

What's the BIG idea?

- An **average** is a single number used to represent a set of data. The **mean**, **median** and **mode** are averages. The **range** is a measure of spread. Level 6
- Using an **assumed mean** can make calculating the mean easier.
  mean = assumed mean + the mean of the differences from the assumed mean Level 6
- A **stem-and-leaf diagram** shows numerical data that is split into a stem and a leaf. For example, you could write 46 and 48 as 4 | 6, 8. The diagram should be ordered and have a key. Level 6
- A **modal class** is the class interval with the highest frequency. Level 6 & Level 7

Did you know?

The average person will eat 10 866 carrots in their lifetime.

(Source: Human Footprint, Channel 4)

Practice, practice, practice!

Level 6

6C I can recognise when to use the range, mean, median or mode

1 Nine friends spend these amounts at a coffee shop.

£1.88 £3.01 £2.46 £1.88 £1.97 £3.10 £2.65 £1.88 £1.35

a Jasper would like to know if he spent more than most.
Which average should he use? What is the value of this average?

b Kelly suggests that next time they pay 'equal shares'.
Which average gives 'equal shares'? How much would each friend have paid this time if they had paid 'equal shares'?

6C I can calculate the mean from a small set of simple discrete data using an assumed mean

2 Find the mean cost of a laptop computer using an assumed mean of £300.

Cost of laptop (£)	349	299	327	320	280
Difference from assumed mean of £300	49	−1	27	20	−20

Total difference = 49 + 27 + 20 − 1 = £75.
Mean difference = £75 ÷ 5 = £15
Mean = assumed mean + mean difference = £300 + £15 = £315

Find the mean price of a cinema ticket using an assumed mean of £5.

Price of cinema ticket (£)	4.70	6.00	4.50	5.80	5.50
Difference from assumed mean of £5					

average mean median modal class/group

3 Find the mean height of these people using an assumed mean of 1.5 m.

Mr Jones 1.8 m, Mrs Jones 1.6 m, Bethany 1.3 m, Jack 1.4 m, Lauren 1.2 m.

4 The table shows the median, mean and range of the distances, in metres, jumped by two triple jumpers in their last competition.

	Median	Mean	Range
Martin Hopper	14.9	14.9	0.5
Kyle Swift	15.4	14.3	10

Their coach needs to decide which triple jumper to pick to represent their athletics club at the local championships.
Each triple jumper will get six attempts in the championships.
Which triple jumper should the coach pick? Explain your reasons.

5 Find the mean height of these hills using a suitable assumed mean.

500 m, 420 m, 350 m, 380 m, 400 m, 390 m, 310 m, 450 m, 420 m, 390 m

6 These are the masses, in kilograms, of 30 parcels.

1.3	0.4	2.5	1.7	3.5	1.2	2.1	0.1	2.8	2.5
4.1	1.2	1.7	2.6	3.8	1.9	0.9	1.4	2.9	1.7
2.4	4.7	3.7	0.1	3.2	1.2	0.8	1.7	3.6	0.2

a Draw a stem-and-leaf diagram for this data.
Remember to include a key.

b What is the mode of the data?

c What is the range?

d What is the median?

Make sure you order the leaves in your stem-and-leaf diagram.

Level 6

6c I can calculate the mean from a small set of simple discrete data using an assumed mean

6b I can compare two sets of data given only the averages and range

6b I can calculate the mean from a set of discrete data using an assumed mean

6a I can construct a stem-and-leaf diagram and use it to find the median

7 The table shows the projected UK population (by age group) for 2010.

Age (years)	0–9	10–19	20–29	30–39	40–49	50–59	60–69	70–79	80–89	90+
Population (thousands)	7280	7457	8663	8127	9197	7551	6669	4477	2432	458

(Source: Office for National Statistics)

a What is the modal class of this data?

b Re-group the data in 20-year intervals (0–19, 20–39, ...).

c What is the modal class now?

8 Find three numbers with

a mean 5 and mode 6

b mean 11, median 10 and range 7.

7c I can find the modal class of a large set of data

7c I can find data values given different averages and the range

Now try this!

A Mean, median or mode?

Write some guidelines to help someone decide whether to use the mean, median or mode to represent a set of data.

B Median and mean

The median of three numbers is 50, and the mean is lower than the median. What could the numbers be?
Add two different numbers to the set so the median is still 50 and the mean is still lower than the median.
Keep adding two numbers to the set. What do you notice?

mode range stem-and-leaf diagram

5.4 Pie charts

- Draw pie charts
- Interpret pie charts

What's the BIG idea?

→ A **pie chart** uses **sectors** of a circle to show different categories of data. Level 6

→ Each sector of the circle represents a different category. For example, if the sector is a quarter of the circle then one quarter of the data is in this category. Level 6

→ There are 360° in a whole circle. Use this formula to work out the angle for each category:

Angle of sector = $\frac{\textbf{frequency}}{\text{total frequency}} \times 360°$ or $\frac{360°}{\text{total frequency}} \times$ frequency

Level 6

Why learn this?

Pie charts are often used in presentations because they help people compare proportions.

Practice, practice, practice!

Level 6

6c I can draw pie charts with more than three categories

1 The table shows the colours of boots worn by the 30 players in a rugby match.

	Black	White	Red	Gold	Total
Number of players	15	3	4	8	30
Angle	$\frac{15}{30} \times 360° = 180°$				

a Copy and complete the table.

b Draw a pie chart to show the data.

Tip

You don't need to draw the last sector in a pie chart, but you should measure its angle to check you haven't made a mistake.

2 The table shows how Kieran spent his Saturday.

Activity	Sleeping	Eating	Cycling	Fishing	Watching TV	Other
Time spent (hours)	8	2	2	4	3	5

Draw a pie chart to represent the data.

Watch out!

Don't draw your pie chart too small. It's easier to label the sectors in a bigger pie chart.

6b I can answer questions about pie charts

3 The pie chart shows the results of a survey of people's favourite holiday destinations. Eleven people chose France as their favourite holiday destination.

a How many people were surveyed altogether?

b How many people chose Spain?

Holiday destinations

frequency interpret

4 This pie chart shows the nutritional content of peanut butter.

a How much protein is in 100 g of peanut butter?

b How much carbohydrate is in 100 g of peanut butter?

Hint: Use your protractor

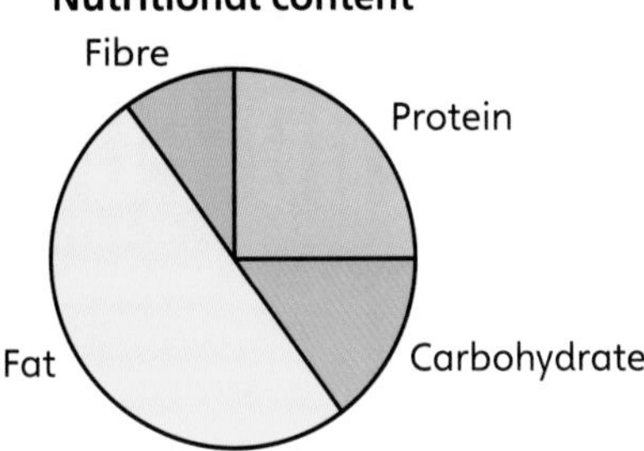

Level 6

6b I can answer questions about pie charts

5 The pie charts show the top five birds seen in Lincolnshire and Merseyside during a birdwatch survey.

Lincolnshire

Blue tit
House sparrow
Collared dove
Blackbird
Starling

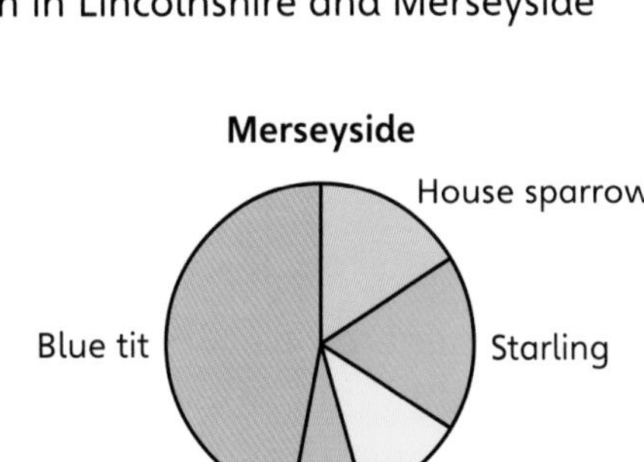

Compare the proportions of birds seen in the two counties.

6a I can compare two pie charts

6 The pie charts show the times that army and navy cadets took to complete an assault course.

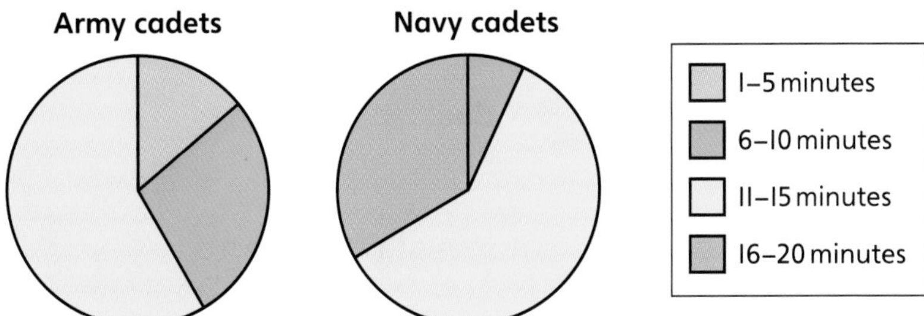

Use the pie charts to decide whether each statement is true or false, or whether there is not enough information to decide.
Explain each of your answers.

a The slowest cadet was an army cadet.

b The fastest cadet was an army cadet.

c More navy cadets than army cadets took at least 6 minutes.

Now try this!

A Nutrition pie chart

Draw a pie chart to show the nutritional content of an item of packaged food.

B Spreadsheet pie charts

Use spreadsheet software to draw the pie chart in Q2.
Draw another pie chart to show how you spent your Saturday.
Combine the two charts using a 'doughnut' graph option.

pie chart proportion sector

maths!

Power for the future

As non-renewable energy sources begin to run out, alternative energy sources are becoming more widely used. In March 2007, in order to help cut CO_2 emissions, the European Union set itself the target of sourcing 20% of energy from renewable resources by 2020.

Sources of energy

This 3-D pie chart shows the energy consumption of the UK in 2007. Figures are given in millions of tonnes of oil equivalent.

- The pie chart has been drawn to exaggerate the size of the oil sector. Explain how this has been done.
- What percentage of energy consumption came from oil?
- What percentage of energy consumption came from renewable sources?

The total energy consumption in 2007 was equivalent to 154.8 million tonnes of oil, which could be broken down as follows:

Sector	Consumption
Industry	31.7
Domestic	44.0
Transport	59.8
Services	19.3

- Represent this data with a pie chart.

A typical wind turbine generates 4.7 million units of electricity every year - this is enough to power 1000 homes, or make 170 million cups of tea!

Energy changes

Renewable resources

The line graphs show the electricity generated using three different renewable energy sources in the UK, between 2004 and 2007.

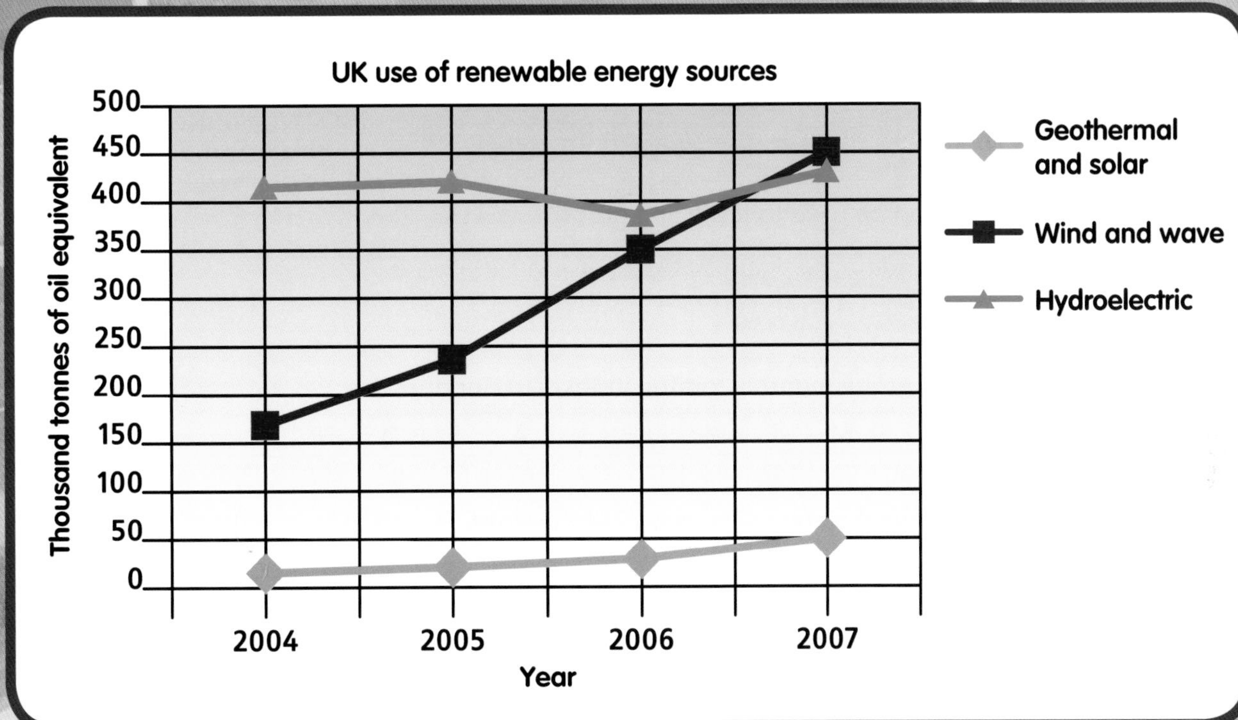

- Write a paragraph that comments briefly on each of these time series.
- What would you expect to see happen next?

Non-renewable resources

This table gives data about the amounts of petroleum and natural gas produced in the UK.

	2004	2005	2006	2007
Petroleum	104.5	92.9	84.0	84.2
Natural gas	96.4	88.2	80.0	72.1

- Plot this data as two line graphs on the same set of axes. Figures are given in millions of tonnes of oil equivalent.
- Comment on the shape of your line graphs.
- Why might it be unhelpful to plot these line graphs on the same axes as the ones above?

5.5 Line graphs

- Draw line graphs on paper or using ICT
- Interpret line graphs

What's the BIG idea?

- A **line graph** is a series of points joined by straight lines. Level 6
- Line graphs are often used to show **trends** over periods of time. These are called **time-series** graphs. Time goes on the horizontal axis. Level 6
- **Spreadsheet** software can be used to draw line graphs. Level 6

Why learn this?

An ECG (electrocardiogram) test measures the electrical activity of the heart. This produces a line graph.

Practice, practice, practice!

Level 6

6b I can construct simple line graphs using ICT

1 The table shows the average monthly rainfall in Jakarta, Indonesia.

Month	J	F	M	A	M	J	J	A	S	O	N	D
Rainfall (mm)	300	300	211	147	114	97	64	43	66	112	142	203

Use spreadsheet software to draw a line graph of the data.

2 The table shows the takings for a second-hand bookshop over a two-week period.

	Mon	Tue	Wed	Thu	Fri	Sat	Sun
Week 1	£197	£218	£154	£235	£278	£369	£342
Week 2	£182	£196	£173	£224	£301	£352	£285

Use spreadsheet software to draw a line graph of the data.

6b I can construct and use line graphs

3 The table shows the quarterly electricity bills for a household over a two-year period.

	Jan–Mar	Apr–Jun	Jul–Sep	Oct–Dec
2008	£90	£75	£60	£85
2009	£110	£90	£75	£100

Tip The year is divided into four quarters with three months in each quarter.

a Draw a line graph of the data.

b Use your graph to estimate the electricity bill for the first quarter of 2010.

6b I can answer questions about line graphs

4 Alice used a datalogger to measure her core body temperature over a 24-hour period. The results are shown in the graph.

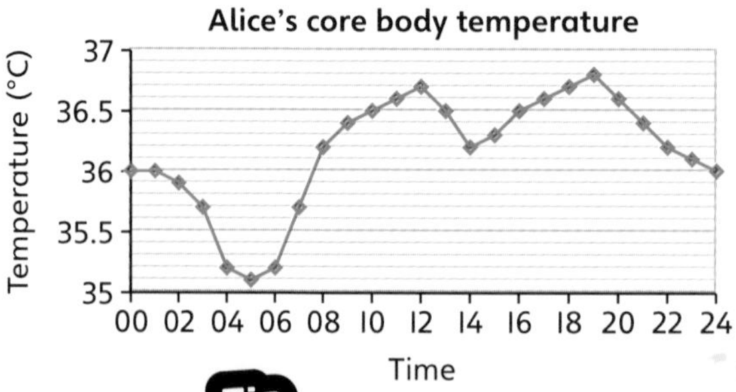

a What was Alice's temperature at midday?

b At what time was her temperature lowest?

c At what time was it highest?

Tip Use a ruler to help you read points off a line graph accurately.

Super fact! Between 13:00 and 16:00 your core body temperature dips, leading to tiredness. People in some countries have a 'siesta' during this time.

line graph time series

5 The graph shows Leroy's and Neil's heart rates over a period during which they exercised.

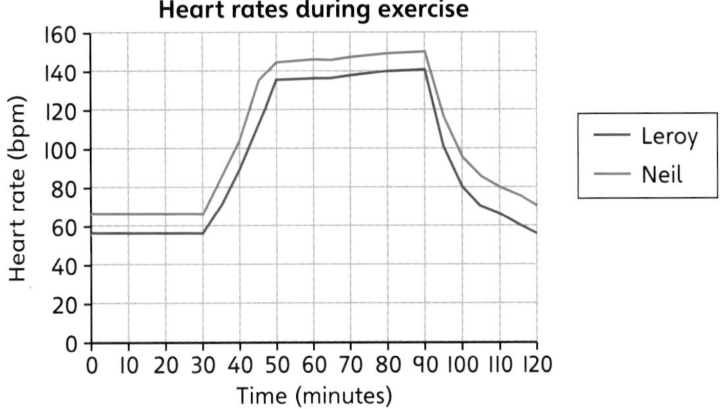

a When did Leroy and Neil start exercising?
b What was Neil's heart rate before he started exercising?
c What was Leroy's maximum heart rate?

6 The graph shows the temperatures of two cups of coffee as they cool. Cold milk was added to each cup at different times.

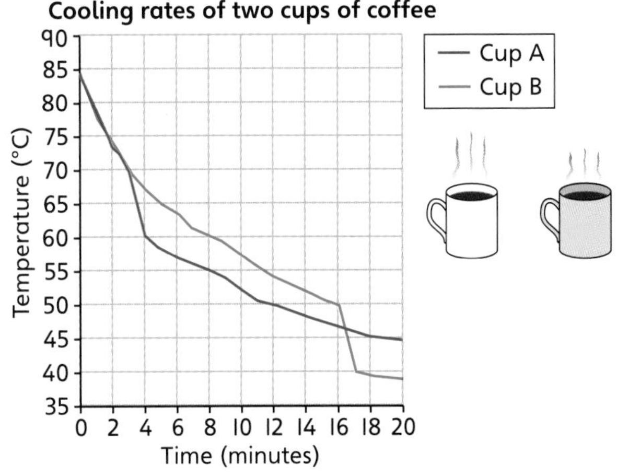

a What was the temperature in both cups after 3 minutes?
b When was cold milk added to cup A? Explain your reason.
c When was milk added to cup B? Explain your reason.
d If you want the hottest coffee possible after 20 minutes, when does the graph suggest you should add the milk, at the start or after 20 minutes?

Level 6

6a I can pick out information from different graphs and use these facts to answer questions

Now try this!

A Weather

Choose a foreign city.
Search websites for graphs showing the weather in your chosen city.
How does it compare to the weather in the UK?

B Key features

Search for websites that show examples of line graphs.
Write a paragraph explaining the key features of one of the graphs you find.

5.6 Frequency diagrams and frequency polygons

- Read information from frequency diagrams
- Group data in equal class intervals
- Draw frequency diagrams for grouped continuous data

Why learn this?

A good diagram can make data come alive.

What's the BIG idea?

- In a grouped **frequency table**, the groups should be of equal width and they should not overlap. You can use inequalities to show the **class intervals**.
  For example 30 kg ≤ weight < 40 kg includes all the weights from 30 kg up to, but not including, 40 kg. **Level 6**
- **Frequency diagrams** can be used to show **grouped continuous data**. frequency diagram looks a bit like a bar chart, but there are no gaps between the bars. **Level 6 & Level 7**
- You can draw a **frequency polygon** by joining the mid-points of the tops of the bars in a frequency diagram. **Level 6 & Level 7**

Tip
Frequency polygons are really useful for comparing different sets of data on the same diagram.

Practice, practice, practice!

Level 6

6c I can construct a simple frequency diagram for continuous data (choosing class intervals)

6b I can compare two distributions using the shape of the distributions

1 One morning, a vet weighed 12 dogs. The mass of each is given in kilograms.

21 18 27 24 23 16 12 29 23 19 22 24

Draw a frequency diagram to show the data. Use four groups.

2 A market research company has drawn these frequency diagrams to show the average times spent food shopping each week by households in the UK and Spain.

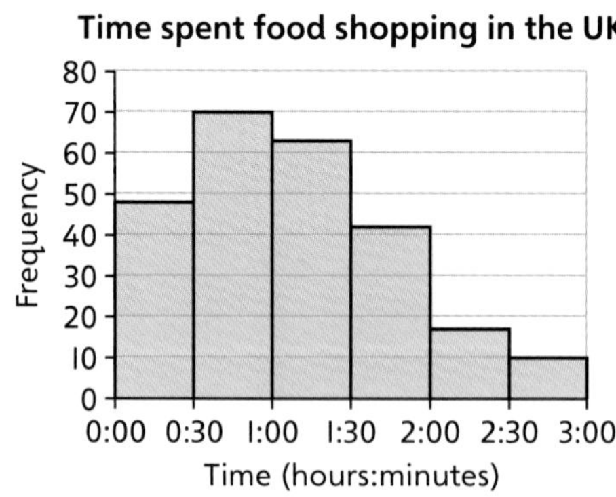

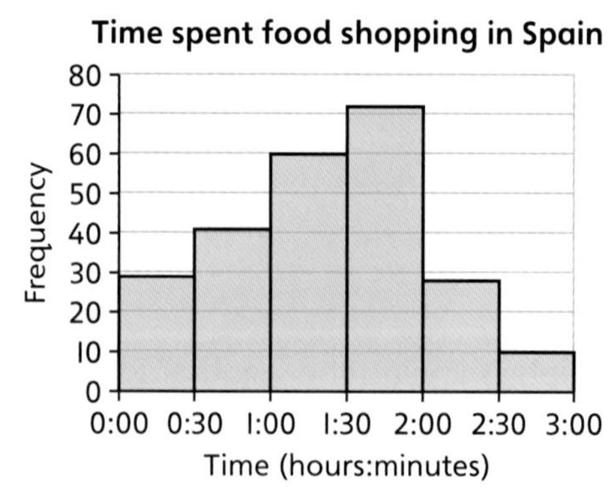

In which country do households spend the least time on their food shopping? Use the frequency diagrams to justify your answer.

class interval continuous discrete frequency diagram

3 At an airport check-in desk, 24 suitcases were weighed. The mass of each is given in kilograms.

20.4	15.3	17.6	17.0	19.4	18.5
21.0	22.5	19.7	20.1	23.1	16.4
21.9	18.2	21.7	22.5	20.5	23.2
15.8	20.7	16.9	19.7	22.7	18.1

Construct a grouped frequency table for the data.

Watch out!
It's easy to miss out a piece of data when constructing a frequency table. Check that the total frequency is the same as the number of pieces of data.

Level 6

6a I can construct more complex frequency tables for continuous data

4 This frequency polygon shows the lengths of time cars stayed in a car park one day.

a How many cars stayed for between 30 and 60 minutes?

b How many cars stayed for longer than 2 hours?

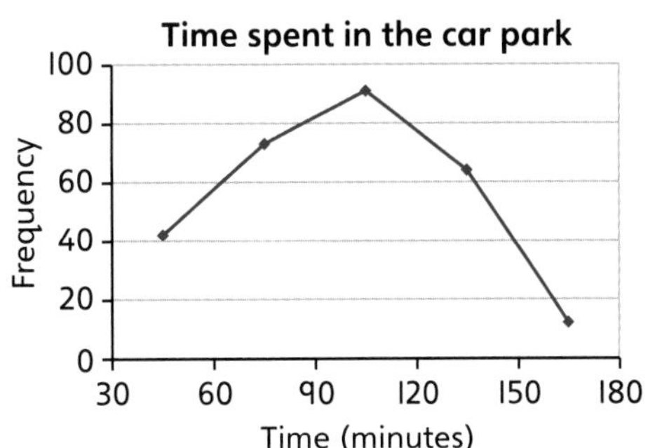

5 These are the distances, in light years, of the 20 nearest stars (excluding the Sun) to the Earth.

11.9	10.5	11.3	7.8	4.4	8.6	9.7	11.4	10.9	11.8
11.8	12.0	4.2	6.0	10.3	8.7	11.4	11.6	11.5	8.3

a Draw a frequency polygon for this data.

b Write a sentence to describe what your frequency polygon shows.

6 The table shows the ages of members of a brass band in 2008 and 2009.

Age	10–19	20–29	30–39	40–49	50–59
2008	17	23	30	19	10
2009	24	28	22	11	8

a Draw a frequency polygon for the 2008 data.

b On the same axes, draw a frequency polygon for the 2009 data.

c Compare the distribution of band members' ages in the two years.

Level 7

7c I can identify key features in frequency polygons

7b I can construct and use frequency polygons

7a I can construct and use superimposed frequency polygons to compare results

Now try this!

A Think of a question

Think of a question that you could answer using a graph on this spread. Write your question on one side of a piece of paper and the answer on the other. Give your question to a partner. Does their answer agree with yours?

B Frequency polygon

Try using the XY (scatter) option in your spreadsheet software to reproduce the frequency polygon you drew for Q6.

frequency polygon **frequency table** **grouped data**

Data on disease

For a humanities project, Reshma is preparing a presentation on world health issues.
She has decided to compare causes of death in developed and developing countries.

1 Suggest possible sources of data on causes of death in different countries.
Are these primary or secondary sources? **Level 6**

2 Reshma gets some data from the World Health Organization.
This table, from the *WHO Health Report 2002*, gives the leading causes of death in 2001.

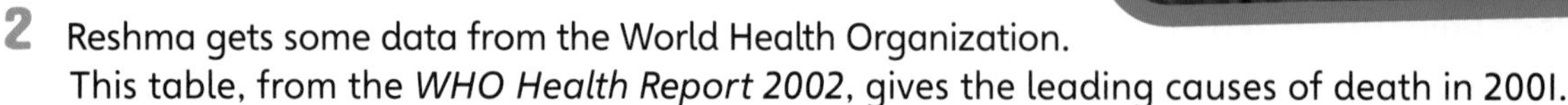

Developing countries	Number of deaths	Developed countries	Number of deaths
1. HIV/AIDS	2 678 000	1. Ischaemic heart disease	3 512 000
2. Lower respiratory infections	2 643 000	2. Cerebrovascular disease	3 346 000
3. Ischaemic heart disease	2 484 000	3. Chronic obstructive pulmonary disease	1 829 000
4. Diarrhoeal diseases	1 793 000	4. Lower respiratory infections	1 180 000
5. Cerebrovascular disease	1 381 000	5. Trachea/bronchus/lung cancers	938 000
6. Childhood diseases	1 217 000	6. Road traffic accidents	669 000
7. Malaria	1 103 000	7. Stomach cancer	657 000
8. Tuberculosis	1 021 000	8. Hypertensive heart disease	635 000
9. Chronic obstructive pulmonary disease	748 000	9. Tuberculosis	571 000
10. Measles	674 000	10. Self-inflicted	499 000

a What was the leading cause of death in developed countries?

b What was the leading cause of death in developing countries?

c Which causes of death were among the ten most common for both types of country? **Level 6**

3 In the table, the causes of death highlighted in green are infectious or communicable diseases – people can 'catch' them. The ones highlighted in blue are non-communicable conditions – they result from lifestyle or environmental factors.
The rest are classified as 'injuries'.
Construct a two-way table to show the number of deaths for each of these three categories for developing and developed countries. **Level 6**

MAKE MATHS FUNCTIONAL!

4 Draw two pie charts to show the data from your table in Q3.
Make one pie chart for developing countries and one for developed countries. **Level 6**

5 People with chronic obstructive pulmonary disease have permanently damaged lungs and find it difficult to breathe. The most common cause is smoking. Reshma decides to find out how many people in her local town smoke. She waits outside her local gym and asks people as they come out.

a Give a reason why this method of sampling may be biased.

b Suggest another way that she could obtain this data. **Level 7**

The BIG ideas

- If you carry out a **survey**, the data you collect is called **primary** data. Alternatively, you can use data that is already available in a newspaper or on a website. This is called **secondary** data. **Level 6**
- A **frequency table** can be used to show a number of observations and their frequency. **Level 6**
- A **two-way table** shows data sorted according to two sets of categories. **Level 6**
- A **pie chart** uses **sectors** of a circle to show different categories of data. **Level 6**
- A **line graph** is a series of points joined by straight lines. **Level 6**
- In a grouped **frequency table**, the groups should be of equal width and they should not overlap. You can use inequalities to show the **class intervals**. **Level 6**
- An **average** is a single number used to represent a set of data. The **mean**, **median** and **mode** are averages. The **range** is a measure of spread. **Level 6**
- **Frequency diagrams** can be used to show **grouped continuous data**. A frequency diagram looks a bit like a bar chart, but there are no gaps between the bars. **Level 6 & Level 7**
- You can draw a **frequency polygon** by joining the mid-points of the tops of the bars in a frequency diagram. **Level 6 & Level 7**
- A survey is carried out on a **sample** of people. A sample should **represent** a whole population. A sample might be **biased** if it only represents part of the population, for example, if you only ask one age group or gender. **Level 7**

Find your level

Level 6

Q1 There are 120 trainee accountants at Moneybagz. 45 of the trainees are women. The pie chart is not drawn accurately. What should the angles be?

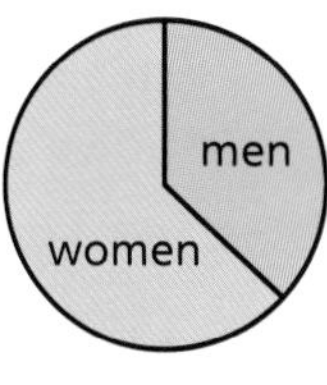

Q2 The graph shows the temperature of a liquid as it cools.

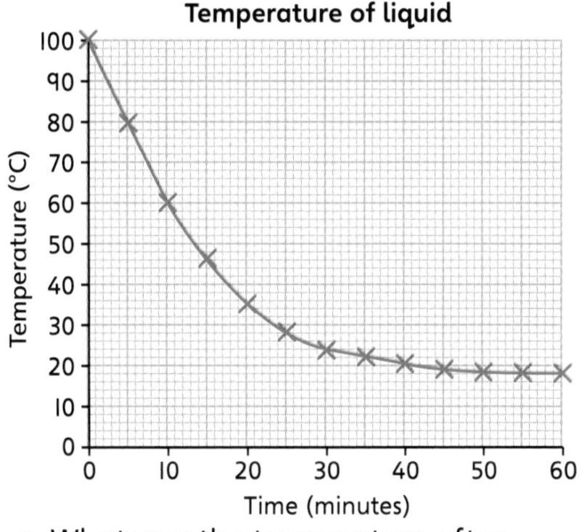

a What was the temperature after 5 minutes?

b How long did it take to reach 35°C?

c What was the lowest temperature reached and when did the liquid reach this temperature?

Level 7

Q3 Find four numbers with

a mode 3, median 6 and range 7

b mean 6, mode 6, median 6 and range 6.

Q4 Kelly is doing a survey on people's favourite foods. She has written this question:

> *Do you think cheese is*
> - *the best topping for jacket potatoes?*
> - *quite a good topping for jacket potatoes?*

What is wrong with her question?

Q5 Doctors advise that people should drink around 1200 m*l* of liquids each day. Alex recorded how much he drank each day for a fortnight, to see if he drinks the recommended amount.

1030	1310	1210	1220	1130	1280	1130
1360	1220	1190	1070	1420	1250	1330

a Construct a frequency table for this data.

b Draw a frequency polygon for the data.

Revision 1

Quick Quiz

Q1 Work out the first four terms of this sequence.
$T(n) = 2n + 5$
→ *See 1.1*

Q2 Decide whether each of these is an expression, an equation or an identity.
a $2p + 2p + 5 - 2 = 4p + 3$
b $6n + 7 = 3n + 14$
c $9t + 8 - 2(t - 3) + 5$
→ *See 2.1*

Q3 Estimate the height of this cube to 1 d.p.
Do not use a calculator.
→ *See 3.7*

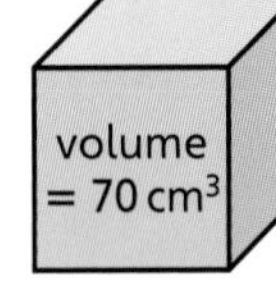

Q4 Calculate the missing angles in this shape.
→ *See 4.2*

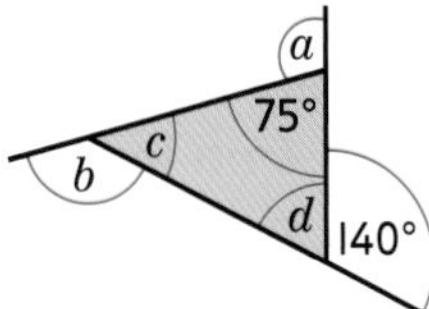

Q5 What is the 5th term in this sequence?
$T(n) = 3n^2 + 4n$
→ *See 1.3*

Q6 Solve the equation $4(x + 2) = 2x + 26$.
→ *See 2.3*

Q7 Work out $\frac{17}{20} + 2\frac{5}{12}$
→ *See 3.2*

Q8 Which of these is the correct formula for the lengths of the sides of this triangle?
a $a^2 + b^2 = c^2$
b $a^2 + c^2 = b^2$
c $b^2 + c^2 = a^2$
→ *See 5.5*

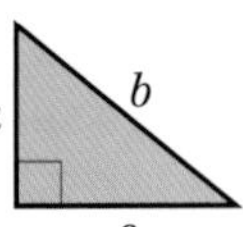

Q9 Find three numbers that have a mode of 4 and a mean of 6.
→ *See 5.3*

Q10 The first four terms of a sequence are
$\frac{3}{1}, \frac{5}{4}, \frac{7}{9}, \frac{9}{16}$.
a Write down the next two terms in the sequence.
b Work out the rule for the nth term.
→ *See 1.3*

Q11 **a** Write down a pair of simultaneous equations using this rectangle.
b Solve your equations to find x and y.
→ *See 2.6*

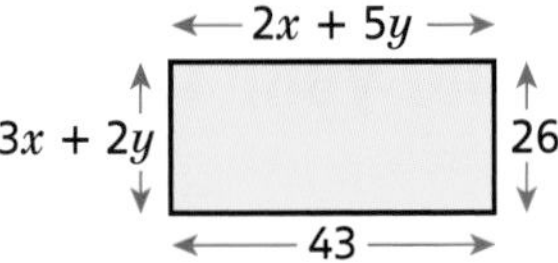

Q12 Sam invests £1800 at an interest rate of 4.5%. What is her investment worth after three years?
→ *See 3.4*

Activity

Imagine you are a dairy farmer. At the end of the year you analyse the milk production figures from your farm.
This table shows the amount of milk your cows produced each month in 2008.

Month	Jan	Feb	Mar	Apr	May	Jun
Milk produced in thousands of litres	56	58	61	62	69	77

Month	Jul	Aug	Sep	Oct	Nov	Dec
Milk produced in thousands of litres	76	80	69	69	66	63

LEVEL 6

Q1 Use an assumed mean of 50 000 litres per month to calculate the mean monthly milk production in 2008.

Q2 The mean and median of the milk produced by two cows during the year is shown below.

	Mean (litres/month)	Median (litres/month)
Daisy	518	510
Flower	495	520

Which cow is the better milk producer? Give a reason for your answer.

Q3 a Draw a line graph to show the monthly milk production in 2008.

b The line graph below shows the average monthly maximum temperature and rainfall at your farm.

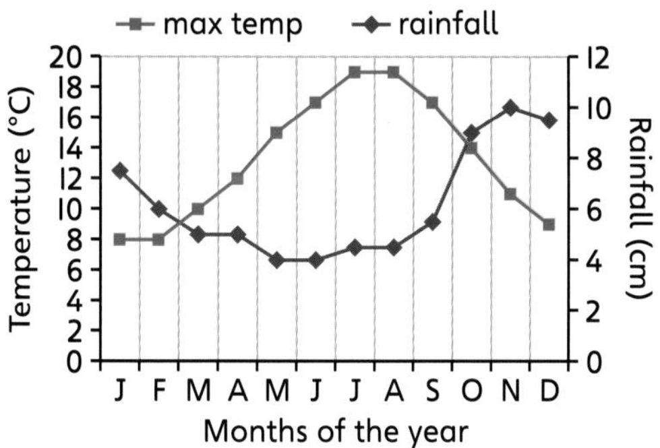

Make two comparisons between your graph in part **a** and the temperature and rainfall line graph above.

LEVEL 7

Q4 The milk company wants more milk to be produced in October, so they pay 140% of the normal milk price.
In October you received a total of £17 388.
What was the normal milk price in pence per litre?

Q5 In 2009, you want to increase your milk production from 2008 by 12%. How much milk do you want to produce in 2009? Give your answer to 1 s.f.

LEVEL 8

Q6 On your farm you have two rectangular fields that are similar. The ratio of the lengths of the sides is 1 : 2. The area of the smaller field is 1500 m^2. You allow your cows to graze in the smaller field for six days before moving them to the larger field. How many days do you expect to graze your cows in the larger field?

Find your level

LEVEL 6

Q1 Look at these pairs of number sequences.
The second sequence is formed from the first sequence by adding a number or dividing by a number.
Work out the missing nth terms.

a 5, 8, 11, 14, … nth term is $3n + 2$
7, 10, 13, 16, … nth term is ……………

b 10, 12, 14, 16, … nth term is $2n + 8$
5, 6, 7, 8, … nth term is ……………

Q2 The area of one face of this cube is $4y^2$.
Write an expression for the **total surface area** of the cube.
Write your answer as simply as possible.

Area = $4y^2$

Q3 Write the missing numbers in these fraction sums.

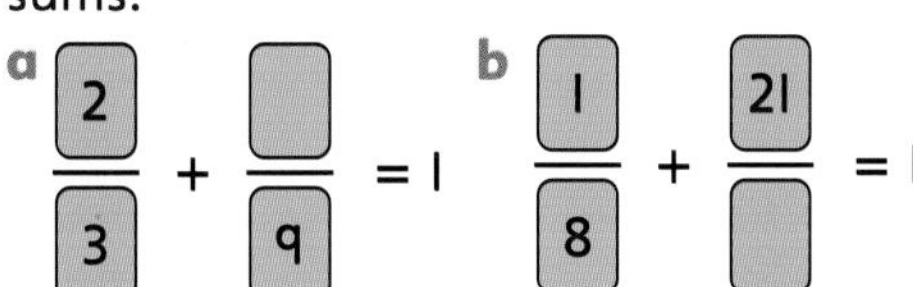

a $\frac{2}{3} + \frac{\square}{9} = 1$ **b** $\frac{1}{8} + \frac{21}{\square} = 1$

LEVEL 7

Q4 The diagram shows a square with side length 4 cm.
The length of the diagonal is x cm.
Show that the value of x is $\sqrt{32}$.

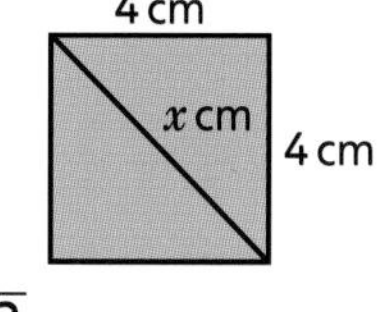

Q5 Lisa is doing a survey on healthy eating.
She has written this question.

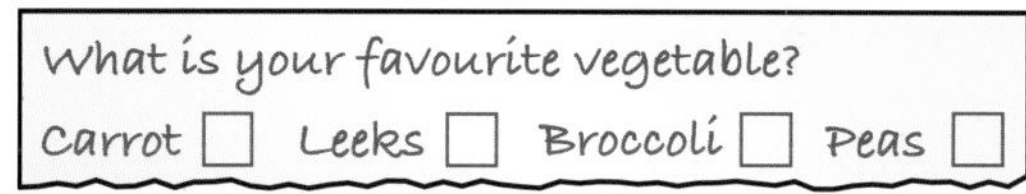

Write down two reasons why this question is not a good question.

LEVEL 8

Q6 Match each sequence card with the correct nth rule card.

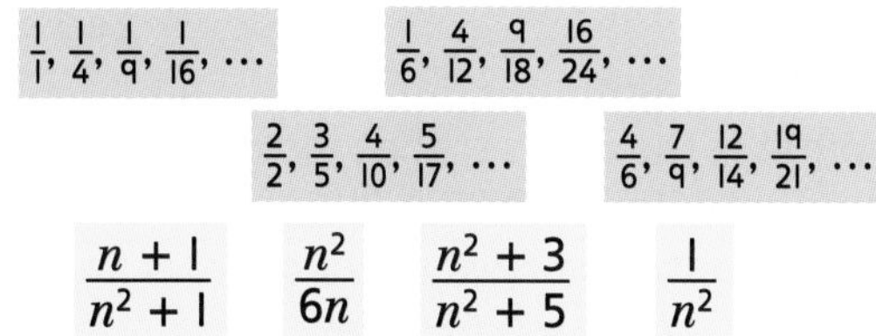

$\frac{1}{1}, \frac{1}{4}, \frac{1}{9}, \frac{1}{16}, \ldots$
$\frac{1}{6}, \frac{4}{12}, \frac{9}{18}, \frac{16}{24}, \ldots$
$\frac{2}{2}, \frac{3}{5}, \frac{4}{10}, \frac{5}{17}, \ldots$
$\frac{4}{6}, \frac{7}{9}, \frac{12}{14}, \frac{19}{21}, \ldots$
$\frac{n+1}{n^2+1}$
$\frac{n^2}{6n}$
$\frac{n^2+3}{n^2+5}$
$\frac{1}{n^2}$

Q7 In this diagram PR is a tangent to the circle with centre O.

a Explain why $\angle OQR = 90°$.

b Calculate the size of $\angle ROQ$.

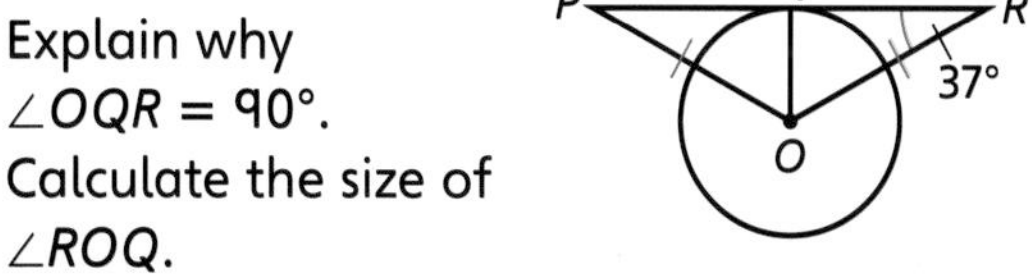

6 Extreme measures

This unit is about units of measurement and calculating areas and volumes.

Mike Fossum describes himself as an astronaut, engineer and space-walker. His work involves very precise measuring and calculating of distances, shapes and speeds. Mistakes in space can spell disaster.

Mike says, 'Coming down from space we have to fire the engines from the other side of the Earth to ensure that we hit the runway in Florida precisely. The Shuttle travels at 17 000 miles per hour so you have to get all your measurements and calculations right, or you'll come down swimming in the Gulf of Mexico, or in Disneyland!'

'My most exciting moment in space was the first time I stepped out to do a space walk. It is an amazing feeling.'

Astronauts need to be careful to use the right units. In the USA, astronauts use imperial measurements (pounds and miles) whereas in Europe they use metric (kg and km). This has caused problems in the past. In 1999 an American Mars Lander was lost because of a mistake converting between units. This error cost hundreds of millions of dollars.

Activities

A Planets do not orbit the Sun in perfect circles and their distances from the Sun vary.

Planet	Distance from Sun	
	Minimum	Maximum
Mars	207 million km	249 million km
Earth	147 million km	152 million km

- What is the nearest that Mars could be to Earth in million km?
- What is the shortest distance from Earth to Mars in miles? 1 km = 0.62 miles.
- A spacecraft takes 8 months to travel to Mars. What is the average speed **a** in km per hour and **b** in miles per hour?

B A craft has 3 astronauts. Each astronaut needs 3 packs of food per day.

- How many packs are needed for a journey lasting 8 months?
- Each pack is a cuboid measuring 5 cm × 10 cm × 10 cm. What volume of space is needed on board the craft to store the food?

The missing letter
Not only do Americans use different units, they say 'math' instead of 'maths'!

Before you start this unit...

1 Copy and complete these.

a 20 cm = ■ mm
b 6.5 m = ■ cm
c 3200 m = ■ km
d 5200 g = ■ kg
e 6.1 kg = ■ g
f 250 cl = ■ l
g 76 ml = ■ cl

2 Work out the area of these rectangles.

Level Up Maths 5-7 page 92

a

b

c

3 Round these numbers to one decimal place.

a 6.449 **b** 4.009
c 15.983 **d** 0.908

6.1 Converting between measures
6.2 Compound measures and bounds
6.3 Circles and perimeter
maths! Roman mosaics
6.4 Circles, sectors and area
6.5 Area and volume
6.6 Volume and surface area
Unit plenary: All wrapped up

Plus digital resources

6.1 Converting between measures

- Convert between area measures
- Convert between volume measures
- Identify the lower and upper bounds of a measurement
- Calculate the lower and upper bounds of an area

What's the BIG idea?

→ You need to know these metric units for area, volume and capacity.

Area	Volume	Capacity
$10\,000\ m^2 = 1$ **hectare** (ha) $10\,000\ cm^2 = 1\ m^2$ $1\,000\,000\ mm^2 = 1\ m^2$ $100\ mm^2 = 1\ cm^2$	$1\,000\,000\ cm^3 = 1\ m^3$ $1000\ mm^3 = 1\ cm^3$	1000 litres (l) $= 1\ m^3$ $1000\ cm^3 = 1$ litre
	$1\ cm^3 = 1$ millilitre (ml) $10\ ml = 1$ centilitre (cl)	

Level 6

→ Measurements are usually given to the nearest whole number or decimal place. A true value lies in a **range** from half a unit below to half a unit above the measurement. Level 7

→ The lower and upper bounds are the minimum and maximum possible values of a measurement. Level 7 & level 8

Why learn this?

The rent of office buildings is often given as an amount per unit area. Different units of area can be used, so you need to be able to convert between units to compare the rents.

Practice, practice, practice!

1 Convert each of these areas to square millimetres (mm^2).

$5\ cm^2$

$5 \times 100 = 500\ mm^2$

a $8\ cm^2$ **b** $4.5\ cm^2$ **c** $19\ cm^2$ **d** $0.6\ cm^2$

2 Convert each of these areas to square centimetres (cm^2).

a $7\ m^2$ **b** $6\ m^2$ **c** $17\ m^2$ **d** $4.2\ m^2$ **e** $0.3\ m^2$

3 Convert each of these areas to square centimetres (cm^2).

a $600\ mm^2$ **b** $3200\ mm^2$ **c** $6940\ mm^2$ **d** $390\ mm^2$
e $41\ mm^2$ **f** $745\,300\ mm^2$ **g** $45\,224\ mm^2$ **h** $2\ mm^2$

4 Convert each of these areas to square metres (m^2).

a $50\,000\ cm^2$ **b** $65\,000\ cm^2$ **c** $590\,000\ cm^2$ **d** $19\,400\ cm^2$
e $4275\ cm^2$ **f** $563\ cm^2$ **g** $16\,574\ cm^2$ **h** $26\ cm^2$

5 Convert each of these volumes to cubic millimetres (mm^3).

a $2\ cm^3$ **b** $12\ cm^3$ **c** $9.6\ cm^3$ **d** $0.82\ cm^3$ **e** $8.3\ cm^3$

Level 6

6b I can convert to smaller metric units used for area

Tip To change large units to small units multiply by the conversion factor.

6b I can convert metric units of area both ways

Tip To change small units to larger units divide by the conversion factor.

6a I can convert to smaller metric units for volume

cubic centimetre cubic metre cubic millimetre hectare lower bound

6 Convert each of these volumes to cubic metres (m^3).

a 9 000 000 cm^3 **b** 2 500 000 cm^3 **c** 42 000 cm^3 **d** 7100 cm^3

7 Convert each of these to litres.

a 7000 cm^3

b 37 000 cm^3

c 257 cm^3

Tip

The unit symbol for litres is the letter *l*. To avoid confusion with the digit 1 (one) the full unit name is often used instead of the letter.

Level 6

6a I can convert metric units of volume both ways

8 Write down the smallest and largest possible values for each measurement.

8.4 cm *range 8.35–8.45 cm*

a 4.2 cm **b** 13.56 m **c** 58 km

d 14 mm **e** 38.2 seconds **f** 725 m*l*

9 These quantities have been measured to the accuracy shown.
Give the upper and lower bounds for each measurement in the form

lower bound $\leqslant x <$ upper bound.

50 m (to the nearest 10 m) *45 m $\leqslant x <$ 55 m*

a 80 km (to the nearest 10 km) **b** 620 cm (to the nearest 10 cm)

c 2300 cm (to the nearest 100 cm) **d** 62 m (to the nearest metre)

e 2.7 km (to 1 d.p.) **f** 3.62 km (to 2 d.p.)

g 47 km (to the nearest kilometre) **h** 8.8 *l* (to 1 d.p.)

i 6.94 km (to 2 d.p.) **j** 370 mm (to the nearest 10)

Level 7

7a I can identify the upper and lower bounds for each measurement

7a I can identify the upper and lower bounds for a given degree of accuracy

10 Each dimension of this rectangle is given to the nearest centimetre.

9 cm

17 cm

a Write down the range of the smallest and largest possible values for each measurement.

b Work out the smallest possible value for the area of the rectangle

c Work out the largest possible value for the area of the rectangle.

d What is the range of the area of the rectangle?

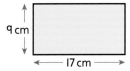

11 Each dimension of this rectangle is given to the nearest millimetre.
What is the range for the area of the rectangle?

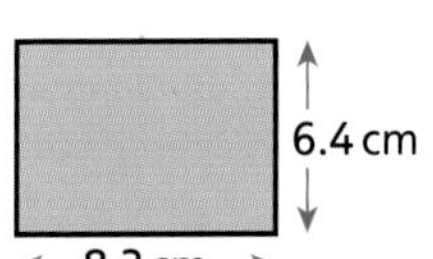

Level 8

8b I can calculate the lower and upper bounds of area measurement

Now try this!

A Conversions poster

Make a poster to show the conversions between different units of area.

B Volume matching

Work with a partner to make a matching game.
You need 16 blank cards. Write different volumes on eight cards and the same volumes but in different units on eight more. Shuffle the cards and challenge another pair to match them up.

range square centimetre square metre square millimetre upper bound

6.2 Compound measures and bounds

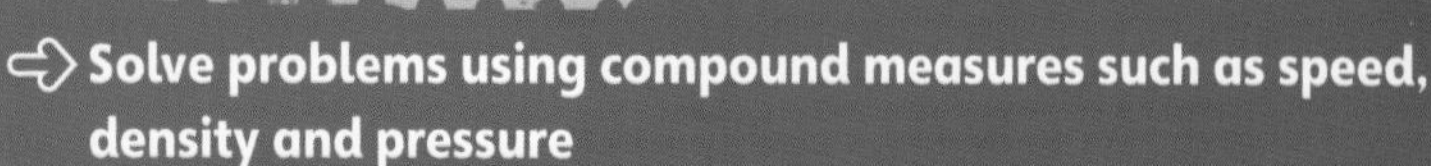

- Solve problems using compound measures such as speed, density and pressure
- Do simple conversions of units of compound measures
- Recognise graphs which show constant, average or variable rates of change
- Calculate the upper and lower bounds of compound measures

Why learn this?

There are legal bounds on the accuracy of a car speedometer. If you are travelling at 60 mph, the the speedometer must show between 60 mph and 72.25 mph.

What's the BIG idea?

- Compound measures combine measures of two different types. For example, speed measures distance and the time taken to travel it.
  For example,
  speed $= \frac{\text{distance}}{\text{time}}$
  density $= \frac{\text{mass}}{\text{volume}}$
  pressure $= \frac{\text{force}}{\text{area}}$ (area on which the force is applied) **Level 7**
- If a rate of change is constant, the two variables are in direct proportion and are connected by a simple formula. **Level 7**
- **average** speed $= \frac{\text{total distance travelled}}{\text{total time taken}}$ **Level 8**
- To calculate a **range** for compound measures, use the least possible values and greatest possible values for each measurement. **Level 8**

Did you know?

Andy Roddick holds the record for the fastest tennis serve. The ball travelled at a speed of 155 mph or 249.4 kph. It happened during a Davis Cup match in 2004.

Watch out!

Check the units required by the question. For speed calculations, rewrite any minutes as a fraction of an hour if you need to.

Practice, practice, practice!

Level 7

7b I can recognise which graphs show constant, average or variable rates of change

1 Which of these graphs is most likely to show

A

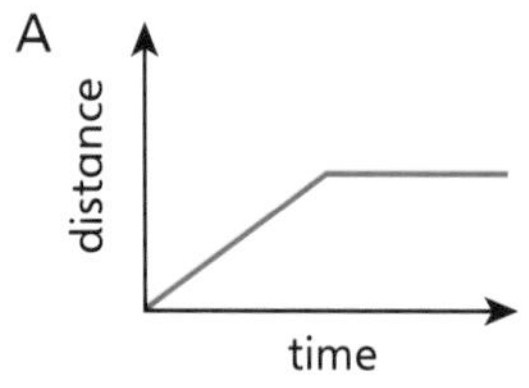

B

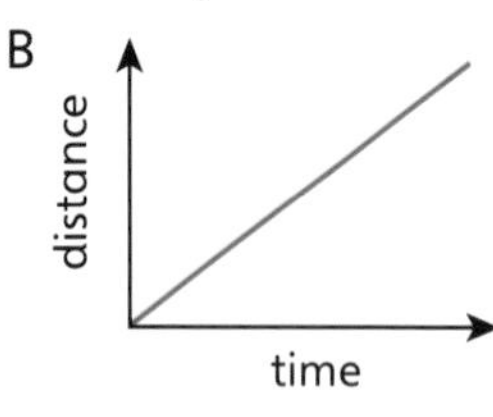

C

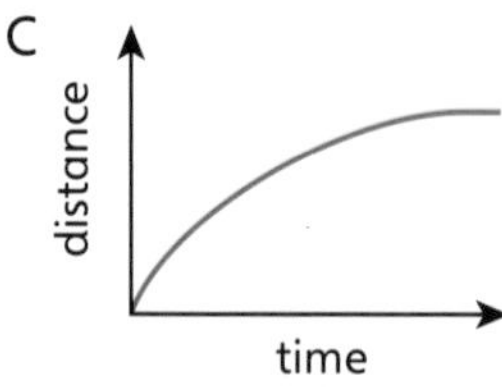

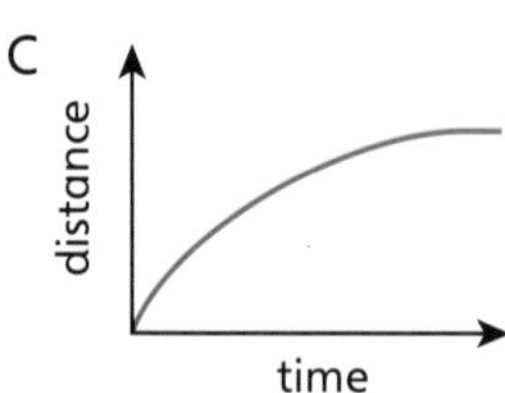

a a car travelling at a constant speed on a motorway

b a car slowing down as it goes up a hill

c a car travelling at a constant speed and then stopping in an emergency?

Super fact!

In a government survey on vehicle speeds published in 2007, 54% of drivers on a motorway exceeded the speed limit of 70 mph.

average density pressure

2 The graphs show how the depth of water in a container varies with time as water is poured into it at a steady rate.
Match each graph to a container.

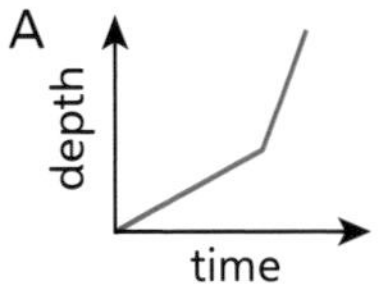

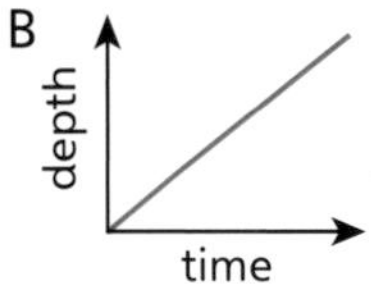

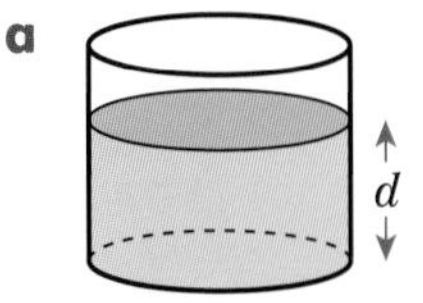

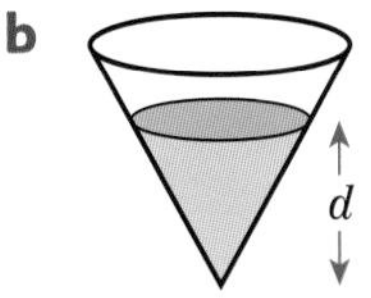

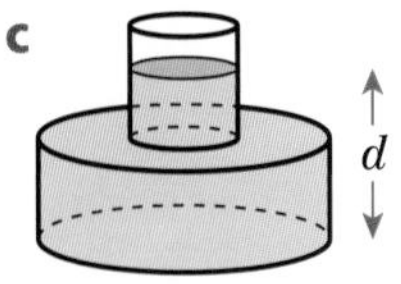

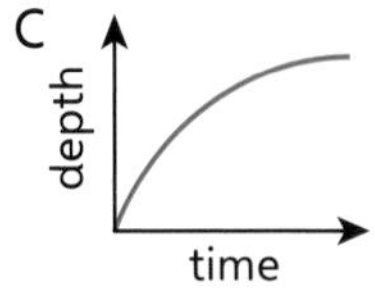

3 Change each of these speeds to metres per second.

72 km/h   1 km = 1000 m
1 hour = 60 × 60 = 3600 seconds $72 \times \frac{1000}{3600} = 20$ m/s

a 90 km/h **b** 150 km/h **c** 2560 km/h **d** 384 km/h

4 Change 15 m/s to kilometres per hour (km/h).

5 A liquid has a mass of 12 g and a volume of 2.5 cm^3. Work out its density.

6 The density of iron is 7870 kg/m^3. Work out the mass of 4.5 m^3 of iron.

7 A force of 20 N pushes on an area of 2.5 cm^2. Work out the pressure.

Level 7

7b I can recognise which graphs show constant, average or variable rates of change

7a I can convert simple units of compound measures

7a I can solve problems using constant rates and related formulae

8 Look at this distance–time graph for three journeys.
Which journey had
a the highest speed?
b the highest average speed over the whole journey?

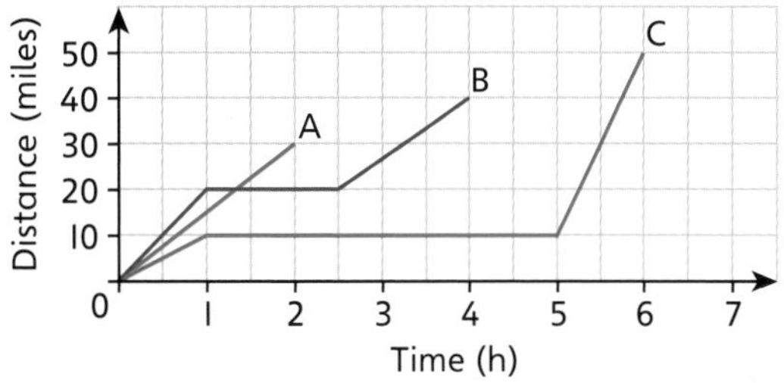

9 Katherine drove 160 miles, to the nearest mile, in exactly 3 hours.
Work out
a Katherine's minimum possible average speed
b her maximum possible average speed.

10 The mass of an object is 56.2 g, correct to 1 d.p.
The volume of the object is 12 cm^3, correct to the nearest whole number.
Work out
a its minimum possible density **b** its maximum possible density.

Level 8

8c I can solve problems using average rates of change and their formulae

8a I can calculate the upper and lower bounds of compound measures

Now try this!

A Speed bingo

Work in groups of four. Three players draw 3 × 3 grids and fill each space with a speed in kph or metres per second. They write down the converted amounts, e.g. kph → m/s amd m/s → kph and give them to the fourth played, the Caller. Players swap grids and the Caller calls out the converted amounts randomly until one player has crossed off all their spaces. Check answers.

B Animal sprint

Use websites to research the speeds of different animals, for example lions, cheetahs, tigers, zebras etc. Compile a list of the fastest animals in ascending order.

range speed

6.3 Circles and perimeter

- Know and use the formula for the circumference of a circle
- Find the radius or diameter of a circle given the circumference
- Find the perimeter of compound shapes involving circles
- Use the formula for finding the length of an arc in a circle

Why learn this?

An odometer measures the distance travelled by a vehicle. On a bicycle it counts the number of times the wheel rotates and multiplies this by the circumference of the wheel to give the distance travelled.

What's the BIG idea?

- **Circumference** is another word for the **perimeter** of a circle. It is found using the formula $C = 2\pi r$ which is sometimes written as $C = \pi d$.

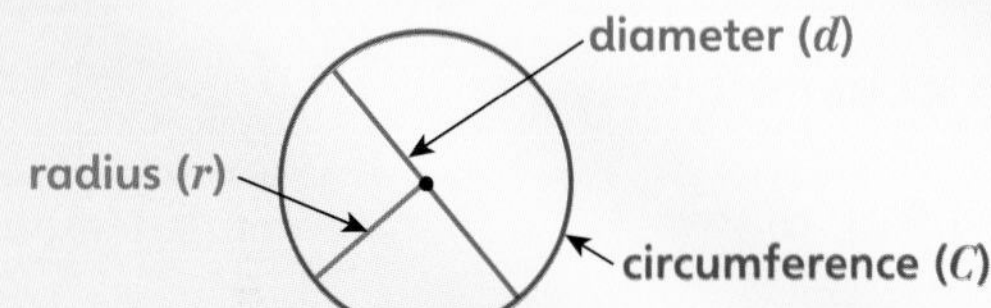

Level 6

- You can rearrange the formulae for the circumference to find the **radius** $r = \frac{C}{2\pi}$ or the **diameter** $d = \frac{C}{\pi}$. Level 7
- The length of an **arc** can be expressed as a fraction of the circumference of a circle.

Length of arc, $l = \frac{\theta}{360°} \times 2\pi r$ or $\frac{\theta}{360°} \times \pi d$. Level 8

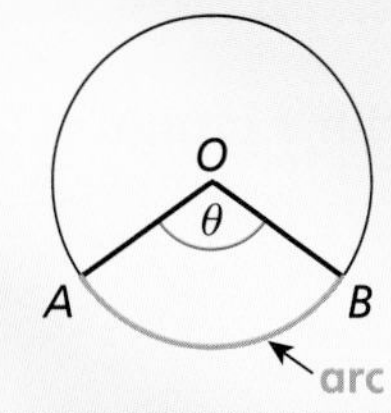

Did you know?

The circumference of the Earth around the equator is 40 075 km.

Tip

Remember that the length of an arc is a fraction of the circumference. This should help you work out the formula if you cannot remember it.

Practice, practice, practice!

In all the questions, use the π key on your calculator.

1 Write down the formula for the circumference of a circle.
Explain what each letter and symbol represents.

2 Find the circumference of each of these circles. Give your answers to 2 d.p.

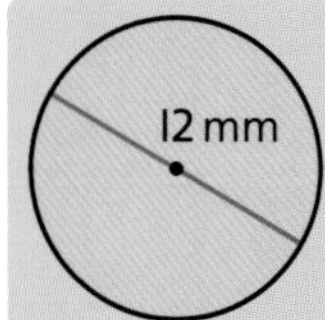

a

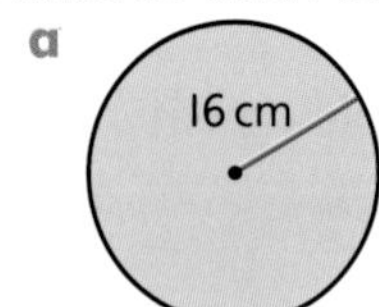

b

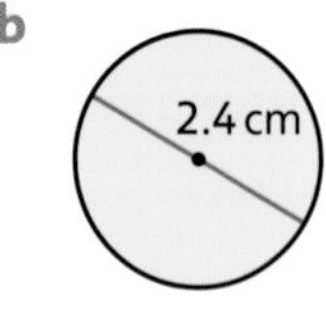

c

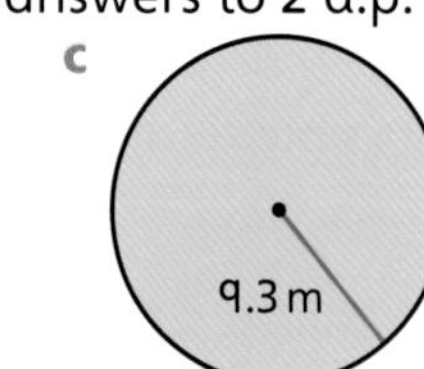

$C = \pi d$, so $C = 3.14 \times 12 = 37.68$ mm

3 The top of a flower pot has a diameter of 8.6 cm.
What is the distance around the edge of the top of the pot?

4 Sam's bicycle has 0.6 m diameter wheels.
On the way to school, the wheels rotate 420 times.
How far does Sam live from the school?

Level 6

6c I can remember the formula for the circumference of a circle

6b I can use the formula for the circumference of a circle

arc circumference diameter

5 The circumference of a circle is 200 cm.
What is the radius of the circle? Give your answer to 1 d.p.

6 The circumference of a circle is 429 cm.
What is the diameter of the circle? Give your answer to 1 d.p.

7 The diagram shows a running track with semicircular curved sections.
What is the distance around the running track?

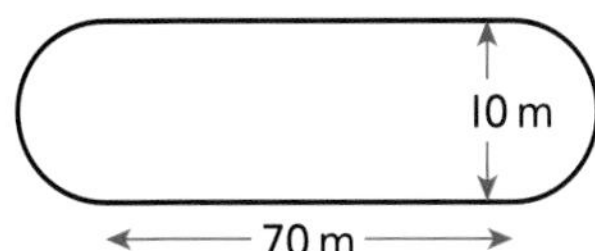

8 Work out the perimeter of this shape.
The curved section is semicircular.

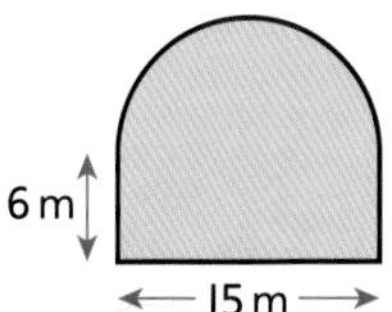

Level 7

7c I can use the formula for the circumference of a circle to calculate the radius or diameter

7b I can find the perimeter of compound shapes involving circles

9 Calculate the length of the minor arc *AB* in each of these circles.
Give your answers to 2 d.p.

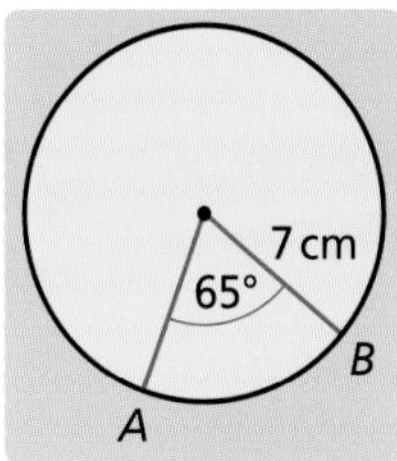

$$AB = \frac{65°}{360°} \times 2\pi r$$
$$= \frac{65°}{360°} \times 2 \times 3.14 \times 7$$
$$= 7.94 \text{ cm}$$

a

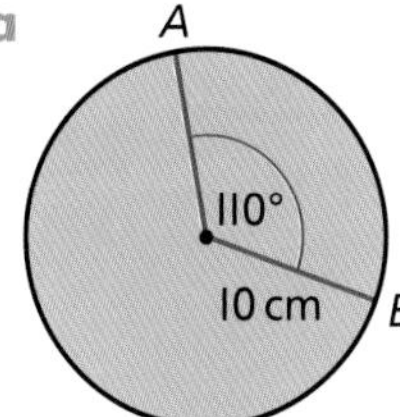

b

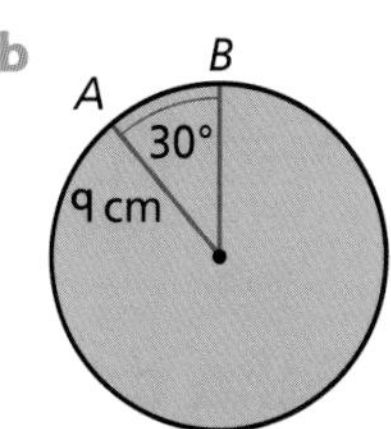

c

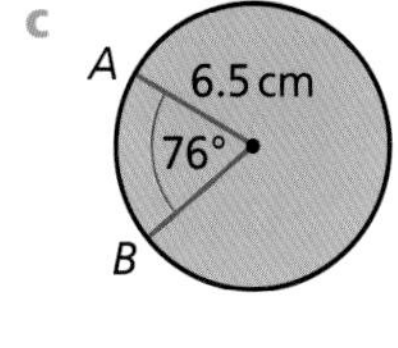

d

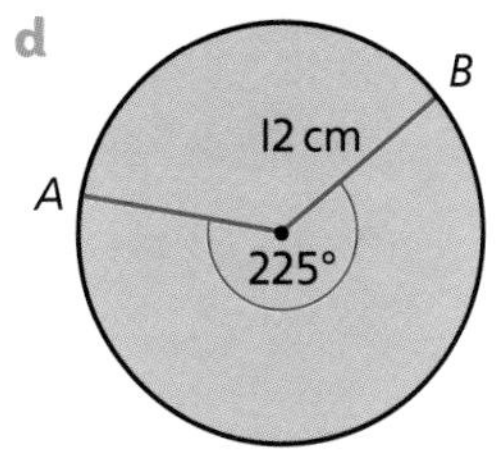

10 A flower bed is in the shape of a sector of a circle.
Calculate the total perimeter of the flower bed.

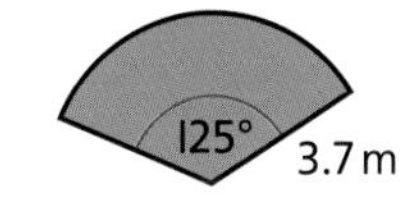

11 The diagram shows a company's logo.
Calculate the total perimeter of the logo.

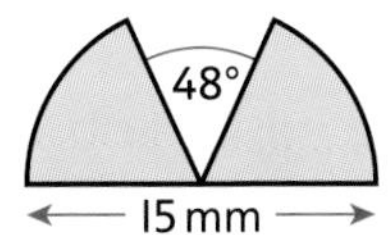

Level 8

8c I can remember and use the formula for the length of an arc

8b I can use the formula for the length of an arc to solve problems

Now try this!

A Circle patterns

Make up your own compound shape out of circles and sectors of circles. Calculate the perimeter of your shape.

B Circle matching

Work with a partner. You need two sets of eight blank cards. On one set of cards draw some circles and on the other set write their circumferences. Shuffle the cards and challenge another pair to match them up.
Repeat the activity with arc lengths of sectors.

Learn this

Learn the formulae for circumference and arc length.

perimeter pi (π) radius

maths!

Roman mosaics

The ancient Romans used small pieces of clay to make up intricate floor patterns called mosaics. Many Roman mosaics have survived in good condition and can be found throughout Europe and North Africa.

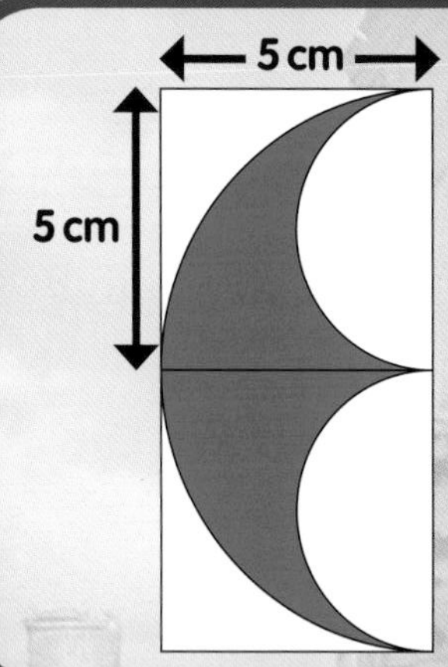

Pelta designs

Roman mosaic designers demonstrated a good grasp of symmetry and rotation. They used a variety of geometric shapes to make the mosaics unique and interesting.

The green shape is called a pelta. It can be found in many Roman mosaics.

Work out the area of the pelta using the measurements shown.

Latin numbers

The citizens of ancient Rome spoke Latin. Latin is called a 'dead' language because it is not spoken today, although many European languages stem from it.
Here are the numbers 1 to 10 in Latin:

1	2	3	4	5	6	7	8	9	10
unus	duo	tres	quattuor	quinque	sex	septem	octo	novem	decem

- Find other European languages that have number words that are similar to their Latin equivalents. Make a list, e.g.:

French: un, deux, ...
Italian: uno, due, ...

Pelta mosaics

Work out the area of the coloured sections in each of these mosaics.
Each tile is a square of side length 5 cm. Give answers to one decimal place.

a

b

c

d

e

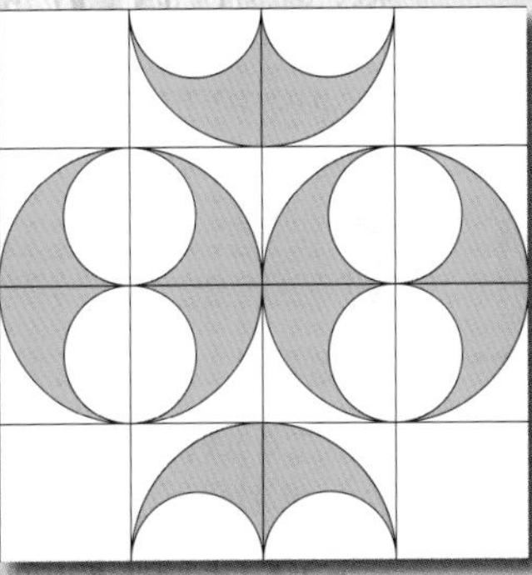

Perimeter activity

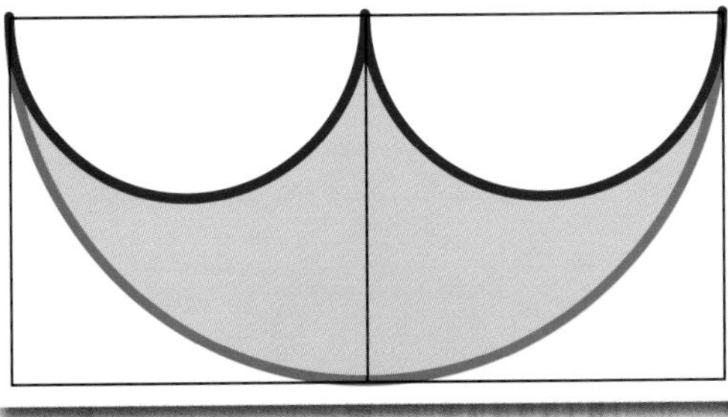

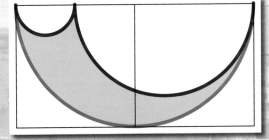

Each tile in these two pelta designs is a square of side length 5 cm. Leave your answers in terms of π

- Work out the length of the red line in the first pelta design.
- Work out the total length of the dark blue line in the first design.
- What do you notice?
- Is this still true if the two smaller semi circles are not the same size (as in the second design)?

6.4 Circles, sectors and area

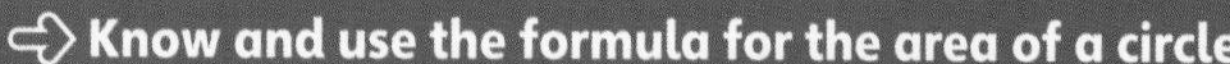

- Know and use the formula for the area of a circle
- Find the area of compound shapes
- Know and use the formula for the area of a sector of a circle

What's the BIG idea?

- **Circumference** is another word for the **perimeter** of a circle. It is found using the formula $C = 2\pi r$ which is sometimes written as $C = \pi d$. Level 6
- The **area** of a circle is found using the formula, $A = \pi r^2$, where r is the **radius**. Level 6
- A compound shape is a shape made up of other shapes. To find its area, find the area of each individual shape and then add all of the areas together. Level 7
- You can rearrange the formula for the area of a circle to find the radius or the **diameter** $r^2 = \frac{A}{\pi}$ so $r = \sqrt{\frac{A}{\pi}}$. Level 7
- The area of a **sector** can be expressed as a fraction of the area of a circle.

  Sector area $= \frac{\theta}{360°} \times \pi r^2$ Level 8

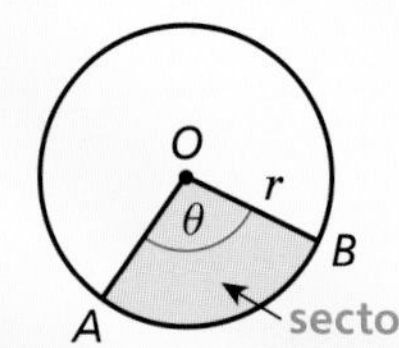

Why learn this?

This lampshade is made from a sector of a circle that has had its centre cut out.

Watch out!

A common mistake is to interpret r^2 as $2 \times r$.

Tip

Remember that the area of a sector is a fraction of the area of the circle. This should help you work out the formula if you cannot remember it.

Practice, practice, practice!

Where necessary, use the π key on your calculator.

1 Find the area of each of these circles. Give your answers to 1 d.p.

a

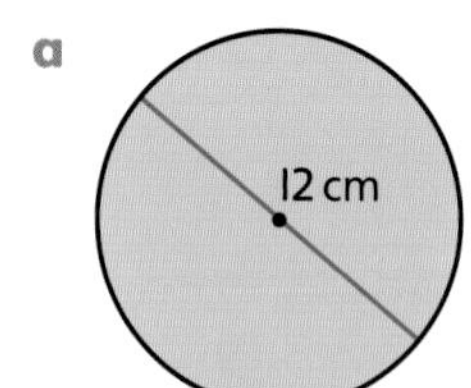

b

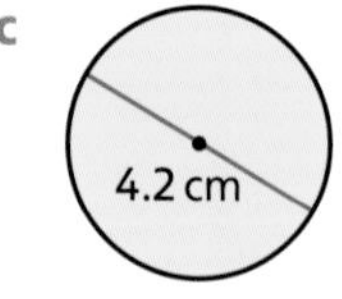

c

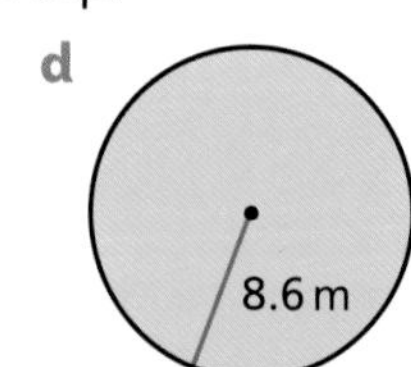

d 8.6 m

Level 6

6a I can use the formula for the area of a circle, if I know the radius or diameter

2 Work out the radius of each of these circles. Give your answers to three significant figures.

a circumference 23 cm
b circumference = 27.5 cm
c area = 500 cm^2
d area = 220 cm^2
e circumference = 60 cm
f area = 42 cm^2

Level 7

7c I can use the circle formulae to find the radius or diameter if I know the circumference or area

area circumference diameter

3 The diagram shows a plan of a garden with semicircular ends.
Work out the area of the garden.

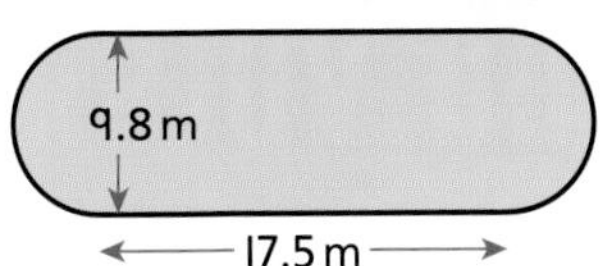

4 Work out the area of the shaded region in this diagram.

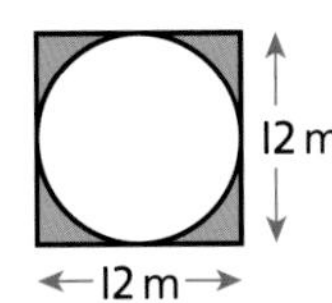

5 In a park there is a circular play space with a diameter of 8.5 m.
At the centre of the play space is a circular sand pit of radius 1.8 m.
What is the area of the play space excluding the sand pit?

Level 7

7b I can find the perimeter of compound shapes involving circles

6 Calculate the area of the sector AOB in each of these diagrams.
Give your answers to 2 d.p.

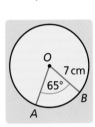

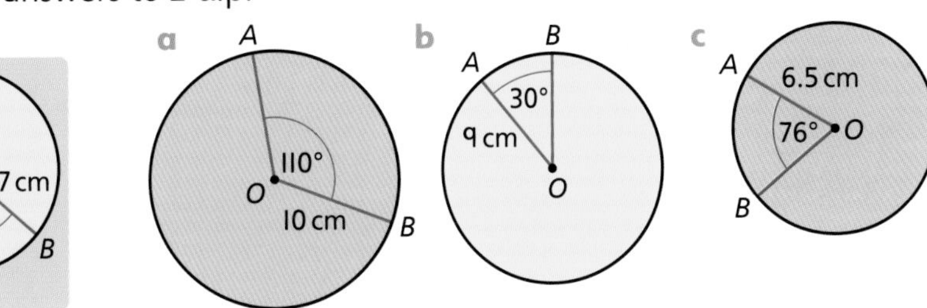

$$\text{Area} = \frac{65°}{360°} \times \pi \times r^2 = \frac{65°}{360°} \times 3.14 \times 7^2 = 27.78 \text{ cm}^2$$

7 A flower bed is in the shape of a sector of a circle.
Work out the area of the flower bed.

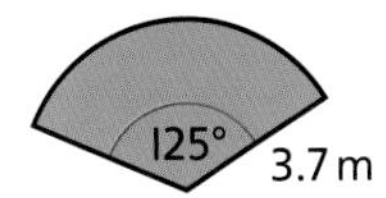

8 Calculate the area of the shaded region in this diagram.

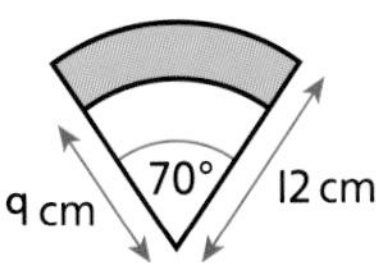

9 A piece of card is in the shape of a rectangle with a sector of a circle on top.
Work out the area of the card.

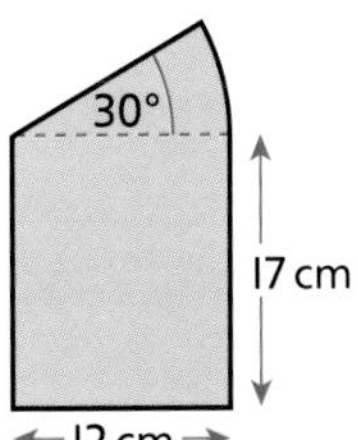

Level 8

8b I can remember the formula for the area of a sector in a circle

8a I can solve problems with areas of sectors of circles

Now try this!

A Circles presentation

Work with a partner. Prepare a presentation on all the key information about circles you have learned so far.

B Circle quiz

Work in a group of four. Prepare five questions about circles. (Make sure you know the answers.) Teams take turns to read out their questions.
The first team to give the correct answer scores 1 point.

Learn this

Learn the formulae for the area of a circle and for sector area.

perimeter pi (π) radius sector

6.5 Area and volume

- Deduce and use the formulae for the areas of a triangles, parallelograms and trapeziums
- Know and use the formula for the volume of a cuboid
- Calculate the surface area of cuboids
- Calculate the volume and surface area of compound shapes made from cuboids

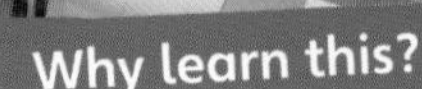

To work out how much wrapping paper you need, you could calculate the surface area of the box.

What's the BIG idea?

- **Area** of a **triangle** = $\frac{1}{2}$ × base × height
  $A = \frac{1}{2}bh$ **Level 6**
- Area of a **parallelogram** = base × height
  $A = bh$ **Level 6**

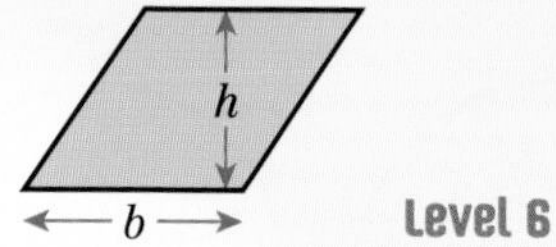

- Area of a **trapezium** = $\frac{1}{2} \times h \times (a + b)$
  $A = \frac{1}{2}h(a + b)$ **Level 6**

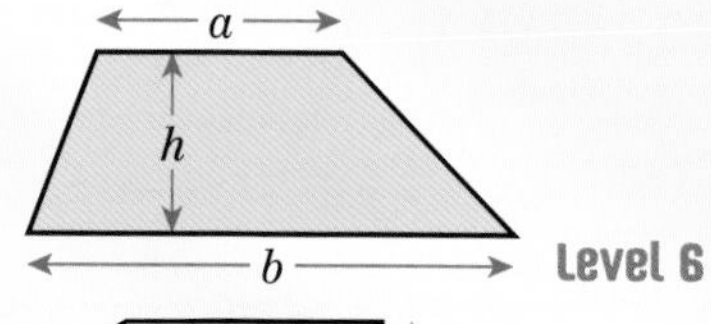

- Multiply the length, height and width of a **cuboid** to find its **volume**.
  $V = l \times h \times w$ **Level 6**

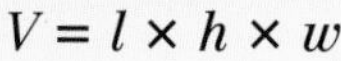

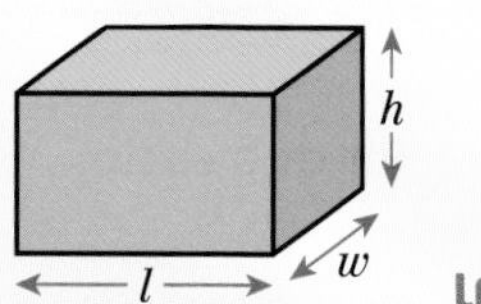

- To calculate the **surface area** of a cuboid, work out the area of each face and add them together.
  Surface area = $2hl + 2hw + 2lw$ **Level 6**

Practice, practice, practice!

Level 6

6C I can explain how to deduce a formula for the area of a triangle

6C I can calculate areas of compound shapes made from rectangles and triangles

6C I can use the formula to calculate the area of a parallelogram

1 **a** Write down the formula for the area of a rectangle.

b Explain how you would use this to work out the formula for the area of a triangle.

2 Calculate the area of each of these shapes.

a

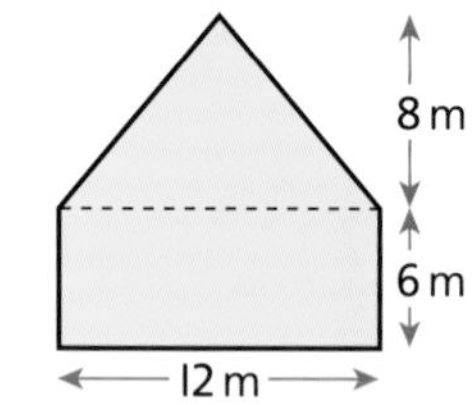

b

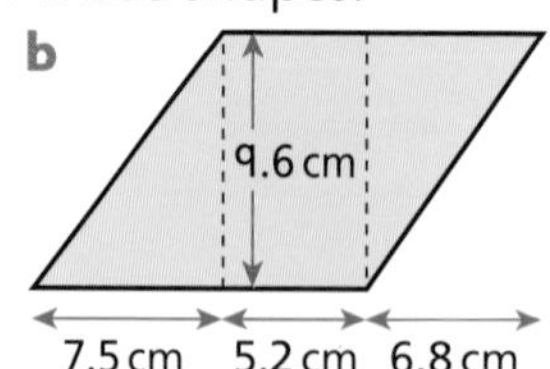

c

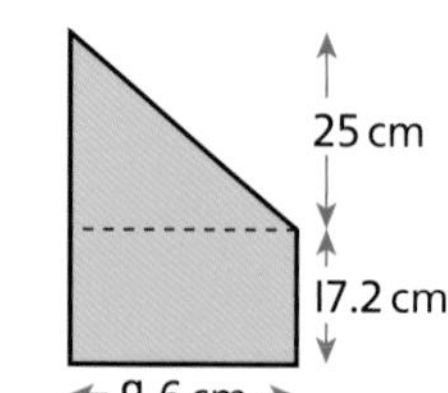

3 Calculate the area of each of these parallelograms.

a

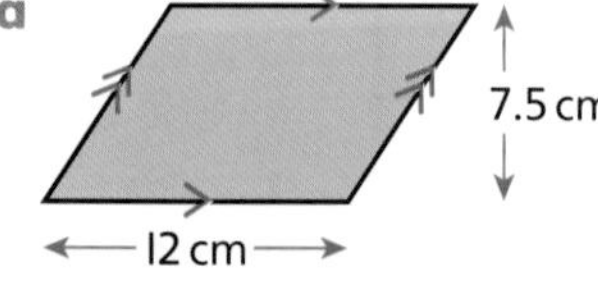

b

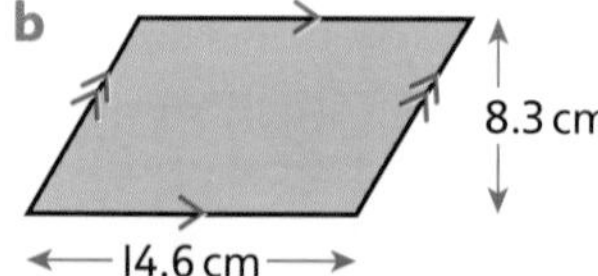

c

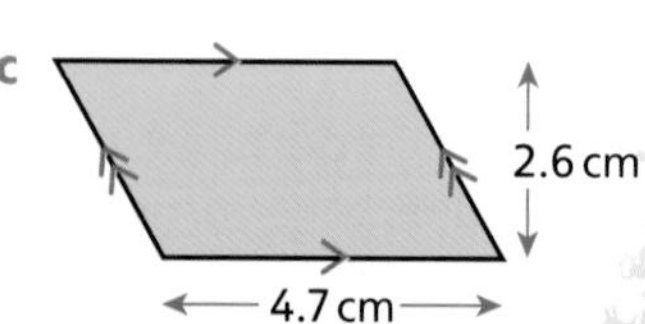

area cuboid parallelogram rectangle

4 Calculate the volume of each of these solid shapes

a
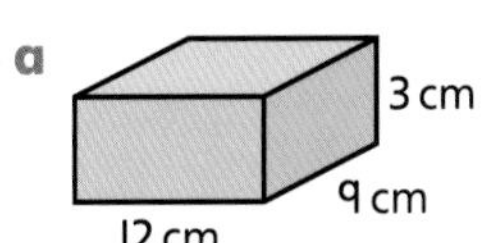

b 2 cm, 3.5 cm, 6.5 cm

c
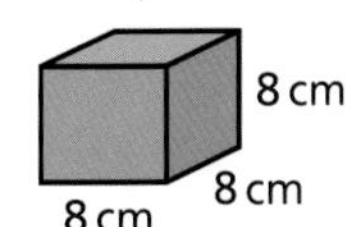

d 12 cm, 16.9 cm, 15.2 cm

5 A cereal box has height and length dimensions of 27.6 cm and 19.2 cm. The width of the box is 5 cm. What surface area of cardboard will be required for the box?

Watch out!
When finding the surface area of cuboids, remember that there are six faces and not three.

6 Calculate the area of each of these trapeziums.

a
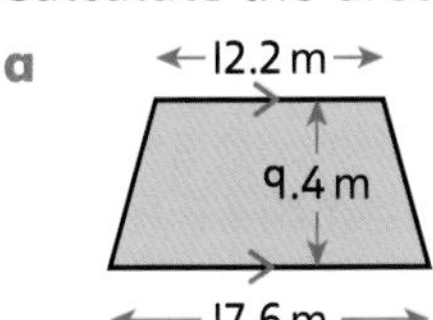

b
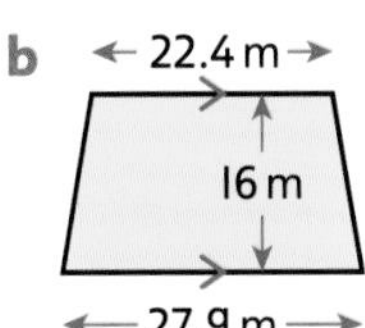

c
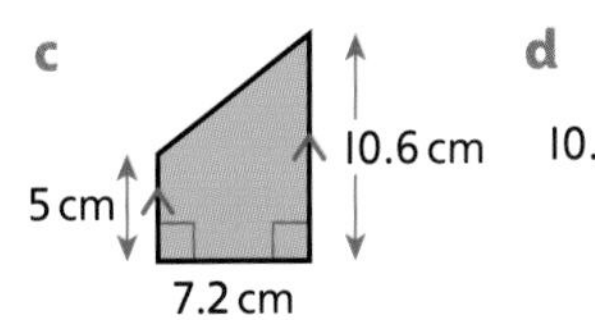

d
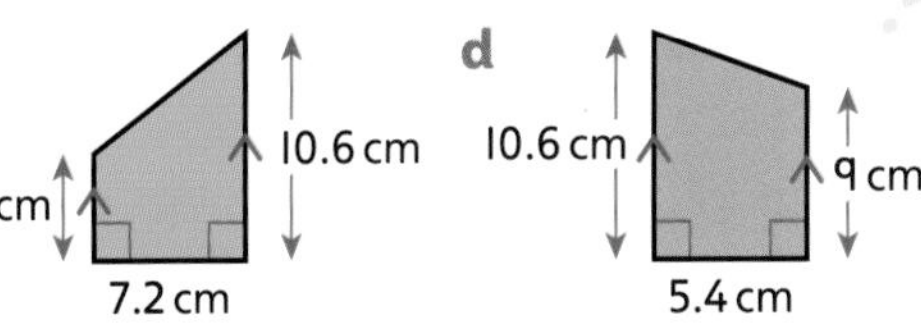

7 Draw a diagram to prove that the formula for the area of a parallelogram = base × vertical height

8 Calculate the volume of each of these solid shapes.

a
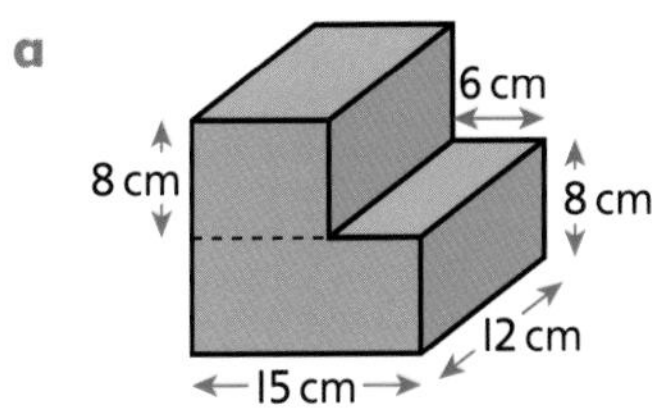

b
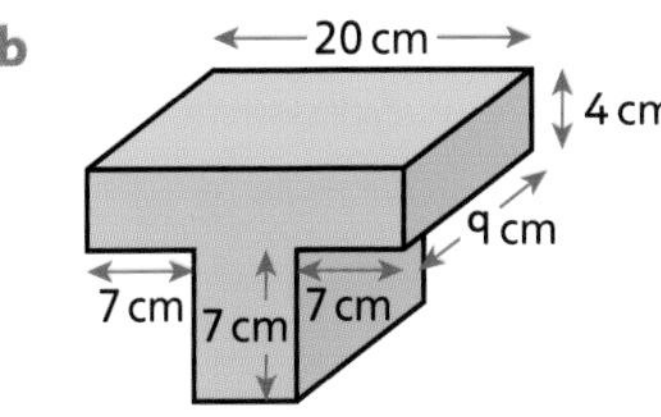

9 Calculate the surface area of the shapes in Q8. **Hint:** Sketch all the different faces first.

10 **a** Draw a diagram to show that a parallelogram can be split into two trapeziums.

b Use your diagram from part **a** to show that the formula for the area of a trapezium is $A = \frac{1}{2}h(a + b)$.

Level 6

6c I can calculate the volumes of cubes and cuboids

6b I can calculate the surface area of cubes and cuboids

6b I can use the formula to calculate the area of a trapezium

6b I can deduce a formula for the area of a parallelogram

6b I can calculate the volume of shapes made from cuboids

6a I can calculate the surface area of compound shapes made from cuboids

6a I can deduce a formula for the area of a trapezium

Now try this!

A Maximum volume 1

Find the maximum volume of an open box that can be made by cutting and folding a 20 cm by 20 cm piece of card.

B Maximum volume 2

Repeat Activity A for an 18 cm by 18 cm piece of card and for a 10 cm by 10 cm piece of card. Find a general rule which connects the height of the box (h) and the size of the square (s), which will always give you the maximum volume (V).

Learn this

Remember to learn the formulae for the volume of a cuboid, and the areas of a triangle and a parallelogram.

surface area trapezium triangle volume

6.6 Volume and surface area

- Calculate the volumes of cuboids, shapes made from cuboids and prisms
- Calculate the surface areas of shapes made from cuboids and prisms
- Calculate lengths and areas from the volumes of prisms and cylinders

Why learn this?

3-D shapes with the same volume can have different surface areas. Package designers try to minimise the cost of packaging by using the 3-D shape which is most practical and has the smallest surface area for the volume that they need.

What's the BIG idea?

- **Volume** of a **cuboid** = length (l) × width (w) × height (h). Level 6
- **Surface area** of a cuboid = total area of all its surfaces $= 2lh + 2hw + 2lw$ Level 6

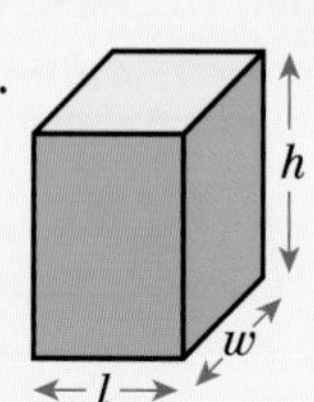

- Volume of a **prism** = area of **cross-section** × length Level 7
- The surface area of the prism is the total area of all surfaces (i.e. the area of all the shapes that make up its net). Level 7
- Given the volume of a prism, you can find its length or the area of its cross-section if you know the other value. Level 7 & Level 8

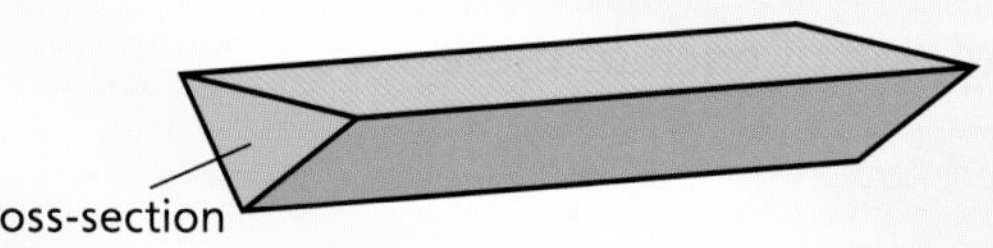

- Volume of a cylinder = $\pi r^2 h$. Level 8
- Surface area of a cylinder = $2\pi rh + 2\pi r^2$. Level 8

Practice, practice, practice!

1 Find the volume of each of these cuboids.

a

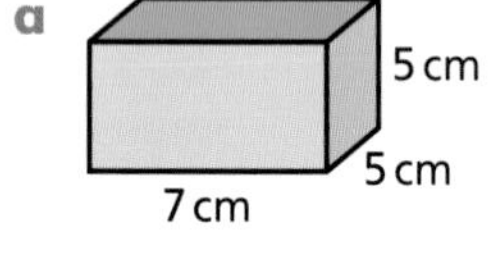

b

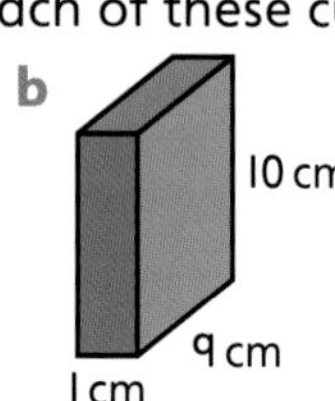

c

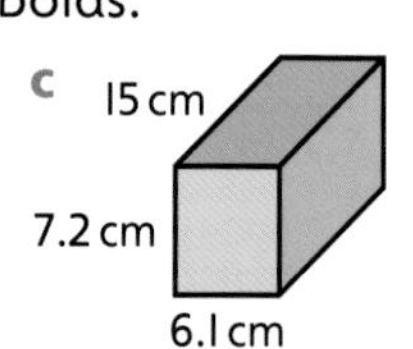

d

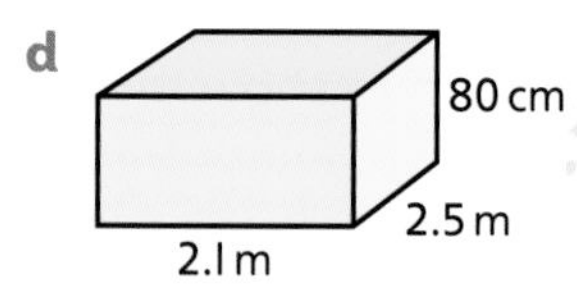

2 A cuboid-shaped package has a length of 3.7 m, a height of 4.1 m and a width of 72 cm. What is the surface area of the package?

3 Find the volume of each of these shapes made from cuboids.

a

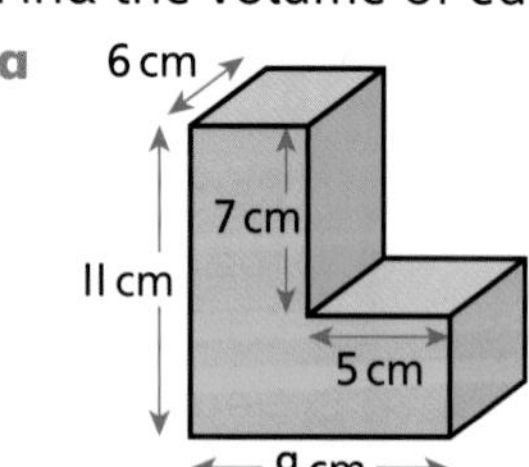

b

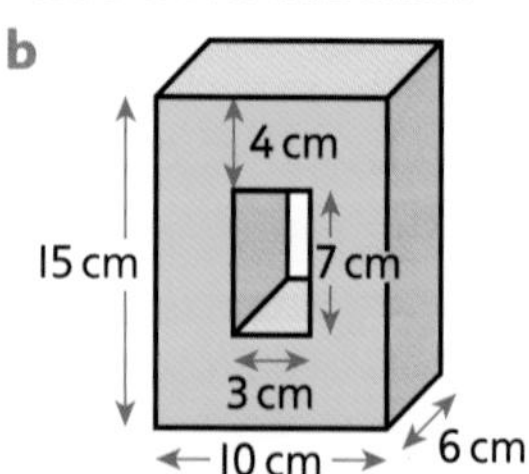

4 Find the surface area of each of the shapes in Q3.

Level 6

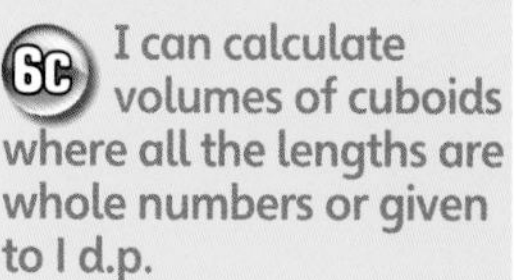

6c I can calculate volumes of cuboids where all the lengths are whole numbers or given to 1 d.p.

6b I can calculate surface areas of cuboids

6b I can calculate volumes of shapes made from cuboids, where all the lengths are whole numbers

6a I can calculate surface areas of shapes made from cuboids, where all the lengths are whole numbers

cross-section cuboid cylinder

5 Find the volume of each of these prisms.

a

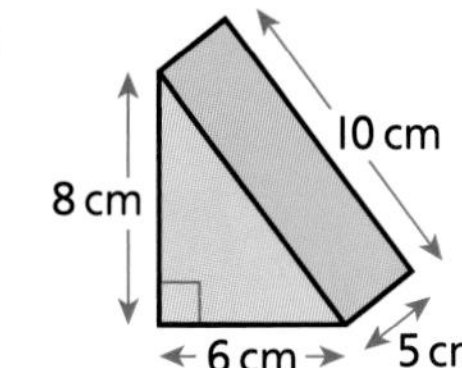

b

c

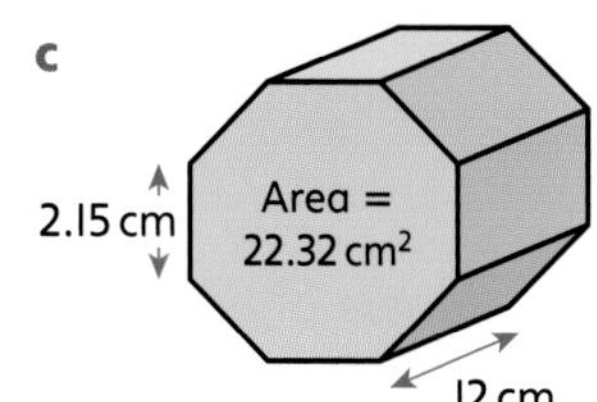

6 Find the surface area of each of the prisms in Q5.

7 A box is 6 cm high and 7.5 cm wide. Its volume is 495 cm³. Find its length and its surface area.

8 Sketch three boxes with a volume of 42 cm³.

Level 7

7c I can calculate the volume of prisms

7b I can calculate the surface area of prisms

7a I can calculate length and surface area of prisms

9 Calculate the missing lengths in each of these prisms.

a

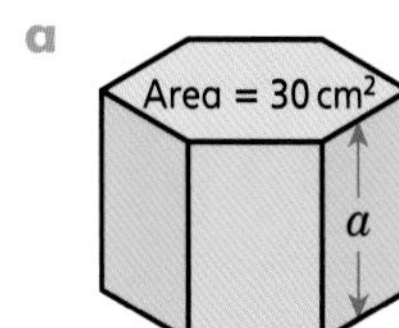
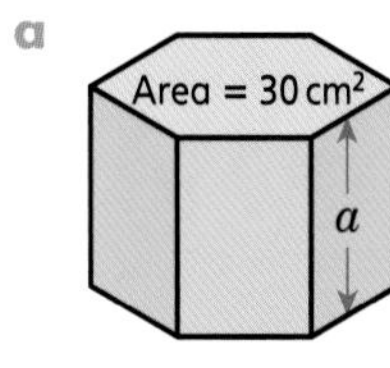

Volume = 255 cm³

b

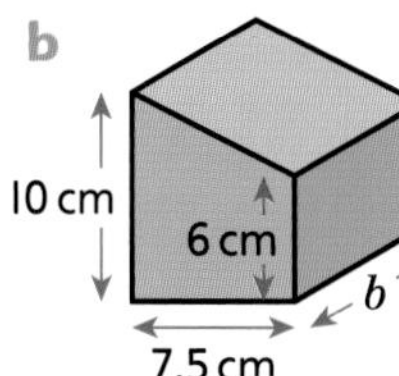

Volume = 375 cm³

c

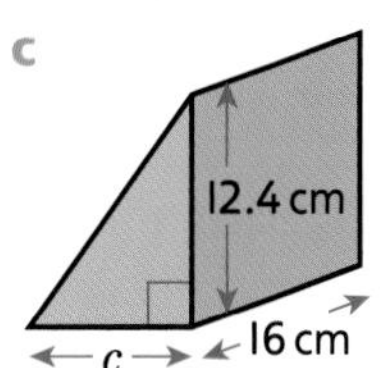

Volume = 813.44 cm³

d

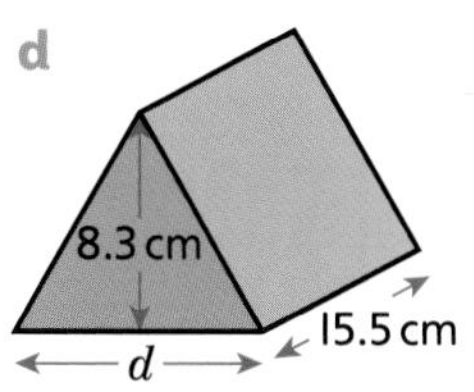

Volume = 771.9 cm³

10 **a** Find the volume of this tin.

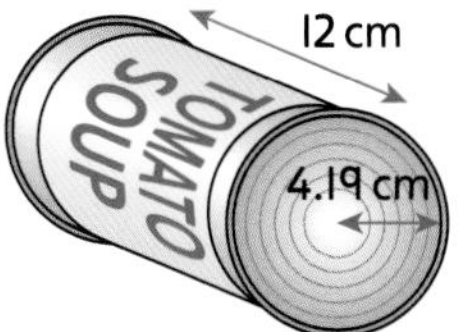

b The volume of this tin is 1690 cm³. Work out the missing length a.

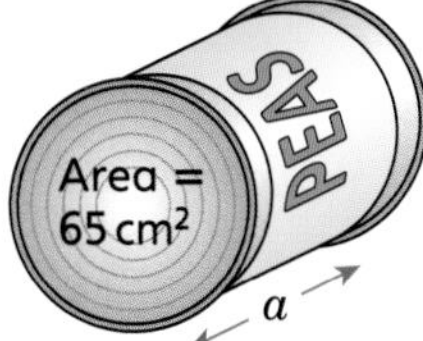

c The volume of this tin is 700 cm³. Work out the missing length b by first finding the area of the end face.

Level 8

8c I can calculate the length or width of a prism given its volume

8a I can calculate the volume, length and surface area of cylinders

Now try this!

A Volume of a cube

The volume of a cube is 216 cm³. Find the dimensions of the cube.

B Tennis ball container

a The diameter of a tennis ball is 70 mm. Design a container to hold three tennis balls.

b Work out the surface area of your container.

prism surface area volume

All wrapped up

Airlines transport large numbers of packages all over the world. Emilio, who lives in the US, has two packages that he needs to send by air to France.

1 The first package is in the shape of a cuboid. Work out the volume of this package. **Level 6**

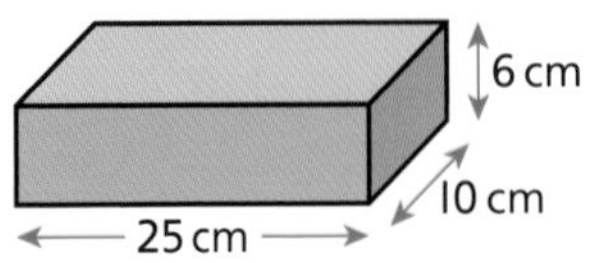

2 Emilio wraps the package in brown paper. There is no overlap. Work out the area of brown paper that he uses. **Level 6**

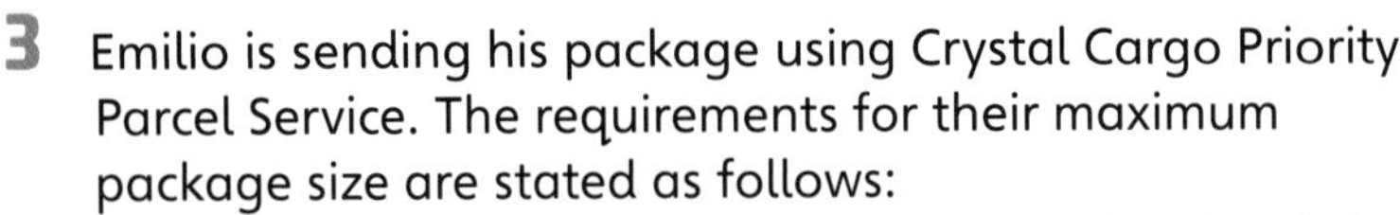

3 Emilio is sending his package using Crystal Cargo Priority Parcel Service. The requirements for their maximum package size are stated as follows:

The combined dimensions (length + width + height) of each package cannot exceed 90 inches, and no single side length may be more than 48 inches.

Does Emilio's package meet these requirements? **Level 6**

Hint: 1 inch ≈ 2.5 cm

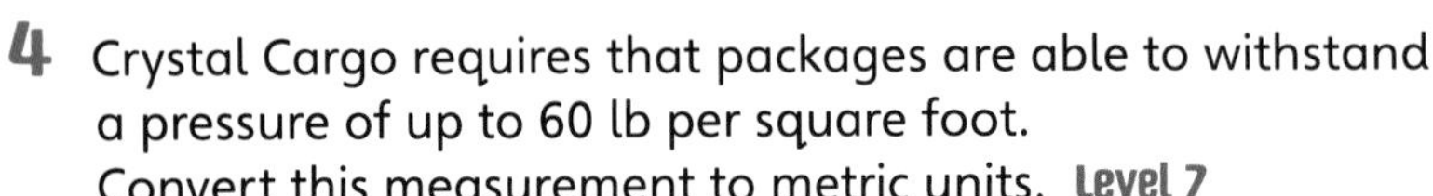

4 Crystal Cargo requires that packages are able to withstand a pressure of up to 60 lb per square foot. Convert this measurement to metric units. **Level 7**

Hint: 1 pound ≈ 0.45 kg
1 foot ≈ 0.3 m

5 Emilio sticks a label on the package. The dimensions of the label are given to the nearest centimetre. Work out the lower bound for the area of the label, in square centimetres. **Level 8**

12 cm | Maria Dupont
87 Rue de Lyon
Paris

15 cm

6 Emilo's second package is in the shape of a sector of a circle. He wraps the package in brown paper and will need adhesive tape to secure it. The adhesive tape will be placed along both curved edges of the package. Emilio has 30 cm of adhesive tape. Does he have enough? **Level 8**

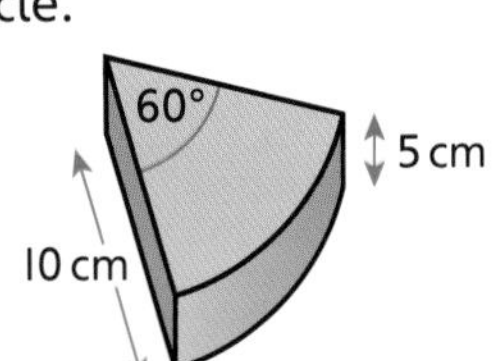

7 Work out the volume of the package in Q6. **Level 8**

8 Both packages are taken by courier to the local airport, which is 45 km away. The courier's journey takes 1 hour 15 minutes. Work out the average speed of the courier in km per hour. **Level 8**

9 The packages are put on a Crystal Cargo aeroplane bound for France. The journey to Charles de Gaulle Airport, Paris takes 7 hours 25 minutes. The average speed of the aeroplane is 783 km/h. Work out the approximate distance that the packages travel on the aeroplane. **Level 8**

The BIG ideas

- Multiply the length, height and width of a **cuboid** to find its **volume**.
  $V = l \times h \times w$ Level 6
- To calculate the **surface area** of a cuboid, work out the area of each **face** and add them together.
  Surface area = $2hl + 2hw + lw$ Level 6
- Measurements are usually given to the nearest whole number or decimal place. A true value lies in a **range** from half a unit below to half a unit above the measurement. Level 7
- Compound measures combine measures of two different types. For example, speed measures distance and the time taken to travel it.

  Speed = $\frac{\text{distance}}{\text{time}}$ Level 7
- **average** speed = $\frac{\text{total distance travelled}}{\text{total time taken}}$ Level 8
- The length of an **arc** can be expressed as a fraction of the circumference of a circle. Length of arc, $l = \frac{\theta}{360°} \times 2\pi r = \frac{\theta}{360°} \times \pi d$. Level 8
- The area of a **sector** can be expressed as a fraction of the area of a circle.

  Sector area = $\frac{\theta}{360°} \times \pi r^2$. Level 8

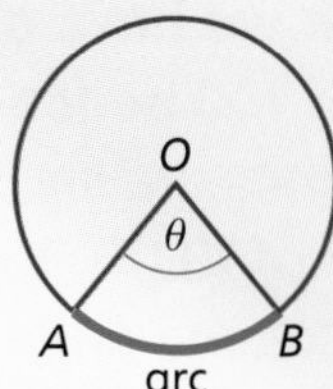

Find your level

Level 6

Q1 The triangle and the rectangle shown have the same area.

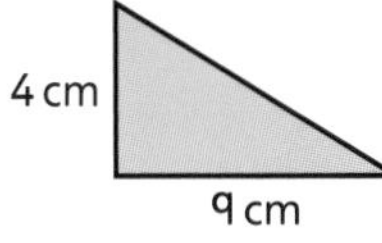

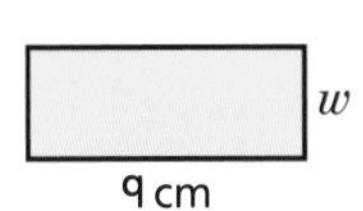

Work out the value of w.

Q2 Jessica says, 'There are 1000 cubic centimetres in a cubic metre.'
Is Jessica correct?
Give a reason for your answer.

Level 7

Q3 a The cross-section of a cylindrical barrel is a circle.
The diameter of the circle is 30 cm.
What is the circumference of the circle?

b A different barrel has a circumference of 84 cm.
What is the radius of this barrel?

Q4 Work out the volume of this triangular prism.

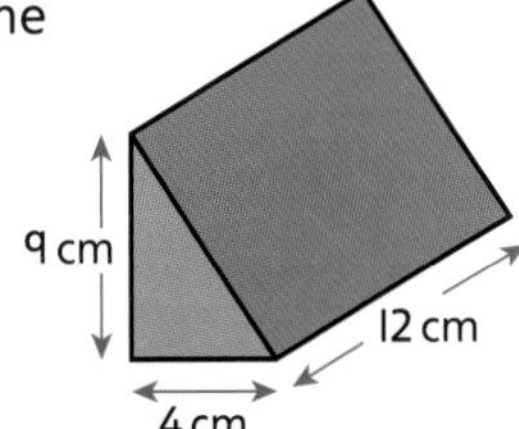

Q5 Two candles with circular cross-sections are placed into a box with height 15 cm. The space around the candles is packed with foam.

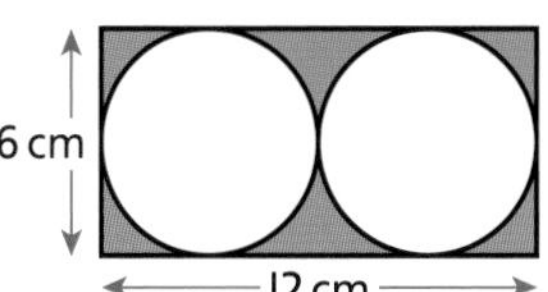

The height of the candles is equal to the height of the box.
What is the volume of foam used?

Level 8

Q6 The diagram shows parts of two circles, sector A and sector B.

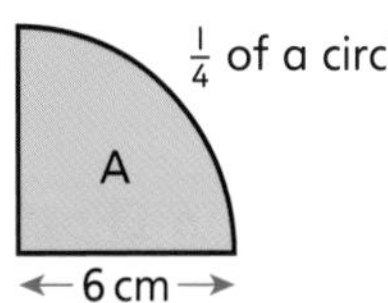

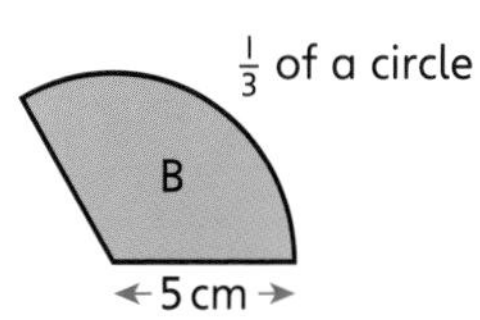

Which sector has the bigger area?
Show your working.

Q7 The cylinder has radius 3 cm.
The volume of the cylinder is $6.3\pi\,\text{cm}^3$.

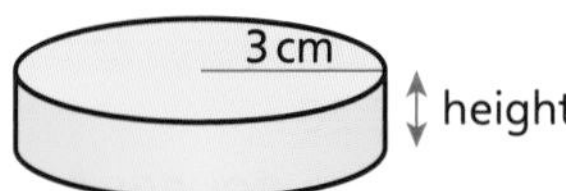

What is the height of the cylinder?

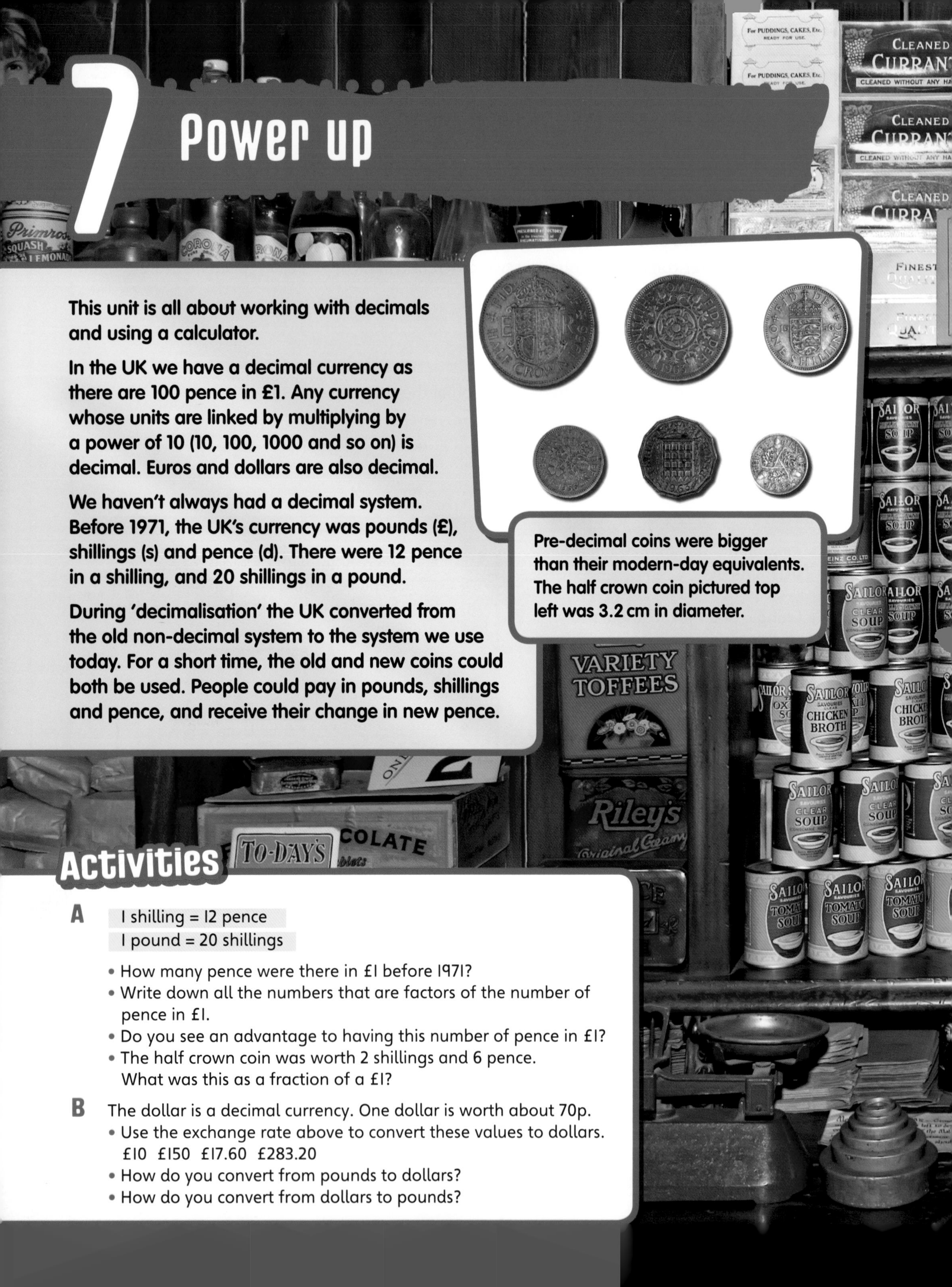

7 Power up

This unit is all about working with decimals and using a calculator.

In the UK we have a decimal currency as there are 100 pence in £1. Any currency whose units are linked by multiplying by a power of 10 (10, 100, 1000 and so on) is decimal. Euros and dollars are also decimal.

We haven't always had a decimal system. Before 1971, the UK's currency was pounds (£), shillings (s) and pence (d). There were 12 pence in a shilling, and 20 shillings in a pound.

During 'decimalisation' the UK converted from the old non-decimal system to the system we use today. For a short time, the old and new coins could both be used. People could pay in pounds, shillings and pence, and receive their change in new pence.

Pre-decimal coins were bigger than their modern-day equivalents. The half crown coin pictured top left was 3.2 cm in diameter.

Activities

A

1 shilling = 12 pence
1 pound = 20 shillings

- How many pence were there in £1 before 1971?
- Write down all the numbers that are factors of the number of pence in £1.
- Do you see an advantage to having this number of pence in £1?
- The half crown coin was worth 2 shillings and 6 pence. What was this as a fraction of a £1?

B The dollar is a decimal currency. One dollar is worth about 70p.

- Use the exchange rate above to convert these values to dollars.
  £10 £150 £17.60 £283.20
- How do you convert from pounds to dollars?
- How do you convert from dollars to pounds?

Before you start this unit...

1 Write these values in order of size, smallest first.

4^2 2^3 5^2 3^2

Level Up Maths 5-7 page 6

2 Round these numbers to the nearest whole number.

a 34.7 **b** 180.9 **c** 1300.4

Level Up Maths 5-7 page 132

3 Work these out without using a calculator.

a $34.7 + 458$ **b** 65×24

c $301 \div 7$

Level Up Maths 5-7 page 134

4 Work out these.

a $4 + 7 \times 2$ **b** $6^2 - 10 \div 5$

c $4 - (3 \times 12)$

Level Up Maths 5-7 page 210

5 Write these values in order of size, smallest first.

100×0.1 0.01×10^2 $10^3 \times 0.1$

Level Up Maths 5-7 page 130

Did you know?

After the French Revolution in 1789, the new government tried to decimalise time. They introduced a system where the day was divided into ten decimal hours and each hour was divided into 100 decimal minutes. Most people didn't want to use the new system and decimal time never caught on.

World's Greatest Maths

7.1 Labyrinth
7.2 To round or not to round?
7.3 Using a calculator
7.4 More rounding
7.5 Roots and standard form on a calculator
7.6 Written methods
7.7 Using a calculator efficiently
7.8 Problem solving
Unit plenary: Decimal Day!

Plus digital resources

World's Greatest Maths!

7.1 LABYRINTH

The evil goblin king has trapped your baby sister in the centre of a dreadful maze. You have one hour to find her before she will be turned into a goblin.

- Find your way through the maze.
- Solve the calculation in each chamber to find the correct exit. Write down each calculation and the correct answer.
- Complete each practice session before moving on to the next section of the maze.

You can use a copy of Resource sheet 7.1.

Start

58 500 00
$5850 \div 10^3$
5.85 58.5 585

3067 0.367
0.03067×10^5
30 670
306.7

Practice session 1

1 Convert these lengths into metres.

- **a** 3 kilometres
- **b** 0.5 millimetres
- **c** 167 micrometres
- **d** 2600 nanometres

2 Work out

- **a** 8×10^2
- **b** 859×10^{-1}
- **c** 62.4×10^3
- **d** 6420×10^{-3}

3 Work out

- **a** $7.3 \div 10^2$
- **b** $386 \div 10^{-4}$
- **c** $170\,000 \div 10^8$
- **d** $0.000857 \div 10^{-6}$

Now complete the first section of the maze.

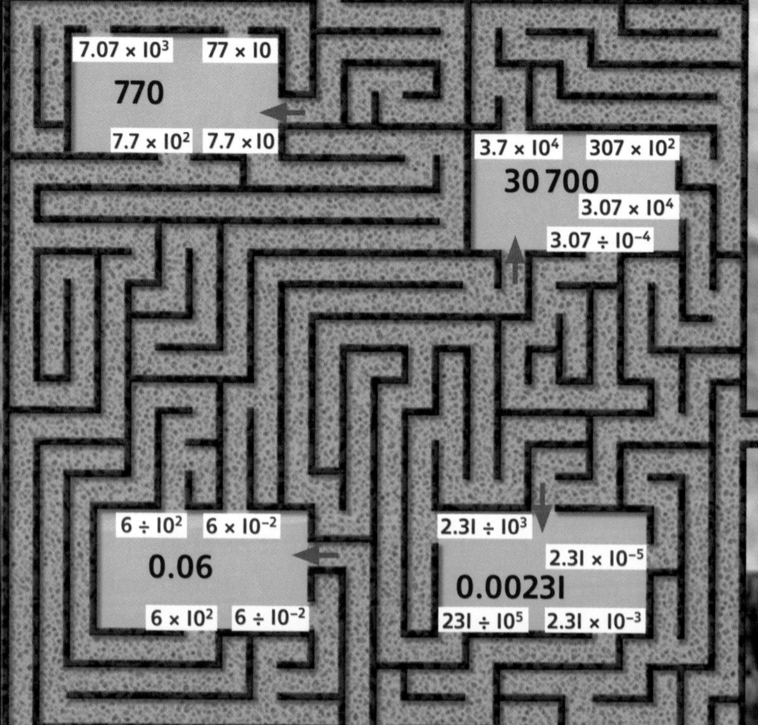

Practice session 2

1 Which of these numbers is in standard form?

A 42×10^3 B 4.2×10^4
C 0.42×10^5 D 42 000

2 Write each number in standard form.

- **a** 3000
- **b** 240
- **c** 80 000
- **d** 609 000

3 Write each number in standard form.

- **a** 0.05
- **b** 0.0049
- **c** 5.62
- **d** 0.0002009

Now complete the second section of the maze. Find the exit which gives the number in standard form.

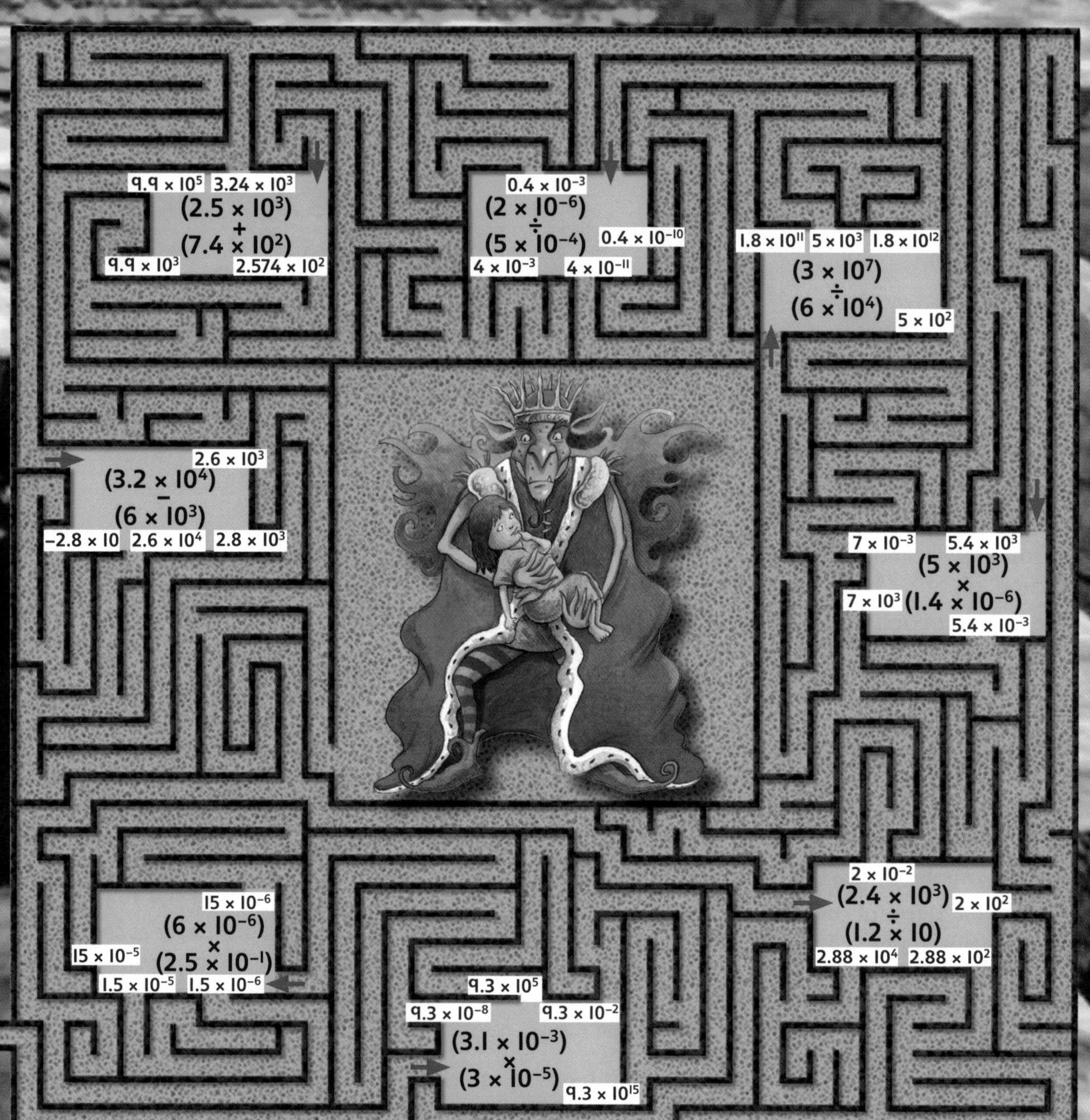

Practice session 3

Work out these, giving your answers in standard form.

1 a $(2 \times 10^3) \times (3.2 \times 10^4)$
b $(4 \times 10^3) \times (1.2 \times 10^{-5})$
c $(1.6 \times 10^2) \times (3 \times 10^6)$
d $(2 \times 10^4) \times (6.2 \times 10^5)$

2 a $(9.6 \times 10^8) \div (3 \times 10^5)$
b $(6 \times 10^7) \div (2 \times 10^4)$
c $(1.6 \times 10^{-12}) \div (4 \times 10^6)$
d $(3 \times 10^{-12}) \div (5 \times 10^{-8})$

3 a $(2.3 \times 10^2) + (1.51 \times 10^3)$
b $(9.8 \times 10^5) + (4.3 \times 10^4)$
c $(7 \times 10^6) - (8.5 \times 10^5)$
d $(1.32 \times 10^3) - (5.1 \times 10^2)$

Now complete the maze to reach your baby sister.

You've rescued your sister from the goblin king, but you need to activate the secret door to escape from the maze. The buttons on each keypad need to be pressed in order of size, smallest first.

For each keypad, write down the numbers on the buttons in the correct order to open the secret door.

1	2	3	4
4.2×10^7	2×10^4	5×10^{-5}	7.25×10^{-2}
6×10^4	1.99×10^4	7×10^{-6}	3.1×10^{-3}
1.99×10^5	9.87×10^3	6×10^{-6}	3.01×10^{-3}
3.8×10^6	3×10^5	4×10^{-5}	8.7×10^{-4}

7.2 To round or not to round?

- Round numbers to one, two or three decimal places
- Write recurring decimals as fractions
- Use rounding to make estimates

Why learn this?

When calculating you need to be able to round decimal numbers to a suitable number of decimal places.

What's the BIG idea?

- To **round** a decimal to two **decimal places**, look at the digit in the third decimal place. If the digit is less than 5, round down. If the digit is 5 or more, round up. **Level 6**
- To round a decimal to three decimal places, look at the digit in the fourth decimal place. **Level 6**
- The degree of accuracy of an answer to a problem must be appropriate to the question.
  For example, the answer to a money problem would be given to two decimal places. **Level 6**
- You can use rounding to make **estimates**.
  For example, $0.47 \times 12.2 \approx 0.5 \times 12 = 6$ **Level 6**
- You can convert a fraction to a decimal by dividing the numerator by the denominator.
  For example, $\frac{1}{3} = 1 \div 3 = 0.33333...$ **Level 6**
- A **recurring decimal** contains a digit, or sequence of digits, which repeats itself forever. A dot over a digit shows that it recurs.
  For example $0.333... = 0.\dot{3}$ **Level 6 & Level 7**
- Recurring decimals are exact fractions.
  For example, $0.1\dot{6} = \frac{1}{6}$ **Level 6 & Level 7**
- You can use an algebraic method to convert a recurring decimal to a fraction. **Level 7**

Learn this

π is an irrational number – it has an infinite number of digits to the right of the decimal point. To 20 decimal places, π is 3.141 592 653 589 793 238 46. The digits never repeat in a pattern, so π cannot be written as a fraction.

Practice, practice, practice!

Level 6

6c I can convert fractions to recurring decimals

6b I can round to one, two and three decimal places

1 Match each recurring decimal to its equivalent fraction from the cloud.

a $0.\dot{6}$ **b** $0.\dot{7}$ **c** $0.\dot{3}$ **d** $0.\dot{1}$ **e** $0.\dot{5}\dot{4}$

2 Copy and complete the table.

Watch out

5.999 to 1 decimal place is 6.0, and to 2 decimal places is 6.00.

	to 1 d.p.	to 2 d.p.	to 3 d.p.
13.4992	13.5	13.50	13.499
157.0384			
0.23456			
61.5972			
0.9999			

approximately equal to (≈) decimal place estimate lower bound

3 Work out these. Give your answers to an appropriate degree of accuracy.

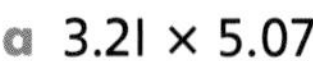

- **a** 3.21 × 5.07
- **b** 9.05 ÷ 17.405
- **c** A box of five recordable CDs costs £12.99. How much does one CD cost?
- **d** What is the volume of a cuboid-shaped box with dimensions 1.2 m by 89.25 cm by 33.05 cm?

Tip Look at the values in the problem to decide whether to give your answer to 1, 2 or 3 decimal places.

Level 6

6a I can give answers to an appropriate degree of accuracy

4 These are the numbers of people using a bus service on seven consecutive days.

34 47 89 32 44 65 9

Calculate the mean number of people on the bus each day.

5 A window measures 1.21 m by 2.34 m. To calculate the area of fabric needed to make curtains, work out the area of the window and multiply it by 2.7.

- **a** By rounding the measurements to the nearest whole number, estimate the area of fabric needed.
- **b** Use a calculator to work out the exact area of fabric needed.
- **c** What is the difference between your estimate and the exact answer?
- **d** Do you think rounding to the nearest whole number is sensible in this case?

6a I can round to make estimates and justify the degree of rounding

6 Use an algebraic method to convert each recurring decimal to a fraction.

$0.\dot{1}$ Let $a = 0.111\,111$ (1)
then $10a = 1.111\,111$ (2)
so $9a = 1$, (2) − (1)
so $a = \frac{1}{9}$

a $0.\dot{6}$ **b** $0.\dot{1}\dot{7}$ **c** $0.\dot{4}\dot{5}$ **d** $\dot{1}.\dot{3}$ **e** $1.\dot{3}1$

7 Rounded to the nearest 10 000, there were 230 000 people at a football match. What are the largest and the smallest numbers of people that could have been at the match?

Level 7

7a I can convert a recurring decimal to a fraction using an algebraic method

7a I can identify the maximum and minimum possible values of a number that has been rounded

Now try this!

A When do they recur?

- **a** Write each of these fractions as a decimal.

  $\frac{1}{2}$ $\frac{1}{3}$ $\frac{1}{4}$ $\frac{1}{5}$ $\frac{1}{6}$ $\frac{1}{7}$ $\frac{1}{8}$ $\frac{1}{9}$ $\frac{1}{10}$
- **b** Which fractions from **a** give recurring decimals? List their denominators.
- **c** What is the link between the denominator and whether the fraction produces a recurring decimal?
- **d** Change the numerators of the fractions identified in part **b**. Do they still produce recurring decimals?

B Recurring conundrum

$\frac{1}{9}$	0.111 111…
$\frac{2}{9}$	0.222 222…
$\frac{3}{9}$	0.333 333…
$\frac{4}{9}$	0.444 444…

- **a** Continue the pattern from the table to write each of these as a recurring decimal.

  i $\frac{5}{9}$ **ii** $\frac{7}{9}$ **iii** $\frac{9}{9}$
- **b** Use the algebraic method from Q6 to prove that 0.999 999… = 1.

recurring decimal round to one decimal place (1 d.p.) upper bound

7.3 Using a calculator

- Do calculations involving fractions and mixed numbers on a calculator
- Enter time on a calculator
- Understand reciprocals and use the reciprocal key on a calculator

Why learn this?

By learning how to enter fractions into a calculator you can save a lot of time, which is important at work.

What's the BIG idea?

→ Most **calculators** have a function **key** for **entering** fractions: $\boxed{a\frac{b}{c}}$

For example, to enter $\frac{2}{3}$, press $\boxed{2}$ $\boxed{a\frac{b}{c}}$ $\boxed{3}$ and to enter $4\frac{3}{8}$, press $\boxed{4}$ $\boxed{a\frac{b}{c}}$ $\boxed{3}$ $\boxed{a\frac{b}{c}}$ $\boxed{8}$. **Level 6**

→ Calculations involving time can be entered into a calculator as a fraction or **mixed number**.

For example, 5 hours 16 minutes is equivalent to $5\frac{16}{60}$ hours because there are 60 minutes in 1 hour. **Level 6**

→ As long as you enter a calculation exactly as it is written, including brackets, your calculator will generally follow the o**rder of operations**. **Level 6**

→ The **reciprocal** of a number is the value given by dividing 1 by that number.

For example, the reciprocal of 4 is $\frac{1}{4}$ or $1 \div 4 = 0.25$. **Level 7**

→ The reciprocal key on a calculator may be $\boxed{\frac{1}{x}}$ or $\boxed{x^{-1}}$. **Level 7**

→ Multiplying a number by its reciprocal gives 1. **Level 7**

Did you know?

Electronic calculators are a relatively recent invention, first appearing in the early 1970s. Before this, mathematicians used abacuses, Napier's bones, mathematical tables, mechanical calculators and slide rules.

Practice, practice, practice!

1 Work out these on your calculator.

a $\frac{3}{5} + \frac{12}{19}$ b $\frac{4}{7} + \frac{15}{17}$ c $\frac{3}{4} - \frac{4}{13}$

d $\left(\frac{5}{7} + \frac{2}{3}\right) \times \frac{8}{11}$ e $\frac{8}{9} - \frac{4}{5} + \frac{2}{7}$ f $\frac{5}{6} - \left(\frac{3}{4} + \frac{6}{7}\right)$

2 Use your calculator to work out

a $3\frac{3}{5} + 2\frac{1}{3}$ b $18\frac{4}{9} - 17\frac{3}{5}$ c $\left(70\frac{2}{3} + 15\frac{7}{8}\right) \div 16\frac{2}{5}$

3 Belinda is on a walking holiday. On Monday she covered $20\frac{2}{3}$ miles. On Tuesday she walked a further $15\frac{7}{8}$ miles. She was tired on Wednesday and walked a third of the distance she walked on Monday.

a How far had she walked by the end of Tuesday?

b How far did she walk on Wednesday?

c The holiday has cost her £850 so far.
How much has it cost her per mile walked?

4 Use your calculator to work out

a 7 × (5 h 18 min) b 14 min 12 s – 3 min 45 s c (18 h 25 min) ÷ 5

Level 6

6b I can use the function key to enter a fraction

6a I can use the function key to enter a mixed number

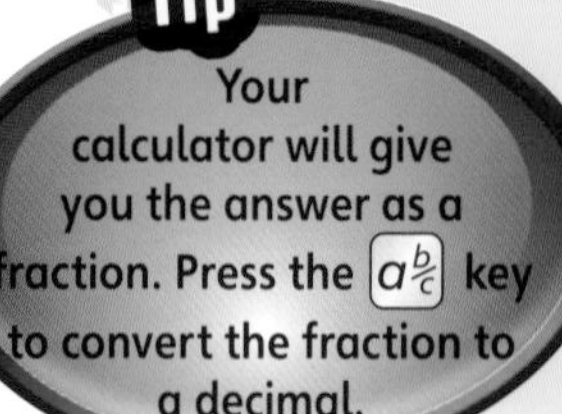

Tip
Your calculator will give you the answer as a fraction. Press the $a\frac{b}{c}$ key to convert the fraction to a decimal.

6a I can enter time on a calculator

calculator decimal place enter inverse key mixed number

5 Mr Zakaria works in a factory where he clocks in and out of his shift.
The table shows the numbers of hours and minutes he worked one week.

Monday	Tuesday	Wednesday	Thursday	Friday
6 h 35 min	7 h 24 min	5 h 42 min	8 h 57 min	9 h 34 min

a For how long did he work in total?
Give your answer in hours and minutes.

b Mr Zakaria is paid £5.26 an hour.
How much did he earn that week?

Level 6

6a I can enter time on a calculator

6 Find the reciprocal of each of these. Give each answer as a fraction.

q $\frac{1}{9}$

a 0.25 **b** 5.5 **c** $\frac{11}{12}$ **d** 0.01

e $\frac{24}{25}$ **f** 12.8 **g** $1\frac{4}{5}$ **h** $12\frac{11}{12}$

7 Use your calculator's reciprocal key to work out these.
Give your answers to two decimal places.

a $\frac{1}{2.5 \times 3.8}$ **b** $\frac{1}{3^2 + 5 \div 2}$ **c** $\frac{1}{2^3 - 4^2 + 3.7}$ **d** $\frac{1}{0.2 + 2 \div 3}$

8 **a** What is the reciprocal of $\frac{11}{19}$?

b What is $\frac{11}{19}$ multiplied by its reciprocal? Why do you get this answer?

c Repeat parts **a** and **b** for **i** 0.89 **ii** $2\frac{3}{4}$ **iii** $\frac{14}{8}$

9 **a** Find the reciprocal of 0.25.

b Find the reciprocal of your answer to part **a**.

c Repeat parts **a** and **b** for **i** $\frac{1}{3}$ **ii** 1.5 **iii** $\frac{8}{9}$
What do you notice?

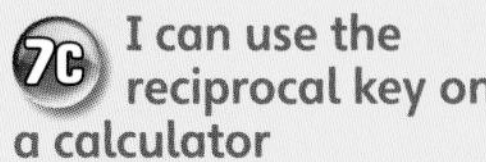

Level 7

7c I can use the reciprocal key on a calculator

7b I can understand what happens when I multiply a reciprocal by the original number

7a I can understand what happens when I find the reciprocal of a reciprocal

Now try this!

A Do I have the time?

If you spent 45 minutes every day of your life watching TV, that would be
45 × 365 × 75 = 1 231 875 minutes (if you live to 75 and if there are 365 days in a year).
This is equivalent to 20 531 hours and 15 minutes, or 855 days, 11 hours and 15 minutes, or 2 years, 125 days, 11 hours and 15 minutes.
Choose one activity and work out how much time you might spend doing it in your lifetime.

B Repeated reciprocals

a Enter the number 1 and press [=].

b Now press [(] [ANS] [+] [1] [)], then press the reciprocal key and [=].
Write down the value you get as a fraction.

c Press [=] five more times, writing down these terms in the sequence.

d Predict the next five terms and check your answers on the calculator.

e Why do you think this sequence is being produced?

Tip

The [ANS] key on a calculator stores the answer to the previous calculation. The [$a\frac{b}{c}$] key can convert a decimal to a fraction.

order of operations reciprocal to two decimal places (2 d.p.)

7.4 More rounding

- Write numbers as decimal multiples of thousands or millions
- Make estimates and approximations
- Round numbers to a given number of significant figures
- Find the upper and lower bounds

Why learn this?

The number of people at a football match is often rounded to the nearest hundred or nearest thousand. In some situations you need to be able to identify the largest and smallest values that the original number could have been.

What's the BIG idea?

- Numbers can be written as **decimal multiples** of thousands or millions.
  For example, the population of the UK is 60 943 912.
  This could be expressed as 60.9 million, rounded to the nearest hundred thousand. Level 6
- You can round numbers to a given number of **significant figures**. The first significant figure is the first non-zero digit in the number, counting from the left. Level 7
- You can use rounding to one significant figure (1 s.f.) to calculate an **approximate** answer to a calculation. Level 7
- When a number has been rounded, the largest and smallest values it could have been can be given. These are called the **upper** and **lower bounds**. Level 8

Did you know?

In the UK a billion traditionally has 12 zeros but in the USA it has nine zeros – so US billionaires are not nearly as rich as billionaires in the UK! However, the US billion is becoming more widely used in the UK.

Practice, practice, practice!

Level 6

6b I can write numbers as decimal multiples of thousands or millions

1 Copy and complete.

a 23 600 000 = ______ million b 35 600 = ______ thousand

c ______ = 4.9 thousand d ______ = 0.5 million

2 The table shows the estimated populations of various countries in 2008.
Round each figure to the nearest hundred thousand and write your answer as a decimal number of millions.

Denmark 5 484 723 ≈ 5 500 000
= 5.5 million
Look at the digit in the ten thousands column.

Country	Population
Denmark	5 484 723
France	64 057 790
Germany	82 369 548
Greece	10 722 816
Norway	4 644 457
Portugal	10 676 910
Spain	40 491 051
Sweden	9 045 389

6b I can use rounding to make estimates

3 The population of the UK is 61 million, rounded to the nearest million.
The total land mass is calculated to be 244 820 million m^2.
There are 30.5 million televisions and 69.657 million mobile phones in the UK.

a Estimate the mean number of televisions per person.

b Estimate the mean number of mobile phones per person.

c Estimate the mean area of land available per person.

approximate approximately equal to ≈ decimal multiples estimate

4 The O_2 arena has a capacity of 23 000. Given that tickets for a concert are £49.50 each, estimate the total revenue from ticket sales when the arena is full.

5 Round these numbers to the given number of significant figures.

a 5724 (1 s.f.) b 8933 (2 s.f.)
c 4 506 789 (3 s.f.) d 23 000 425 (2 s.f.)
e 17 425 909 (3 s.f.) f 9830 (1 s.f.)
g 8 970 235 (3 s.f.) h 999 999 (2 s.f.)

Watch out!
Once you've rounded, don't forget to replace the other digits with zeros.

6 Round these numbers to the given level of accuracy.

a 0.0379 (1 s.f.) b 0.059 34 (2 s.f.) c 0.082 078 (3 s.f.)
d 0.000 300 68 (1 s.f.) e 2.050 437 (2 s.f.) f 1.234 567890 (3 s.f.)

Level 7
7c I can round to one significant figure to make estimates
7b I can round to a given number of significant figures

7 A company estimates its profits as £2.3 million.

a What are the largest and the smallest profits the company could have made?
b Use inequality signs to identify the upper and lower limits of the possible profit.

8 The total revenue taken for one night at the O_2 arena was £1.5 million, to the nearest hundred thousand.
Exactly 23 000 tickets were sold for that night.
Work out the upper and lower bounds of the ticket price.

Level 8
8c I can identify upper and lower bounds for discrete data
8a I can determine upper and lower bounds in complex problems

Now try this!

A Crack the code

Each letter of the alphabet is assigned a different number.

A	B	C	D	E	F	G	H	I	J	K	L	M
0.10	0.11	0.12	0.13	0.14	0.15	0.16	0.17	0.18	0.19	0.20	0.21	0.22
N	O	P	Q	R	S	T	U	V	W	X	Y	Z
0.23	0.24	0.25	0.26	0.27	0.28	0.29	0.30	0.31	0.32	0.33	0.34	0.35

Round each number to two significant figures to reveal the message.

0.234, 0.239, 0.315
0.321, 0.274, 0.176, 0.288, 0.137
0.098
0.223, 0.144, 0.284, 0.276, 0.101, 0.164, 0.141
0.239, 0.145
0.339, 0.239, 0.295, 0.266
0.238, 0.324, 0.225
0.184, 0.225
0.115, 0.237, 0.128, 0.1395

B Who am I?

I am 6 million to the nearest million. I am 6.4 million to the nearest hundred thousand. The last three of my digits are zero but none of the others are. All my digits are even. Apart from the three zeros, none of my digits are repeated. Who am I?

lower bound significant figures upper bound

7.5 Roots and standard form on a calculator

- Use a calculator to find square roots and cube roots
- Use a calculator to find other roots
- Enter numbers expressed in standard form into a calculator
- Interpret answers given in standard form on a calculator

Why learn this?

Scientists often work with very large or very small numbers.

What's the BIG idea?

- You can use a calculator to find square and cube roots. **Level 6**
- The inverse of raising a number to the **power** of 4 is finding the fourth root $\sqrt[4]{\ }$, the inverse of raising a number to the power of 5 is finding the fifth root $\sqrt[5]{\ }$, and so on. **Level 7**
- Very large and very small numbers can be written in **standard form**.
  For example, the mass of an electron is approximately
  0.000 000 000 000 000 000 000 000 000 000 910 9 kg.
  In standard form this is 9.109×10^{-31} kg. This is read as 'nine point one zero nine times ten to the power of minus thirty-one'. **Level 7 & Level 8**
- Numbers written in standard form can be entered into a calculator.
  For example, to enter 3.6×10^5, press [3] [•] [6] [EXP] [5].
  The [EXP] key does the '×10' operation. **Level 7 & Level 8**
- Calculators give very large or very small numbers in standard form.
  For example, if a calculator gives the answer 7.9×10^{-5}, it means 0.000 079.
  Some calculators show this as 7.9E^{-5}, and this means the same thing. **Level 7 & Level 8**

Learn this

Standard form

8.1×10^4

number between 1 and 10 (8.1); power of 10 (10^4)

Did you know?

A googol is the number 10^{100}.
A googolplex is $10^{\text{googol}} = 10^{10^{100}}$.

Practice, practice, practice!

Level 6

1 Use the square roots you know to estimate the square roots of these numbers.
Use the square root key on your calculator to check your answers.

$\sqrt{6.25}$ $2^2 = 4$ and $3^2 = 9$
So $\sqrt{6.25}$ must lie between 2 and 3. It will be about halfway between.
Estimate: $\sqrt{6.25} = 2.5$ Calculator check: 2.5 ✓

a $\sqrt{21}$ **b** $\sqrt{105}$ **c** $\sqrt{77.44}$ **d** $\sqrt{132.25}$

6c I can estimate square roots and use the square root key to check answers

2 Work out these using your calculator.
Give your answers to one decimal place where appropriate.

a $\sqrt[3]{1728}$ **b** $\sqrt[3]{0.01}$ **c** $\sqrt[3]{166.375}$ **d** $\sqrt[3]{\frac{1}{3}}$

e The volume of a cube is 2744 m^3. What is its side length in centimetres?

6a I can use the function key for cube roots

3 A solid cube has a volume of 250 cm^3. Work out these using your calculator.
Give all your answers to two decimal places.

a What is the length of one of its sides?
b What is the area of one of its faces?
c What is the surface area of the cube?

6a I can use the function key for cube roots

cube root inverse power

4 Work out these using your calculator.
Give your answers to one decimal place.

a $\sqrt[4]{81}$ **b** $\sqrt[6]{36}$ **c** $\sqrt[10]{1000}$

d Given that $x^5 = 1024$, what is the value of x?

Tip
The root function may be shown as $\sqrt[x]{\ }$ on your calculator.

5 Write these numbers in standard form:

a 2 400 000 **b** 4500 **c** 18 **d** 524

6 Write these numbers in standard form:

a 0.15 **b** 0.032 **c** 0.00041 **d** 0.08

7 Use your calculator to work out these.
Give your answers in standard form to one decimal place.

a $3.1 \times (4.2 \times 10^{17})$ **b** $5 \div (9.45 \times 10^{-12})$

c $(5.2 \times 10^6) \times (1.3 \times 10^{10})$ **d** $(9.8 \times 10^{-3}) \div (7.9 \times 10^7)$

e $(5.2 \times 10^4) \times (3.843 \times 10^4)$ **f** $(1.1 \times 10^{-3}) \times (2.2 \times 10^{-4}) \div (3.3 \times 10^5)$

8 Write your answers to Q7 as ordinary numbers.

Level 7

7c I can use a calculator to find any root

7b I can write numbers greater than 10 in standard form

7a I can write numbers less than 10 in standard form

7a I can enter numbers in standard form into a calculator

7a I can interpret numbers given in standard form on a calculator

9 The width of an average human hair is approximately the same width as 1 million carbon atoms.
Given that the width of a hair is 0.1 mm, what is the width of one carbon atom?

10 A single drop of water contains about 2 sextillion (21 zeros!) oxygen atoms.
Given that a glass contains 1.3×10^3 drops of water, how many oxygen atoms does it contain?

11 Hair grows at approximately 4.8×10^{-9} metres per second.

a How much will your hair grow in one day?

b Assuming that there are 365 days in the year, how much will your hair grow in a year?

12 The Earth rotates at approximately 1.67×10^3 km/h, measured at the equator.
How many kilometres does the Earth rotate at the equator in 100 years?

Hint: Assume that there are 365 days in a year.

Level 8

8c I can solve problems involving standard form

A The root of the matter

When you square root a number there are always two answers: a positive and a negative answer. For example, the square roots of 16 are +4 and −4. This is because $4 \times 4 = 16$ and $-4 \times -4 = 16$.

a How many solutions will there be when you cube root a number?

b How many solutions will there be for the 4th root? … the 5th root? … the 6th, 7th, 8th root?

c Write down a rule to link the number of solutions to the root being found.

B When you don't have a calculator

a Use your calculator to work out $(3.2 \times 10^{15}) \times (1.5 \times 10^6)$.

b Without using your calculator, work out

i 3.2×1.5 **ii** $10^{15} \times 10^6$

iii $(3.2 \times 1.5) \times (10^{15} \times 10^6)$

c What do you notice about **iii** and the answer given by the calculator?

d Using the same process, write down ten questions, involving standard form, that you could work out without a calculator.

square root standard (index) form

7.6 Written methods

- Use standard column procedures to add and subtract integers and decimals
- Multiply and divide integers and decimals using written methods

Why learn this?

Money calculations involve addition, subtraction, multiplication and division with answers to two decimal places.

What's the BIG idea?

- Line up the decimal points when adding and subtracting decimal numbers. You can use a **zero place holder** to replace any 'missing' digits. Level 6
- You can use an **equivalent** calculation to **multiply** a decimal by a decimal. For example, 4.2 × 2.3 = (42 ÷ 10) × (23 ÷ 10) = 42 × 23 ÷ 100 Level 6
- You can also use an equivalent calculation to **divide** a decimal by a decimal. For example, 4.25 ÷ 0.12 = 425 ÷ 12 (both values have been multiplied by 100) Level 6
- Look at your answers in the context of the problem. You should give the answer to a suitable number of **decimal places** or **significant figures**. Level 6 & Level 7

Practice, practice, practice!

Level 6

6c I can multiply decimals with up to two decimal places

1 Use a mental or written method to work out

a 0.5 × 0.2 b 0.04 × 0.5 c 0.3 × 0.9
d 0.06 × 0.6 e 0.04 × 0.08 f 0.9 × 0.09

2 Use a standard written method to work out

a 4.2 × 1.8 b 12.7 × 5.2 c 1.25 × 3.4
d 18.44 × 1.1 e 9.05 × 3.2 f 1.88 × 0.22

3 Ryan is buying a new carpet for his living room. The room measures 9.4 m by 6.75 m.

a What area of carpet will he need?
b The carpet costs £2.20 per square metre. Work out the total price of the carpet.

6c I can divide an integer by a decimal

4 Use a standard written method to work out these divisions.

225 ÷ 1.5 *Multiply both values by 10. This is equivalent to 2250 ÷ 15*

```
15 ) 2 2 5 0
     1 5 0 0     (15 × 100)
     -------
       7 5 0
       7 5 0     (15 × 50)
     -------
       0 0 0
```

225 ÷ 15 = 100 + 50 = 150

a 182 ÷ 2.8
b 114 ÷ 1.2
c 19 ÷ 0.25
d 9 ÷ 0.18

6b I can divide a decimal by a decimal

5 Use a mental or written method to work out

a 0.10 ÷ 0.5 b 0.09 ÷ 0.3
c 0.04 ÷ 0.05 d 0.5 ÷ 0.01

6 Work out these using a written method.

a 26.18 ÷ 1.7 b 134.4 ÷ 2.4
c 3.42 ÷ 0.12 d 87.21 ÷ 1.9

Watch out!

You cannot divide by zero! Dividing means 'sharing', so dividing by zero does not make sense. If you try to divide by 0 on a calculator you will get an error message.

addition decimal place divide division divisor equivalent integ

7 The price of a local paper is £0.35. One newsagent's bill, for the local paper, is £92.75. How many local papers did the newsagent buy?

8 Amelie wants to buy some gold to create a piece of jewellery. Gold is selling at £14.47 per gram so she buys 4.75 grams. The jewellery designer will charge £38.48 per hour for her work and she estimates the work will take her 14.25 though if she has problems it will take her 16.75 hours.

a How much will the gold cost Amelie in total?

b What is the minimum amount the designer will charge?

c Work out the maximum possible cost to Amelie of the piece of jewellery.

9 Work out these using a written method.

a 0.003 04 + 12.05 + 0.208 + 3.9 **b** 9.067 + 3268 + 4.5259 + 0.78

c 6.5 − 0.06 − 0.0079 **d** 1847 − 0.010 05 + 35.67 − 0.0099

10 In an experiment, 1.2785 g of material A is combined with 0.229 g of material B. Following the reaction, samples C (0.1076 g) and D (0.9 g) are removed. The remainder is divided in half to produce samples E and F. 0.014 g of impurities are removed from sample F before it is stored in a container with mass 3 g.

a What is the mass of sample E?

b What is the total mass of the purified sample F and the container?

Level 6

6b I can divide a decimal by a decimal

6a I can multiply with decimals to solve problems, giving answers to an appropriate degree of accuracy

6a I can add and subtract integers and decimals of any size

11 Work out these.
Give your answers to a suitable number of significant figures.

a 0.075 ÷ 0.056 **b** 0.0145 ÷ 0.24 **c** 0.18 ÷ 0.039

12 **a** James is hanging lining paper. The area of the wall to be papered is 11.5 m^2.
The height is 2.7 m.
How wide is the wall (to the nearest cm)?

b Lining paper comes in rolls with width 50 cm and length 10 m.
How many rolls will James need?

Level 7

7b I can divide decimals and give answers to a suitable number of significant figures

Now try this!

A Counting the digits

a **i** Work out 3.2 × 4.07.

ii How many digits in total are there after the decimal points in the question?

iii How many digits are there after the decimal point in the answer?

iv What do you notice?

b Work out 2.29 × 3.25. Does the same thing happen?

c Work out 1.5 × 2.2. Does it still work?
Explain your answer fully.

B Exchange rates

Find the exchange rates between £ and each of these currencies. Then use a written method to convert each amount into the given currency.

a £52.34 (euros)

b £199 (US dollars)

c £352.74 (yen)

Think carefully about the level of accuracy needed.

ultiplication multiply significant figures subtraction zero place holder

7.7 Using a calculator efficiently

- Use brackets and the memory on a calculator
- Use the power, root and reciprocal keys on a calculator
- Carry out calculations with more than one step
- Understand why it is important not to round during the intermediate steps of a calculation

Why learn this?

A calculator is only as efficient as the person operating it.

What's the BIG idea?

- Scientific **calculators** follow the correct **order of operations**, so you can use brackets to tell them which values to calculate first.
  For example, $\frac{25 - 2.5}{12 - 6}$ can be entered as (25 − 2.5) ÷ (12 − 6). **Level 6 & Level 7**
- The fraction **key** can be used to **enter** calculations in the form of fractions.
  For example, $\frac{4.5}{7.1 + 8.24}$ can be entered as 4.5 $a\frac{b}{c}$ (7.1 + 8.24). **Level 6 & Level 7**
- The **memory** on a calculator can store values so that you can use them again later in the calculation. **Level 6 & Level 7**
- The ANS key on a calculator stores the answer to the previous calculation. **Level 6 & Level 7**
- Try to avoid **rounding** during an intermediate step of a calculation. **Level 7**

Practice, practice, practice!

Watch out! Don't forget to clear the memory between calculations.

Level 6

6b I can use the brackets and memory keys to carry out calculations expressed as a fraction

1 Use the brackets or memory keys on your calculator to work out these.
Give your answers to two decimal places.

a $\frac{9.5 \times 7.2}{3.5 + 7.7}$ **b** $\frac{(25.4 + 3.09)^2}{2^2 \div 0.5}$ **c** $\frac{18.45^3 - 3.2}{\sqrt{165} \times 0.4}$ **d** $\frac{(3.32 - 0.9)^2}{0.2 \times 8^2}$

6b I can use the brackets and memory keys to carry out calculations with more than one step

2 Use the brackets and memory keys on your calculator for this problem.

Carrots	74p per kg
New potatoes	£1.49 per kg
Onions	69p per kg
Broccoli	£1.38 per kg
Courgettes	£2.03 per kg

a Mr Flakowska buys 2.5 kg of carrots, 1 kg of new potatoes, 750 g of onions, 1 kg of broccoli and 350 g of courgettes. Work out the cost of his shopping.

b The prices are reduced by 10 per cent at the checkout, how much would Mr Flakowska pay for these items?

c The original prices are increased by 25 per cent because of high delivery charges. How much would he pay for these items?

6a I can use a calculator to work out calculations with three or more steps

3 Work out these. Give your answers to two decimal places where appropriate.

a $[3.24 + (5.8 \times 7.24) - 3.65]^2$ **b** $7.6^2 - (4.23 + 8.43 - 9)^2$

c $\sqrt{76} + (8.3 \div 7)^3$ **d** $9.3^3 - [(3.4 + 6.7) \div 12.5]$

e $3(\sqrt[3]{2} + 9)$ **f** $\left[\frac{1}{5} + \left(\frac{2}{3} - \frac{1}{4}\right)\right]^3$

g $\left(\frac{3}{7}\right)^2 \div \sqrt{\frac{5}{4}}$ **h** $\left(\frac{3}{8} - 0.006\right)^3 \times 45.26$

brackets calculator clear display enter key memory

4 In a sale, Adrian buys 12 books priced £7.99 and five books priced £4.99. The total price is reduced by 25 per cent at the checkout.

a Write down one way of calculating the total price with a single calculation.

b Find the final price using a calculator.

Level 6

6a I can use a calculator to solve problems involving more than one operation

5 **a** Calculate $\sqrt[3]{100} \times (3.21 + 5.67)$

b Use your answers to part **a** to help you work out the value of

i $\dfrac{1}{\sqrt[3]{100} \times (3.21 + 5.67)}$ **ii** $\left[\sqrt[3]{100} \times (3.21 + 5.67)\right]^2$

iii $\dfrac{\sqrt[3]{100} \times (3.21 + 5.67)}{(18^4 - 12) \div 3}$

Level 7

7c I can use the power, reciprocal, root, fraction, brackets and memory keys on a calculator

6 Here are the masses (in kilograms) of babies born in one hospital in April.

Baby	A	B	C	D	E	F	G	H	I	J
Mass (kg)	2.50	3.60	3.95	3.28	2.98	2.61	4.01	2.02	3.45	3.99

7c I can use the memory key on the calculator appropriately

a Work out the mean mass of the babies.

b The hospital has decided not to include baby H's mass in the mean. Work out the new mean.

c The mean mass of the 12 babies born in May was 3.2 kg. Work out the overall mean mass of the 21 babies born in April (with baby H excluded) and May.

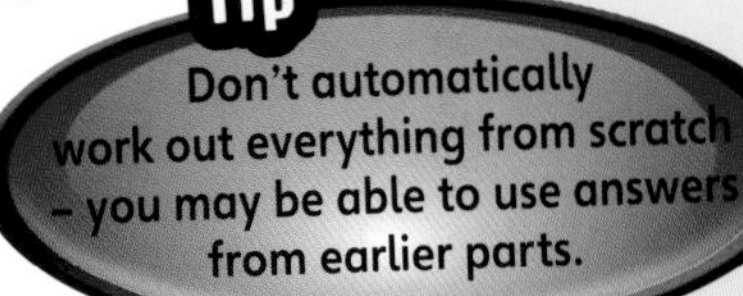
Tip
Don't automatically work out everything from scratch – you may be able to use answers from earlier parts.

7 Here is the plan of an allotment.

a Find the length of AB.

b Find the total area.

Give your answers to an appropriate degree of accuracy.

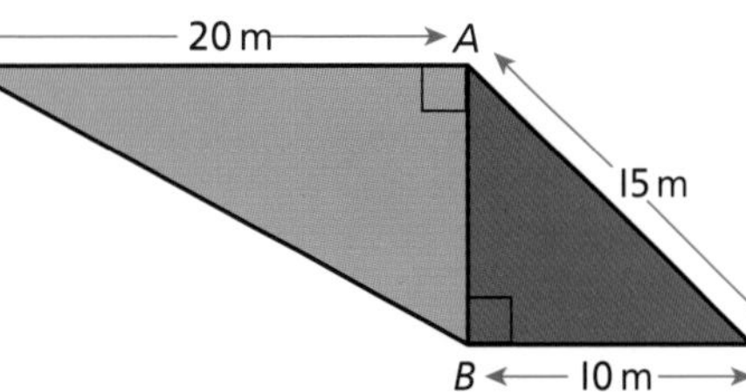
20 m
A
15 m
B
10 m

7b I can apply Pythagoras' theorem and know not to round during intermediate steps of a calculation

Now try this!

A Oscillating sequences

a Find the value of 1 × −1 on your calculator. Write down the value you get.

b Multiply your answer to part **a** by −1 by entering [ANS] [×] −1. Write down the value you get.

c Press [=] again. Write down the value you get.

d Repeat part **c**, several times.

e What happens each time you press [=]? Why do you think this is?

f This is an oscillating sequence since it bounces between positive and negative numbers. Use your calculator to generate another oscillating sequence.

B Interesting account

A bank offers an annual interest rate of 12 per cent on deposits of more than £100 if the money is left in the account for at least 10 years.

a £100 is invested. How much is in the account at the end of the first year?

b How much is in the account at the end of the tenth year?

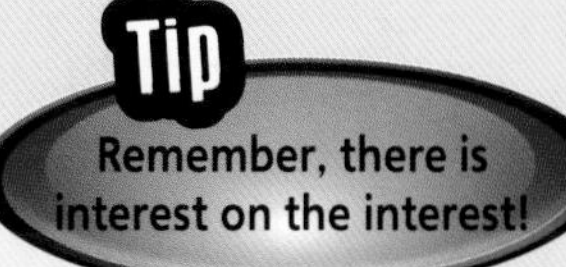
Tip
Remember, there is interest on the interest!

c How many years would the money have to be left in the account for there to be over £500?

order of operations power reciprocal root round

7.8 Problem solving

- Use a range of calculation techniques, including trial and improvement methods
- Solve substantial problems by breaking them down into small steps
- Use a calculator efficiently

Why learn this?

Can't see the wood for the trees? Solving problems often involves 'stepping back' from the whole problem and breaking it down into a number of smaller steps.

What's the BIG idea?

- When solving big problems, break them down into simpler questions that are easier to answer. Level 6, Level 7 & Level 8
- If you can, draw a **diagram** to represent the problem. Level 6, Level 7 & Level 8
- Decide which methods you are going to use.
  - **Trial and improvement** – **estimate** an answer and test it to see how close it is to the solution you want. Improve your estimate repeatedly.
  - Algebraic – form an **equation** and solve it to find the unknown.
  - Accurate diagram – construct an accurate diagram or appropriate graph. Level 6, Level 7 & Level 8
- Decide what resources you will need – calculator, spreadsheet package, drawing implements. Level 6, Level 7 & Level 8

Tip

When problem solving, always write down exactly what you are doing. Then, if you make a mistake, you can retrace your steps and correct it.

Practice, practice, practice!

1 The sum of two whole numbers is 63. Their product is 882.
- **a** Write three different pairs of whole numbers whose sum is 63.
- **b** Test the pairs from part **a** to see if their product is 882.
- **c** If you haven't found the correct pair of numbers yet, try some other pairs.

2 I think of a number x, add 15 and multiply by 4. The answer is 108.
Form an equation and solve it to find the value of x.

3 A teacher is organising a school trip to France for a week.
The price of the flight is £72.99 per person plus airport tax of 10 per cent.
The price of the accommodation is £15 per person per night.
Work out the total cost of the trip for 35 pupils.

Level 6

6c I can use trial and improvement techniques to solve problems

6b I can solve problems by forming and solving equations

6b I can solve problems by breaking them down into simpler steps

diagram equation estimate lower bound product reciprocal

4 A gardener has decided to plant a semicircular flower bed with bulbs.
The recommendation is for no more than one bulb per 100 cm^2.
The radius of the flower bed is 5 m.

- **a** Draw a diagram of the flower bed.
- **b** Work out the area of the flower bed.
- **c** How many bulbs will be needed for the flower bed?

Hint: Convert the area to square centimetres.

Level 6

6a I can use diagrams to help visualise problems

5 A rectangular plot is fenced on two adjacent sides.
The width is given by x and the length is the square of the width.
The total length of fencing for the two sides is 375 m.
Form an equation and use trial and improvement to find the value of x to two decimal places.

6 A company offers two different mobile phone tariffs.
Tariff A: £12.99 a month line rental, calls 5p per minute, texts 10p each
Tariff B: no monthly line rental, calls 7p per minute, texts 10p each

- **a** Kate makes 3 hours of calls a month. Tariq makes 15 hours of calls a month. Both want the cheapest option.
Which tariff should each choose? **Hint:** Use a graph to compare the tariffs.
- **b** How many minutes of calls a month would someone need to make for tariff A to be cheaper than tariff B?

Level 7

7c I can solve problems by forming equations involving x^2 and using trial and improvement techniques

7b I can solve a problem by using graphs

7 Airlines allow you to carry up to 100 ml of fluid in hand luggage.
A company produces cylindrical bottles for carrying fluid on aircraft.
Given that the height of the bottle is 5 cm, what is the maximum diameter? Give your answer to one significant figure.

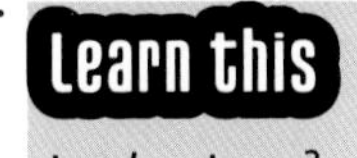

8 I think of a number x, find its reciprocal, square it and add 0.5. The answer is $\frac{3}{4}$.
What was the number I was originally thinking of?

9 The dimensions of a rectangular garden are 5 m by 8 m, to the nearest metre.

- **a** What are the smallest and largest possible areas of the garden?
- **b** Use inequality signs to state the limits of the area.

Level 8

8a I can solve problems by finding a variable that is not the subject of a formula

8c I can solve problems by forming an equation to solve

8a I can solve problems by understanding upper and lower bounds

Now try this!

A Imagine that!

Work with a partner.
Each use the numbers 36, 15 and 3 to write a problem that can be solved by using trial and improvement, by drawing a diagram and/or by forming and solving an equation.
Make sure you can solve the problem and then swap with your partner.
Whose problem is the better one?

B How many?

Research the rate at which the population of the UK is changing.
Estimate what the population of the UK in 2020 will be.

Tip
Include birth rates, death rates, immigration and emigration.

significant figures   sum   trial and improvement   upper bound

Decimal Day!

It's Decimal Day and Mr Masus has changed all the prices in his restaurant into decimal prices.

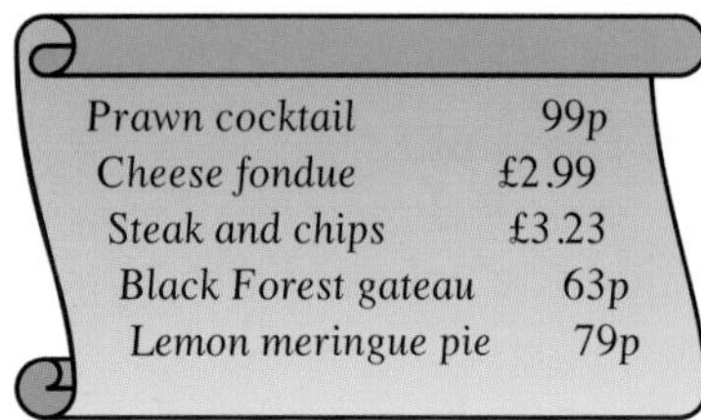

Prawn cocktail	99p
Cheese fondue	£2.99
Steak and chips	£3.23
Black Forest gateau	63p
Lemon meringue pie	79p

1 What is the cost of three cheese fondues, two steak and chips, one Black Forest gateau and five lemon meringue pies? **Level 6**

2 Mr Masus works out the total bill for a group of teenagers to be £4.41. Each teenager had one portion of Black Forest gateau only. How many teenagers were there? **Level 6**

3 Before decimalisation in 1971, there were 240 pence in £1.

a Mr Masus used to charge $\frac{1}{3}$ of a pound for a tomato soup. How much was this in old pence?

b Could Mr Masus charge $\frac{1}{3}$ of a pound in decimal pence? Explain your answer. **Level 6**

4 Mr Masus wishes to carry out this calculation.

$$\frac{2 \times (2.99 + 0.63) + 3 \times 0.79 + 0.99}{4}$$

a Use a mental method to estimate the answer to this calculation.

b Use a written method to work out the exact value.

c Use your calculator to check your answer to part **b**. **Level 7**

5 Before decimalisation a farthing was $\frac{1}{4}$ of a penny.
A farthing is now worth 1.04×10^{-3} decimal pounds.
Work out how much a farthing is worth in decimal pence. **Level 8**

6 Mr Masus rounds one of the customer's bills to the nearest pound.
He then charges the customer £24.
What are the largest and smallest amounts the bill could have been? **Level 8**

7 Mr Masus charged one customer £12.84.
He knows the customer ordered several steak and chips and several Black Forest gateaux (and nothing else), but he can't remember exactly how many of each.
Work out how many steak and chips and how many Black Forest gateaux the customer ordered. **Level 8**

8 Mr Masus held a special event on Decimal Day. Customers could order a special meal at a set price.
He sold 40 set-price meals that day (and nothing else).
The revenue for that event was £900, to the nearest £100.
What are the upper and lower bounds of the set price? **Level 8**

The BIG ideas

- Line up the decimal points when adding or subtracting decimal numbers. You can use a **zero place holder** to replace any 'missing' digits. **Level 6**
- You can use an **equivalent** calculation to **divide** a decimal by a decimal. For example, $4.25 \div 0.12 = 425 \div 12$ (both values have been multiplied by 100). **Level 6**
- You can use rounding to make **estimates**. For example, $0.47 \times 12.2 \approx 0.5 \times 12 = 6$ **Level 6**
- A **recurring decimal** contains a digit, or sequence of digits, which repeats itself forever. A dot over a digit shows that it recurs. For example $0.333... = 0.\dot{3}$. **Level 6 & Level 7**
- Scientific **calculators** follow the correct **order of operations**, so you can use brackets to tell them which values to calculate first. **Level 6 & Level 7**
- You can round numbers to a given number of **significant figures**. The first significant figure is the first non-zero digit in the number, counting from the left. **Level 7**
- Very large and very small numbers can be written in **standard form**. For example, 34 000 in standard form is 3.4×10^4 **Level 7 & Level 8**
- When a number has been rounded, the largest and smallest values it could have been can be given. These are called the **upper** and **lower bounds**. **Level 8**
- When solving big problems:
  - break them down into simpler questions that are easier to answer
  - draw a **diagram** if appropriate
  - decide which methods you are going to use, e.g. algebraic, accurate diagrams or trial and improvement techniques. **Level 6, Level 7 & Level 8**

Find your level

Level 6

Q1 Write these numbers in order of size, smallest first.

3^2 $\sqrt{100}$ 2^4 2^3

Q2 Use a written method to work out

a 0.0045 + 17.627 **b** 43 − 7.005

Level 7

Q3 Copy and complete these multiplication grids.

a

×	9	
8	72	
−7		42

b

×	0.3	
7		2.8
		8

Q4 During her holidays from university, Laura works part-time.
One week she worked for 26 hours.
During that week, she worked for the same length of time each day from Monday to Thursday. On Friday, Saturday and Sunday, she worked for twice as long as on Monday.
For how long did Laura work on Saturday?

Level 8

Q5 The number of people at a concert was recorded as 250 000, rounded to two significant figures.

What is the largest number of people who could have attended?

Q6

Country	Population	Area (km²)
Australia	2.1×10^7	7.7×10^6
China	1.3×10^9	9.6×10^6
South Africa	4.9×10^7	1.2×10^6
UK	6.1×10^7	2.4×10^5
USA	3.0×10^8	9.8×10^6

a Which country has the smallest population?

b How many people per km² are there in China?

c How many more people per km² are there in the UK than in Australia?

8 Graphic detail

This unit is about solving problems using graphs and properties of numbers.

Marcus du Sautoy is a professor of maths at Oxford University. He says, 'Prime numbers are nature's gift to the mathematician. For me, the most interesting thing is that primes seem to underpin the universe and the way nature works. Here's an example.

'Cicadas spend the first years of their lives underground and then emerge for a few weeks to eat, mate and lay eggs. Different species stay underground for different lengths of time. The North American species stay underground for 7, 13 or 17 years. These are all prime numbers. Why?

'One reason for them being primes is that it might have helped them to avoid a predator that also appeared periodically. For example, if a predator appears every six years then the cicada that appears every seven years won't meet it until year 42. A non-prime number cicada that appears every eight years will get eaten in year 24, the lowest common multiple of six and eight. So primes are literally the key to the evolutionary survival of cicadas.'

A keen football fan, Marcus du Sautoy plays in a prime number shirt for his Sunday League team.

Activities

A
- Write out the numbers from 1 to 50 to represent 50 years.
- Write a 'P' next to every 6th number to represent a predator that appears every 6 years.
- Use different colours to represent three different species of cicada. Red appears every 7 years, blue every 8 years and yellow every 9 years.
- How many times does each species meet the predator over the 50 years?

B Mathematicians think that every even number can be made from the sum of two primes but no one has been able to prove that this is always true.
- Write down all the even numbers from 10 to 20.
- Show that each of these can be made from a sum of two primes.
- Choose an even number with three digits. Find two prime numbers that add up to your number.

An unsolved mystery

Why are the primes scattered along the number line in what looks like a random pattern? Anyone who can work out the pattern will become rich and very famous.

Before you start this unit...

1 **a** Write down all the factors of 12 and 18.
b Which numbers are factors of both 12 and 18?

Level Up Maths 5-7 page 8

2 Work these out without using a calculator.
a $\sqrt{144}$ **b** $\sqrt{0.16}$
c $\sqrt[3]{64}$ **d** $\sqrt[3]{0.008}$

Level Up Maths 5-7 page 138

3 **a** Draw a grid with x- and y-axes from -4 to 4.
b Plot the graphs of $y = x$ and $y = -x$.
c What size is the angle between the two lines?

Level Up Maths 5-7 page 76

8.1 Prime factors, HCF and LCM
8.2 Using factors and multiples
8.3 Multiplying and dividing with indices
8.4 Index laws with negative and fractional powers
8.5 Reviewing straight-line graphs
8.6 Investigating the properties of straight-line graphs
8.7 Graphs of quadratic and cubic functions
maths! Parabolas
8.8 Functions and graphs in real life
8.9 Investigating patterns
Unit plenary: Primed for action
Plus digital resources

Did you know?

- A prime number has only two factors, itself and the number 1.
- Every natural number (1, 2, 3, ...) can be made by multiplying prime numbers together.

8.1 Prime factors, HCF and LCM

- Find the HCF and LCM of three numbers less than 100
- Find the prime factor decomposition of a whole number
- Find the HCF and LCM of a group of numbers from their prime factor decompositions

Why learn this?

Prime factors are important because they are the 'building blocks' of the natural numbers (1, 2, 3, ...). Every single natural number greater than 1 can be written as a product of its prime factors.

What's the BIG idea?

- A **factor** divides into a number exactly. The **highest common factor (HCF)** of a group of numbers is the highest number that is a factor of all the numbers.
  For example, 9 is the HCF of 27, 45 and 54. **Level 6**
- A **multiple** is a member of the multiplication table for that number. The **lowest common multiple (LCM)** of a group of numbers is the lowest number that is a multiple of all the numbers.
  For example, the LCM of 2, 4 and 13 is 52. **Level 6**
- A **prime number** has exactly two factors, 1 and itself. Any other number can be written as a **product** of its prime factors. This is called the **prime factor decomposition**.
  For example, the prime factor decomposition of 60 is $2 \times 2 \times 3 \times 5$. **Level 6**
- You can write a number as a product of its prime factors using powers.
  For example, $60 = 2 \times 2 \times 3 \times 5 = 2^2 \times 3 \times 5$
  $1323 = 3 \times 3 \times 3 \times 7 \times 7 = 3^3 \times 7^2$ **Level 6**
- You can work out the HCF and LCM of a group of numbers from the prime factor decomposition of each number. **Level 6 & Level 7**

Did you know?

Common land is land that is shared by everybody. Common factors are shared factors.

Practice, practice, practice!

Level 6

6c I can find the HCF of three whole numbers less than 100

6c I can find the LCM of three whole numbers less than 100

6b I can find the prime factor decomposition of a whole number

1 Find the highest common factor (HCF) of these numbers.

a 6, 8 and 12 **b** 14, 28 and 49 **c** 18, 27 and 81 **d** 42, 63 and 84

2 Find the lowest common multiple (LCM) of these numbers.

8, 40, 24 Multiples of 8: 8, 16, 24, 32, 40, 48, 56, 64, 72, 80, 88, 96, 104, 112, (120), 128, ...
Multiples of 24: 24, 48, 72, 96, (120), 144, ...
Multiples of 40: 40, 80, (120), 160, ...
LCM = 120.

a 45, 18 and 24 **b** 9, 10 and 15 **c** 10, 30 and 48 **d** 15, 20 and 24

3 **a** List the factors of 48.

b List the factors of 48 that are prime numbers.

c Write 48 as a product of its prime factors. This is called its prime factor decomposition.

d Repeat parts **a** to **c** to find the prime factor decomposition of
i 36 **ii** 45 **iii** 72 **iii** 96.

Tip
When identifying factor pairs, start at 1 (1×48) and then move on to 2, and so on.

4 Find the prime factor decomposition of each of these numbers.

a 32 **b** 54 **c** 64 **d** 100

Tip
Always check that a prime factorisation is correct by multiplying out.

factor highest common factor (HCF) lowest common multiple (LCM) multiple

5 Use prime factor decomposition to find the HCF and LCM of these numbers.

40 and 56 *Prime factor decompositions* $40 = 2 \times 2 \times 2 \times 5$
$56 = 2 \times 2 \times 2 \times 7$
Identify the common factors and multiply them:
$HCF = 2 \times 2 \times 2 = 8$
Multiply together all the factors but only include overlaps once.
$LCM = 2 \times 2 \times 2 \times 5 \times 7 = 280$

a 8 and 12 b 9 and 12 c 8 and 9
d 16 and 24 e 36 and 96 f 32 and 48
g 96 and 144 h 120 and 225 i 180 and 240
j 216 and 240 k 256 and 480 l 330 and 396

6 a Use prime factor decomposition to find the LCM and HCF of these numbers.
i 30 and 77 ii 25 and 49 iii 64 and 81 iv 36 and 121
b What do you notice about the HCF and LCM of these pairs of numbers?

7 Mark has two light bulbs, A and B, that turn on and off in a regular pattern.
Bulb A turns on, stays on for 5 s and then turns off and stays off for 5 s.
Bulb B turns on, stays on for 6 s and then turns off and stays off for 6 s.
Mark turns both bulbs on at the same time.
How long will it be before both bulbs turn on again at the same time?

Bulb A	on	off	on
Bulb B	on	off	on

Level 6

6a I can find the LCM and HCF of two numbers using prime factor decomposition

6a I can find the LCM and HCF of two numbers using prime factor decomposition

6a I can use multiples and LCMs to solve problems

8 Use prime factor decomposition to find the HCF and LCM of these numbers.
a 12, 16 and 20 b 24, 36 and 45 c 20, 30 and 72
d 56, 84 and 72 e 120, 144 and 156 f 32, 102 and 136

9 Rhiannon has three sets of Christmas tree lights that flash on and off.
The first set turn on, stay on for 5 s and then turn off and stay off for 5 s.
The second set turn on, stay on for 4 s and then turn off and stay off for 4 s.
The third set turn on, stay on for $\frac{3}{4}$ s and then turn off and stay off for $\frac{3}{4}$ s.
Rhiannon turns all three sets of lights on at the same time.
How long will it be before for all three sets of lights switch on again at the same time?

Level 7

7c I can find the LCM and HCF of three numbers using prime factor decomposition

7a I can use multiples and LCMs to solve problems involving fractions

Did you know?

The first Christmas tree to be lit by electric lights was in New York in 1882. It used 80 bulbs.

Now try this!

A LCM and HCF dash

Work with a partner. Each select a whole number between 50 and 250.
Work separately to
- find the prime factor decomposition of each number
- work out the LCM and HCF for the pair.

Compare and check answers.

B Unique solution?

- Find two numbers which have an LCM of 180 and an HCF of 15.
- Is it possible to fnd a different pair of numbers for which this is true?

prime factor decomposition prime number product

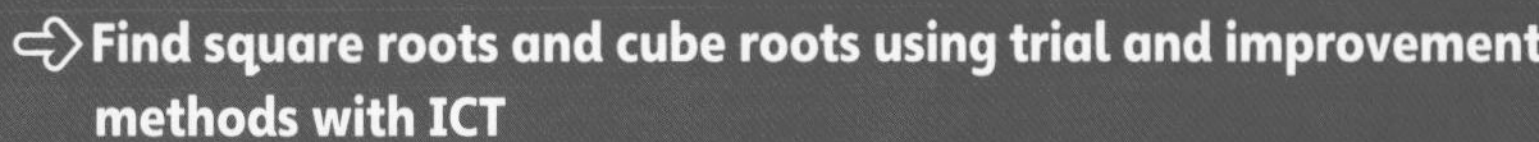

8.2 Using factors and multiples

- Find square roots and cube roots using trial and improvement methods with ICT
- Find square roots and cube roots using factors
- Solve problems using factors and multiples
- Simplify algebraic fractions such as $\frac{a}{b} + \frac{c}{d} = \frac{ad + bc}{bd}$

Why learn this?

Factors and multiples are useful not just in maths, but in other subjects such as science, engineering and economics.

What's the BIG idea?

→ To solve a problem using **trial and improvement**, make an initial 'guess' at the solution, test it to find out whether the guess is too big or too small, and then improve the guess. When you have the solution sandwiched between two guesses of the required level of accuracy, use the 'halfway' test. **Level 6**

→ You can use **factors** to help you find **square roots** and **cube roots**.
For example, to find the square root of 225
$225 = 5 \times 5 \times 3 \times 3 = (5 \times 3) \times (5 \times 3) = 15 \times 15$
So the square root is 15. **Level 7**

→ A surd is a square root that cannot be simplified further.
You can use factors to simplify square roots to answers in surd form.
For example, $\sqrt{12} = \sqrt{4 \times 3} = \sqrt{4} \times \sqrt{3} = 2\sqrt{3}$ **Level 8**

→ To add and subtract fractions, find the **lowest common multiple** (LCM) of the **denominators**. Write all the fractions with this denominator. You can do this with numbers and with algebraic variables. **Level 7 & Level 8**

Practice, practice, practice!

Level 6

6a I can find square roots using trial and improvement

1 Using only the [×] key on your calculator, estimate these square roots. Give your answer to one decimal place.

a $\sqrt{255}$ **b** $\sqrt{1020}$ **c** $\sqrt{720}$
d $\sqrt{905}$ **e** $\sqrt{5620}$ **f** $\sqrt{350}$

6a I can find cube roots using trial and improvement

2 Using only the [×] key on your calculator, estimate these cube roots. Give your answers to one decimal place.

a $\sqrt[3]{210}$ **b** $\sqrt[3]{635}$ **c** $\sqrt[3]{800}$
d $\sqrt[3]{1720}$ **e** $\sqrt[3]{500}$ **f** $\sqrt[3]{3225}$

Level 7

7c I can find square roots using factors

3 Use factors to work out these square roots. (Don't use a calculator.)

$\sqrt{256} = \sqrt{2 \times 2 \times 64} = 2 \times \sqrt{64} = 2 \times 8 = 16$

a $\sqrt{1024}$ **b** $\sqrt{729}$ **c** $\sqrt{900}$
d $\sqrt{5625}$ **e** $\sqrt{324}$ **f** $\sqrt{1764}$

cube root denominator factor lowest common denominator

4 Use factors to work out these cube roots. (Don't use a calculator.)

$\sqrt[3]{216} = \sqrt[3]{2 \times 2 \times 2 \times 27} = 2 \times \sqrt[3]{27} = 2 \times 3 = 6$

a $\sqrt[3]{125}$ **b** $\sqrt[3]{8000}$ **c** $\sqrt[3]{1728}$

d $\sqrt[3]{512}$ **e** $\sqrt[3]{3375}$ **f** $\sqrt[3]{5832}$

Level 7

7a I can find cube roots using factors

5 Write these as simply as possible in surd form.

a $\sqrt{18}$ **b** $\sqrt{45}$ **c** $\sqrt{28}$ **d** $\sqrt{27}$

6 The area of a square is 40 cm². Find the length of one side.
Give your answer in surd form.

7 Here is a rectangle.
Work out its perimeter.
Give your answer in its simplest form.

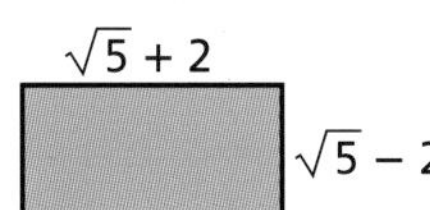

8 Copy and complete to find equivalent fractions.

a $\frac{3}{8} = \frac{\square}{32}$ (×4 above, ×4 below)

b $\frac{1}{x} = \frac{\square}{xy}$ (×y above, ×y below)

c 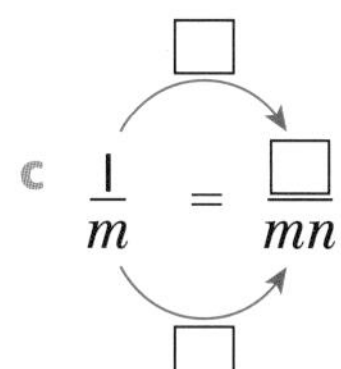 $\frac{1}{m} = \frac{\square}{mn}$ ($\square$ above, $\square$ below)

d $\frac{1}{p} = \frac{\square}{3pq}$ ($\square$ above, $\square$ below)

e $\frac{a}{c} = \frac{\square}{cd}$ (×d below)

f $\frac{m}{2n} = \frac{\square}{2nx}$

9 Simplify these expressions.

$\frac{1}{a} + \frac{1}{b} = \frac{b}{ab} + \frac{a}{ab} = \frac{b+a}{ab} = \frac{a+b}{ab}$

a $\frac{1}{p} + \frac{1}{q}$ **b** $\frac{2}{a} + \frac{3}{b}$ **c** $\frac{1}{x} + \frac{a}{y}$

d $\frac{a}{x} + \frac{1}{y}$ **e** $\frac{1}{q} - \frac{1}{s}$ **f** $\frac{1}{b} - \frac{3c}{a}$

g $\frac{5}{x} + \frac{3}{y}$ **h** $\frac{p}{2q} - \frac{r}{5s}$ **i** $\frac{a}{2b} - \frac{3c}{d}$

Level 8

8c I can simplify expressions that involve surds

8c I can simplify expressions that involve surds

8c I can change algebraic fractions to equivalent fractions

8a I can add and subtract algebraic fractions such as $\frac{a}{b} + \frac{c}{d} = \frac{ad + bc}{bd}$

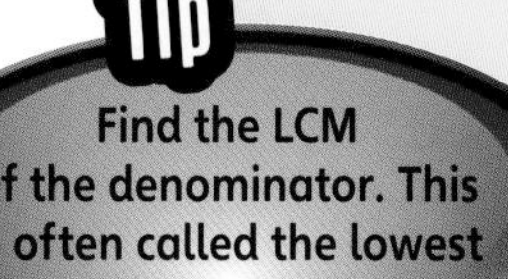

Tip
Find the LCM of the denominator. This is often called the lowest common denominator.

Now try this!

A Rooting for victory

Work with a partner. Each select a whole number between 100 and 999. Swap numbers. Using trial and improvement and ICT, find the square root of each other's number to one decimal place. The winner is the first to work out the correct answer. Check with a calculator.

B Cube root challenge

Play in groups of three. Select a number by rolling a dice four times. So if you roll 5, then 2, then 5 and then 1, your number is 5251. Using trial and improvement and ICT, find the cube root of the number to two decimal places. The winner is the first to work out the correct answer. Check with a calculator.

lowest common multiple square root trial and improvement

8.3 Multiplying and dividing with indices

- Use the index laws for multiplication and division of positive powers of numbers and variables
- Know that any number or variable to the power 0 is 1
- Use the index laws for raising a power to another power

What's the BIG idea?

- When you multiply numbers or variables written as **powers** of the same number, add the powers (or **indices**).
  For example,
  $2^3 \times 2^2 = 2^{3+2} = 2^5$
  $x^3 \times x^2 = x^{3+2} = x^5$
  $x^a \times x^b = x^{a+b}$ **Level 6 & level 7**
- When you divide numbers or variables written as powers of the same number, subtract the powers (or indices).
  For example,
  $2^5 \div 2^2 = 2^{5-2} = 2^3$
  $x^5 \div x^2 = x^{5-2} = x^3$
  $x^a \div x^b = x^{a-b}$ **Level 6 & level 7**
- Any number or variable to the power 0 is 1.
  For example,
  $5^0 = 1$
  $x^0 = 1$ **Level 7 & Level 8**
- When you raise a number or variable written as a power to another power, multiply the powers (or indices).
  For example
  $(2^2)^3 = 2^2 \times 2^2 \times 2^2 = 2^{2+2+2} = 2^6$
  $(x^2)^3 = x^{2 \times 3} = x^6$
  $(x^a)^b = x^{a \times b}$ **Level 8**

Why learn this?

Indices are important when writing out calculations involving very large numbers, such as distances in space.

Super fact!

The Hubble Space Telescope can see galaxies that are 1.1×10^{23} km away.

Practice, practice, practice!

Level 6

6a I can multiply and divide numbers written as powers

1 Simplify each expression. Leave your answer as a power.

$2^2 \times 2^3 = 2 \times 2 \times 2 \times 2 \times 2 = 2^5$ $\quad 3^5 \div 3^3 = \frac{3 \times 3 \times 3 \times 3 \times 3}{3 \times 3 \times 3} = 3^2$

a $5^4 \times 5^3$ **b** $4^2 \times 4^2$ **c** $7^5 \times 7^3$

d $5^6 \div 5^3$ **e** $8^7 \div 8^2$ **f** $10^4 \div 10^3$

g $5^3 \div 5^3$ **h** $2^4 \times 2^3 \times 2^4 \times 2^3$ **i** $\frac{6^5}{6^2}$

j $\frac{5^7}{5^3}$ **k** $\frac{10^4}{10^2}$ **l** $\frac{6^5 \times 6^3}{6^4}$

expression indices

2 Simplify each expression. Leave your answer as a power.

$5^7 \times 5^2 = 5^{7+2} = 5^9$ $\quad 5^7 \div 5^2 = 5^{7-2} = 5^5$

a $2^2 \times 2^3$	**b** $2^3 \times 2^3$	**c** $5^3 \times 5^3$	**d** $3^2 \times 3^4$
e 5×5^3	**f** $2^6 \div 2^4$	**g** $3^4 \div 3^3$	**h** $5^6 \div 5^3$
i $6^3 \div 6^3$	**j** $4^6 \div 4$	**k** $\frac{5^4}{5^2}$	**l** $\frac{2^4 \times 2^6}{2^9}$

Level 6

6a I can use index laws in number

3 Simplify each expression. Leave your answer as a power.

$x^2 \times x^3 = x \times x \times x \times x \times x = x^5$ $\quad x^5 \div x^3 = \frac{x \times x \times x \times x \times x}{x \times x \times x} = x^2$

a $a^4 \times a$	**b** $a^2 \times a^2$	**c** $a^5 \times a^3$	**d** $a^6 \div a^3$
e $a^7 \div a^2$	**f** $a^4 \div a^3$	**g** $a^3 \div a^3$	**h** $a^4 \times a^3 \times a^4 \times a^3$
i $\frac{c^5}{c^2}$	**j** $\frac{t^7}{t^4}$	**k** $\frac{y^8}{y^2}$	**l** $\frac{p^5 \times p^4}{p^7}$

4 Simplify each expression. Leave your answer as a power.

a $a^4 \times a^3$	**b** $b^3 \times b^3$	**c** $x^5 \times x^3$	**d** $p^2 \times p^5$
e $t \times t^3$	**f** $x^6 \div x^4$	**g** $a^8 \div a^3$	**h** $a^3 \div a$
i $a^3 \div a^3$	**j** $a^6 \div a$	**k** $2t^2 \times t^3$	**l** $6b^3 \div 2b$

Level 7

I can use index laws in algebra

5 Work out the value of each expression.

a 5^0	**b** 10^0	**c** $225^0 \times 7^2$	**d** y^0
e p^0	**f** $10^0 \times q^0$	**g** $(cd)^0$	**h** $6a^0$
i $4^5 \div 4^5$	**j** $6^3 \div 6^3 \times 3^2$	**k** $7^2 \div 7^2 \times 2^5$	**l** $x^3 \div x^3$

6 Simplify each expression. Leave your answer as a power.

$(3^4)^2 = 3^{4 \times 2} = 3^8$ $\quad (a^3)^2 = a^{3 \times 2} = a^6$

a $(5^2)^3$	**b** $(7^4)^3$	**c** $(6^5)^2$	**d** $(3^4)^4$	**e** $(8^5)^3$	**f** $(p^2)^3$
g $(x^2)^5$	**h** $(y^5)^2$	**i** $(b^c)^c$	**j** $(s^t)^u$	**k** $(2p^4)^2$	**l** $(q^0)^3$

7 **a** Work out $(3^3)^2$ and $(3^2)^3$. What do you notice?

b Repeat part **a** for $(5^4)^2$ and $(5^2)^4$. What do you notice here?

c Prove that $(x^3)^2 = (x^2)^3$.

Level 8

8c I can work out a number or variable written to the power 0

8c I can understand and use index laws for raising a power to another power

Now try this!

A Power dice

a Roll a dice twice. Raise the number from your first throw to the power of the number from your second throw. So if you roll 4 and then 5, you need to work out 4^5. Now swap the numbers and work out the new value, in this case 5^4. Which gives the higher result? Repeat five times.

b What are the highest and lowest results possible?

c Now play against a partner. Each roll a dice twice and select your higher result. Repeat five times and add your five results together. The winner is the player with the higher total.

B Roll an expression

A game for two players. Each roll a dice three times to create an expression, either $x^a \times x^b$ or $x^a \div x^b$.

- The first roll gives you the value of a.
- The second roll gives you the value of b.
- The third roll decides the operation. If it is even, multiply; if it is odd, divide.

So if you roll 4, then 2 and then 1, your expression is $x^4 \div x^2$. Now each simplify your own expression and work out its value when $x = 5$. The player with the higher value scores 1 point. Repeat five times.

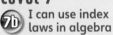

Tip

$2^{-2} = \frac{1}{2^2}$

power(s)

8.4 Index laws with negative and fractional powers

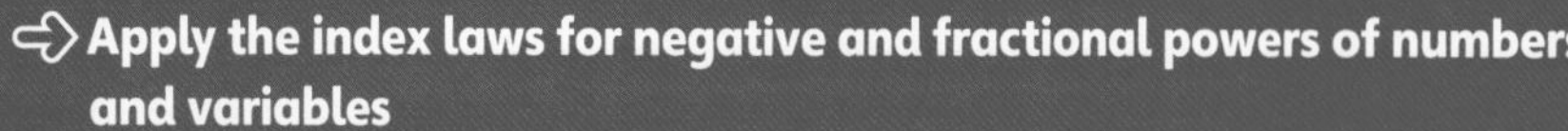

- Apply the index laws for negative and fractional powers of numbers and variables
- Know that when a number is raised to a negative power the result is smaller than 1

What's the BIG idea?

- A negative **power** can be written as a **reciprocal**.
  For example,
  $2^{-3} = \frac{1}{2^3}$ $\quad x^{-3} = \frac{1}{x^3}$ $\quad x^{-a} = \frac{1}{x^a}$ **Level 8+**
- Multiplying and dividing numbers or variables with negative **indices** follow the same rules as for positive indices.
  For example,
  $3^5 \times 3^{-2} = 3^{5 + -2} = 3^3$ $\quad 2^2 \div 2^{-3} = 2^{2 - -3} = 2^5$
  $x^3 \times x^{-5} = x^{3 + -5} = x^{-2} = \frac{1}{x^2}$ $\quad x^3 \div x^{-5} = x^{3 - -5} = x^8$ **Level 8+**
- A fractional power can be written as a root (e.g. as a **square root** or **cube root**).
  For example,
  $16^{\frac{1}{2}} = \sqrt{16}$ $\quad 27^{\frac{1}{3}} = \sqrt[3]{27}$ $\quad 8^{\frac{5}{3}} = (\sqrt[3]{8})^5 = 2^5 = 32$
  $x^{\frac{1}{2}} = \sqrt{x}$ $\quad x^{\frac{1}{3}} = \sqrt[3]{x}$ $\quad x^{\frac{5}{3}} = (\sqrt[3]{x})^5$ **Level 8+**
- Multiplying and dividing numbers or variables with fractional indices follow the same rules as for integer indices.
  For example,
  $8^{\frac{3}{4}} \times 8^{\frac{5}{4}} = 8^{\frac{3}{4} + \frac{5}{4}} = 8^2$ $\quad 2^{\frac{3}{4}} \div 2^{\frac{1}{4}} = 2^{\frac{3}{4} - \frac{1}{4}} = 2^{\frac{1}{2}}$
  $x^{\frac{1}{2}} \times x^{\frac{1}{4}} = x^{\frac{1}{2} + \frac{1}{4}} = x^{\frac{3}{4}}$ $\quad x^{\frac{5}{6}} \div x^{\frac{1}{6}} = x^{\frac{5}{6} - \frac{1}{6}} = x^{\frac{2}{3}}$ **Level 8+**
- Raising a number or variable written as a negative or fractional power to another power follows the same rule as for positive integer indices.
  For example,
  $(2^{\frac{1}{4}})^3 = 2^{\frac{1}{4} \times 3} = 2^{\frac{3}{4}}$ $\quad (x^{\frac{2}{3}})^{\frac{1}{2}} = x^{\frac{2}{3} \times \frac{1}{2}} = x^{\frac{1}{3}}$ **Level 8+**

Why learn this?

Since using indices can speed up calculations, it is logical to extend their use to negative and fractional indices.

Tip

$\sqrt{q} \times \sqrt{q} = q$
$q^{\frac{1}{2}} \times q^{\frac{1}{2}} = q^1$
So $q^{\frac{1}{2}} = \sqrt{q}$

Super fact!

Loudness is proportional to $\frac{1}{r^2}$, where r is the diatance from the source. So if you halve your distance from a loudspeaker, it sounds four times as loud.

Practice, practice, practice!

Level 7

7b I can use index laws for multiplication and division of positive integer powers

1 Simplify each expression.

a $p^2 \times p^3$ **b** $b^4 \times b^3$ **c** $y^3 \times y^3$ **d** $t^2 \times t^5$ **e** $t^2 \times t^3$
f $x^7 \div x^4$ **g** $a^8 \div a^4$ **h** $a^3 \div a^6$ **i** $a \div a^3$ **j** $a^2 \div a^5$

Level 8

8c I can use negative indices and know that a number raised to a negative index is smaller than 1

2 Find the value of each expression without using a calculator. Give your answer as a fraction and as a decimal.

$5^{-2} = \frac{1}{5^2} = \frac{1}{25} = 0.04$

a 2^{-3} **b** 10^{-2} **c** 2^{-1} **d** 10^{-5} **e** 3^{-1}
f 4^{-2} **g** 5^{-1} **h** 10^{-3} **i** 2^{-2} **j** 2^{-4}

cube root expression index (indices)

3 Find the value of each expression without using a calculator.

$16^{\frac{3}{2}} = (16^{\frac{1}{2}})^3 = (\sqrt{16})^3 = 4^3 = 64$

a $8^{\frac{1}{3}}$ b $8^{\frac{2}{3}}$ c $16^{\frac{1}{2}}$ d $64^{\frac{3}{2}}$ e $25^{\frac{1}{2}}$

f $27^{\frac{2}{3}}$ g $64^{\frac{2}{3}}$ h $125^{\frac{1}{3}}$ i $100^{\frac{3}{2}}$ j $16^{\frac{3}{4}}$

k $(6^{\frac{1}{2}})^4$ l $(5^4)^{\frac{1}{2}}$ m $(4^{2.5})^2$ n $(8^{\frac{1}{2}})^{\frac{2}{3}}$ o $(8^2)^{\frac{1}{3}}$

4 Find the value of each expression. Write your answer as a fraction.

$16^{-\frac{3}{2}} = \frac{1}{(\sqrt{16})^3} = \frac{1}{4^3} = \frac{1}{64}$

a $8^{-\frac{1}{3}}$ b $8^{-\frac{2}{3}}$ c $16^{-\frac{1}{2}}$ d $64^{-\frac{5}{2}}$ e $25^{-\frac{1}{2}}$

f $27^{-\frac{1}{3}}$ g $64^{-\frac{1}{3}}$ h $125^{-\frac{1}{3}}$ i $100^{-\frac{5}{2}}$ j $16^{-\frac{3}{4}}$

5 Simplify each expression.

a $(y^4)^2$ b $(p^4)^4$ c $(y^{\frac{1}{2}})^3$ d $(s^5)^{\frac{1}{2}}$ e $(y^{2.5})^{0.5}$

f $(x^{\frac{1}{3}})^{\frac{1}{4}}$ g $(y^{\frac{1}{2}})^4$ h $(s^4)^{\frac{1}{2}}$ i $(y^{2.5})^2$ j $(x^{\frac{1}{2}})^{\frac{2}{3}}$

6 Simplify each expression.

a $p^{-2} \times p^3$ b $b^4 \times b^{-3}$ c $y^{-3} \times y^3$ d $t^2 \times t^{-5}$

e $t^2 \times t^{-3}$ f $x^{-7} \div x^4$ g $a^8 \div a^{-4}$ h $a^3 \div a^{-6}$

i $a \div a^{-3}$ j $a^{-2} \div a^5$ k $(y^4)^{-2}$ l $(p^{-4})^4$

7 Simplify each expression.

a $(y^{\frac{3}{2}})^2$ b $(p^4)^{\frac{1}{2}}$ c $(y^{\frac{1}{2}})^3$ d $(s^5)^{\frac{1}{2}}$

e $(y^2)^{0.5}$ f $(x^{\frac{1}{3}})^{\frac{3}{4}}$ g $x^{\frac{1}{2}} \times x^{\frac{3}{2}}$ h $y^{\frac{1}{3}} \times y^{\frac{1}{2}}$

i $x^2 \div x^{\frac{1}{2}}$ j $x^2 \div x^{\frac{3}{2}}$ k $x \div x^{\frac{1}{2}}$ l $x^4 \div x^{3\frac{3}{4}}$

8 Simplify each expression.

a $p^{-2} \times p^{\frac{3}{2}}$ b $p^{-3} \times p^{-\frac{1}{3}}$ c $p^{-4} \div p^{\frac{3}{4}}$ d $\frac{1}{x^{-2}}$

e $p^{-3} \div p^{\frac{3}{2}}$ f $s^{-\frac{3}{2}} \times s^{\frac{3}{2}}$ g $q^{-4} \div q^{-\frac{3}{2}}$ h $\frac{1}{8^{-\frac{2}{3}}}$

i $(y^{\frac{1}{2}})^{-3}$ j $(s^{-5})^{\frac{1}{2}}$ k $(y^{-2.5})^{-0.5}$ l $(x^{\frac{1}{3}})^{-\frac{3}{4}}$

Level 8

8b I can use fractional indices and write a fractional power as a root

8a I can work out negative fractional powers of numbers

8+ I can use index laws to raise a power to another power

8+ I can use index laws for negative powers

8+ I can use index laws for fractional powers

8+ I can use index laws for fractional and negative powers

Now try this!

A Power throws

Roll a dice twice and record your results. Raise the number from your first throw to a power decided by your second throw.

- If you roll 1, the power is 2.
- If you roll 2, the power is $\frac{1}{2}$.
- If you roll 3, the power is 3.
- If you roll 4, the power is $\frac{1}{3}$.
- If you roll 5, the power is -2.
- If you roll 6, the power is -1.

Now work out the value of this expression – this is your score. So if you roll 5, then 2, you need to work out $5^{\frac{1}{2}}$.
Repeat 10 times. What are the highest and lowest scores possible?

B Roll another expression

A game for two players. Each roll a dice three times to create an expression, either $x^a \times x^{\frac{1}{b}}$ or $x^a \div x^{\frac{1}{b}}$.

- The first roll gives you the value of a.
- The second roll gives you the value of b.
- The third roll decides the operation. If it is even, then multiply; if is odd, divide.

So if your roll 3, then 2 and then 2, your expression is $x^3 \times x^{\frac{1}{2}}$.
Now each simplify your own expression and work out its value when $x = 5$.
The player with the higher value scores 1 point.
Repeat five times.

power reciprocal square root

8.5 Reviewing straight-line graphs

- Plot graphs of linear functions in four quadrants
- Find out if a point lies on a straight line
- Know that all linear functions in x and y can be rearranged into the form $y = mx + c$
- Explain the effect on straight lines of the form $y = mx + c$ when the gradient m or the intercept c changes
- Compare the properties of straight-line graphs of the form $y = mx + c$ without drawing the graphs

What's the BIG idea?

- All **linear functions** in x and y can be written as an equation in the form $y = mx + c$. **Level 6**
- Linear functions can be represented by straight-line graphs. **Level 6**
- You can check whether points lie on a straight line by substituting their coordinates into its equation.
  For example, the point $(-2, -10)$ lies on the line $y = 2x - 6$ because $(2 \times -2) - 6 = -10$. The point $(1, -5)$ does not lie on the line because $(2 \times 1) - 6 \neq -5$. **Level 6**
- For graphs with an equation of the form $y = mx + c$, m is the **gradient** and c is the **intercept** with the y-axis.
  - The steeper the line, the greater the value of m (ignoring the sign in front of m).
  - **Parallel** lines have the same gradient, but different y-axis intercepts. **Level 7 & Level 8**
- You can compare the properties of straight-line graphs without drawing them by looking at their equations in the form $y = mx + c$. **Level 7 & Level 8**

Why learn this?

When you want to find out if there is a relationship between two variables it can be a good idea to plot a graph. The scientist Robert Hooke found that the relationship between the weight attached to a spring and how much its length increased formed a straight-line graph.

Practice, practice, practice!

1 The point (3, 5) lies on the straight-line graph of $y = 2x - 1$.

a Which of these points also lie on this line?

A (1, 1) B (0, 1) C (−1, −3) D (−1, −1) E (0, −1)

b Copy and complete each coordinate pair so that the point lies on the straight line $y = 3x - 2$.

i (1, ☐) **ii** (0, ☐) **iii** (☐, 4) **iv** (☐, −8) **v** (−1, ☐)

2 **a** Copy and complete this table of values for the straight-line graph $y = 2x + 1$.

x	−3	−2	−1	0	1	2
$y = 2x + 1$				1		

b Plot the points on a coordinate grid with x values from −3 to +2 and y values from −5 to +5. Join the points to give a straight line.

Tip
Always use a sharp soft pencil to plot points and draw lines.

Level 6

6c I can find coordinate points that follow the algebraic rule for a straight-line graph

6b I can plot and draw straight-line graphs using coordinates with both positive and negative numbers

gradient intercept

3 Plot these straight-line graphs on a coordinate grid with x values from -4 to $+4$ and y values from -10 to $+10$.

a $y = x + 1$ **b** $y = 2x + 2$ **c** $y = 3x - 1$

d $y = -x + 1$ **e** $y = -2x + 3$ **f** $y = -3x - 2$

4 Look at the graphs you drew in Q3.

a Write the equation of the graph with the steepest gradient.

b Write the equation of the graph that crosses the y-axis at

i the highest point **ii** the lowest point.

Level 6

6b I can plot and draw straight-line graphs using coordinates with both positive and negative numbers

6a I can plot and draw straight-line graphs and compare their features

5 Look at these equations of straight lines.

A $y = x + 1$ B $y = 2x + 3$ C $y = 5x - 2$ D $y = -x + 1$

E $y = 3x + 5$ F $y = 3x - 2$ G $y = 2(x - \frac{1}{2})$

a Which straight line is the steepest?

b Which straight lines pass through (0, 1)?

c Which straight line passes through (0, 3) and has gradient 2?

6 Rearrange these equations of straight lines into the form $y = mx + c$.

$5y - 15x = 20$	Add $15x$ to both sides	$5y = 20 + 15x$
	Divide both sides by 5	$y = 4 + 3x$
	Rearrange into the form $y = mx + c$	$y = 3x + 4$

a $x + y = 1$ **b** $y + \frac{1}{2}x = 3$ **c** $y - 3x = 2$

d $0 = 9 - y - 2x$ **e** $9x - 3y = 6$ **f** $6 - 4x - y = 3$

Level 7

7b I can compare the features of straight-line graphs by looking at their equations

7a I can rearrange a linear function into the form $y = mx + c$

7 Here are the equations of some straight lines.

A $y = x + 1$ B $y = 2x + 3$ C $y + 2x = -2$

D $y = -x + 1$ E $y = 3x + 5$ F $y = 3x - 2$

G $2y = x + 1$ H $x + y = 3$ I $-2x + 2 - y = 0$

a Which straight lines are parallel?

b For each straight line, give the coordinates of where it crosses the y-axis.

Level 8

8c I can recognise when lines are parallel and where a line crosses the y-axis from the equation of the line

Now try this!

A Line chance!

Re-label the faces of a dice as -3, -2, -1, 1, 2 and 3. Roll it twice. The first number you throw is the gradient of the line and the second number is the value of where the line crosses the y-axis. Repeat this ten times. Use ICT to draw all ten lines on one grid. How many of them are parallel?

B Find the line!

A game for two players. Use a dice like the one in Activity A. Each draw a coordinate grid with values of x and y from -3 to $+3$. Each roll the dice twice. The first throw gives the x-coordinate of a point and the second gives the y-coordinate. Both plot the two points on your grids and join them with a straight line. The first to give the correct equation of the line scores 1 point. Repeat five times.

linear function parallel

8.6 Investigating the properties of straight-line graphs

- Understand the meaning of m and c in $y = mx + c$
- Know that for the function $y = mx + c$, increasing the value of x by 1 increases the value of y by m
- Work out the gradient of a straight-line graph
- Identify parallel and perpendicular lines on a graph
- Know that parallel lines have the same gradient
- Know the gradient of a line perpendicular to $y = mx + c$

Why learn this?

It saves time if you can recognise or sketch the graph of a linear function without having to work out coordinates.

What's the BIG idea?

- The **gradient** m of a straight line is the change in the value of y divided by the change in the value of x. **Level 7**
- Each time the value of x increases by 1 on a straight-line graph of the form $y = mx + c$, the value of y changes by m. **Level 7**
- **Parallel** lines have the same gradient. **Level 8**
- For a straight line written in the form $y = mx + c$, any line **perpendicular** to this line will have a gradient of $-\frac{1}{m}$. **Level 8**

Tip

y, m, c, 1, x

Practice, practice, practice!

Level 6

6b I can recognise parallel and perpendicular lines on a graph

1 **a** Which of these lines are parallel to one other?

b Which lines are perpendicular to one other?

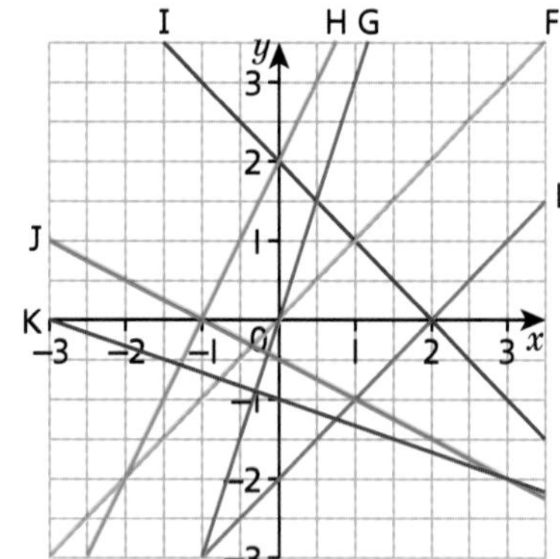

Tip

Make sure you understand the scale on an axis before you read a value from it.

Level 7

7c I know how to calculate the gradient of a straight-line graph and work out its equation

7c I can use ICT to draw a straight-line graph

2 For each line in Q1

a find its gradient

b state the coordinates of the point where it crosses the y-axis

c write down its equation.

3 Use ICT to plot each straight-line graph from Q2c. Compare the lines with the graph in Q1.

gradient intercept

4 Here is table of values for some linear functions. The equation of each graph can be written in the form $y = mx + c$.

x	−3	−2	−1	0	1	2	3
Line A y	−2	−1	0	1	2	3	4
Line B y	−4	−3	−2	−1	0	1	2
Line C y	−6	−4	−2	0	2	4	6
Line D y	−8	−5	−2	1	4	7	10
Line E y	3	2	1	0	−1	−2	−3

a For each function, write down the value of m.

Line A: When x increases by 1, y increases by 1, so $m = 1$.

b Find the equation of each line.

Line A: crosses the y-axis at $y = 1$, so the equation is $y = x + 1$.

Level 7

7c I can work out the gradient of a straight line by finding the increase in y when x increases by 1

5 For each of these linear functions, state the coordinates of the point where its graph crosses the y-axis.

A $y = 2x + 1$ B $y = 2x + 3$ C $y + 2x = -2$
D $y = -x + 1$ E $y = 3x + 5$ F $y = \frac{1}{3}x - 2$
G $2y = 4x + 1$ H $2x + 2y = 3$ I $y = -\frac{1}{2}x + 2$

7a I can find where the graph of a straight line crosses the y-axis from its equation

6 Draw the graphs of these linear functions on a coordinate grid with the x-axis from 0 to 2 and the y-axis from −6 to 10.
For each one
- plot the y-intercept
- plot a second point by moving 1 unit across and m units up or down.

A $y = 2x - 1$ B $y = 2x + 2$ C $y = -2x - 1$
D $y = -x + 2$ E $y = 3x - 3$ F $x + y = 5$
G $y = 4x + 1$ H $2x + 2y = 5$ I $y = -\frac{1}{2}x + 2$

7a I can draw a straight-line graph from the intercept and gradient

7 Without drawing the graphs of the functions in Q5, write down the equations of the lines

a that are parallel to one other

b that are perpendicular to one other.

c Use ICT to plot the graph of the functions in Q5 and check your answers to Q5 and Q6.

Level 8

8c I can recognise when lines are parallel or perpendicular from their equations

Now try this!

A Parallel or perpendicular?

Work with a partner to write down the equations of ten straight lines, including some that are parallel to one another and some that are perpendicular. Swap equations with another pair and, without drawing the lines, identify the parallel and perpendicular lines from each other's set. Check your answers with the other pair.

B Head lines

A game for two players. Roll a dice four times. The first two rolls give you one pair of coordinates. The third and fourth rolls, each multiplied by −1, give you a second pair. So if you roll 1, then 2, then 3, then 4, your points are (1, 2) and (−3, −4). Without plotting them, each work out the equation of a line drawn between these two points, and compare your results. Score 2 points if you get the value of m right, and 2 points if you get the value of c right. Repeat five times.

parallel perpendicular

8.7 Graphs of quadratic and cubic functions

- Construct tables of values, including negative numbers, for quadratic and simple cubic functions and draw their graphs
- Recognise graphs of quadratic functions and know their properties
- Recognise the graph of $y = x^3$
- Use ICT to investigate families of graphs based on $y = ax^2 + bx + c$

Why learn this?

Many objects travel through the air following a curved flight-path. Quadratic functions can be used to work out precise landing points.

What's the BIG idea?

- All **quadratic functions** can be written in the form $y = ax^2 + bx + c$ where a is any number except 0 and b and c are any numbers. **Level 7 & Level 8**
- The functions $y = x^3$ and $y = 3x^3$ are both **cubic functions**. **Level 7 & Level 8**
- You can plot the graphs of quadratic and cubic functions by creating tables of values and plotting them. **Level 7 & Level 8**
- The graph of a quadratic function is a curve that is symmetrical about its turning point. The y-coordinate at the turning point is either a **maximum** or a **minimum**. **Level 7 & Level 8**

Practice, practice, practice!

1 Here is a table of values for the quadratic function $y = 3x^2$.

x	−3	−2	−1	0	1	2	3
$y = 3x^2$	27	12					27

a Copy and complete the table of values.

b Make a coordinate grid with x values from −3 to 3 and y values from 0 to 32.

c Plot and draw the graph of $y = 3x^2$ using your table of values.

Tip
When you work out $3x^2$ make sure you follow the correct order of operations.

2 Here is a table of values for the quadratic function $y = 3x^2 + 4$.

x	−3	−2	−1	0	1	2	3
$3x^2$	27	12					27
+4	+4	+4	+4	+4			+4
$y = 3x^2 + 4$	31	16					31

a Copy and complete the table of values.

b On your coordinate grid from Q1, plot and draw the graph of $y = 3x^2 + 4$.

Level 7

7c I can construct a table of values including negative values for a quadratic function like $y = ax^2$ and plot its graph

7b I can construct a table of values including negative values for a quadratic function like $y = ax^2 + b$ and plot its graph

cubic function maximum point

3 By making a table of values for each function, draw graphs of $y = x^2$ and $y = -x^2$ on the same coordinate grid. What do you notice?

4 Here is a table of values for the cubic function $y = 2x^3$.

x	−3	−2	−1	0	1	2	3
$y = 2x^3$	−54						54

a Copy and complete the table of values.

b On a coordinate grid with x values from −3 to 3 and y values from −60 to 60 plot and draw the graph of $y = 2x^3$.

Learn this

$1^2 = 1$ $(-1)^2 = 1$
$1^3 = 1$ $(-1)^3 = -1$
$2^2 = 4$ $(-2)^2 = 4$
$2^3 = 8$ $(-2)^3 = -8$
$3^2 = 9$ $(-3)^2 = 9$
$3^3 = 27$ $(-3)^3 = -27$

Level 7

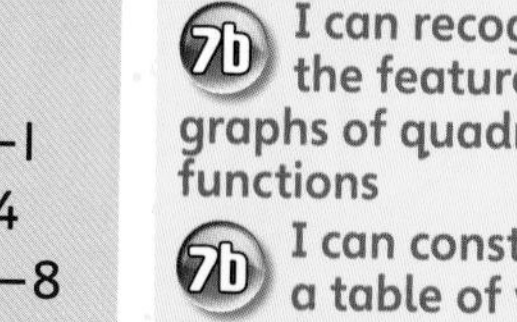

7b I can recognise the features of graphs of quadratic functions

7b I can construct a table of values including negative values for a cubic function like $y = ax^3$ and plot its graph

5 Seven quadratic and cubic graphs are shown on the coordinate grid. Match each of these functions to the correct graph.

a $y = x^2$
b $y = -x^2$
c $y = 5x^2$
d $y = -5x^2$
e $y = 5x^2 + 10$
f $y = 5x^2 - 20$
g $y = x^3$

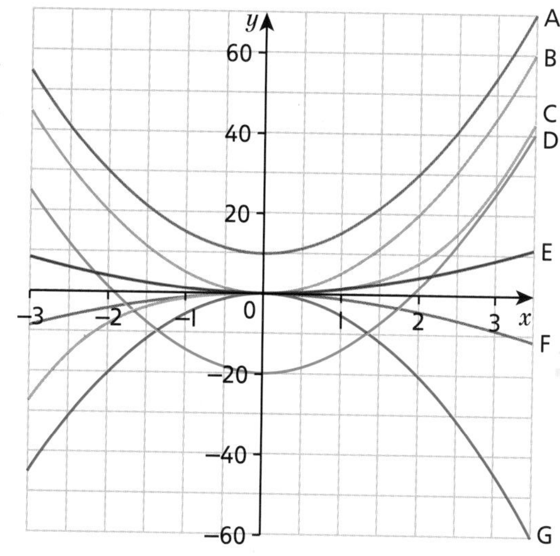

Level 8

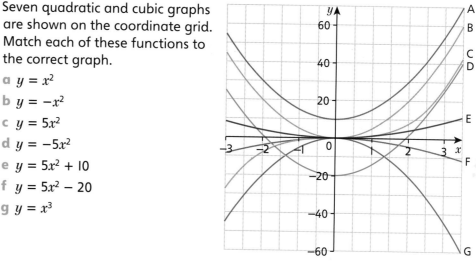

8c I can recognise graphs of functions of the form $y = ax^2 + b$ and $y = ax^3$

6 a Use ICT to plot the graphs of

i $y = x^2$ ii $y = x^2 + 3$ iii $y = x^2 + 6$ iv $y = x^2 - 4$

b What do you notice?

c Where do you think the graph $y = x^2 - 2$ would cross the y-axis? Draw it to check.

8c I can explain the effect on a quadratic graph of changing the parameter

7 a For each curve on the grid in Q5, state whether it has a maximum or a minimum.
Give the coordinates of any maximum or minimum.

b Write down the equations of any lines of symmetry of the curves.

8b I can identify maxima, minima and lines of symmetry on quadratic and cubic graphs

Now try this!

A Quadratic changes

Work with a partner. Use ICT or a graphical calculator to investigate what happens to the graph of $y = x^2 + bx + 2$ when you change the value of b.

B Quadratic effects

Work in groups of three. Start with the quadratic equation $y = ax^2 + bx + c$. As a group, choose values for a, b and c. Use ICT or a graphical calculator to plot the graph of your function. Now work separately. One of you investigates what happens to the graph if you change the value of a, another investigates changing b and the third investigates changing c. When you have finished, discuss your results.

minimum point   quadratic function

maths!

Parabolas

Curves where the highest power is x^2, such as $y = 2x^2 + 3$ or $y = 5x^2 - 3x - 8$, are called parabolas. Parabolas turn up in the physical world in many different situations.

If an object, like a tennis ball, moves unassisted through the air, its path will be an almost perfect parabola. (Air resistance means that the shape is not quite exact.)

Parabolas are used in the design of satellite dishes because the shape collects and focuses signals from a large area into one spot, called the 'focus'. The same principle is used in parabolic microphones. These are used to pick up signals from the side of a football pitch, for example, in sports broadcasting.

What would happen to the shape of this parabola if

a the power of the water jet was changed?

b the angle of the water jet was changed?

How does a parabolic reflector help turn one small light into a wide beam?

Hint: Look at the diagram.

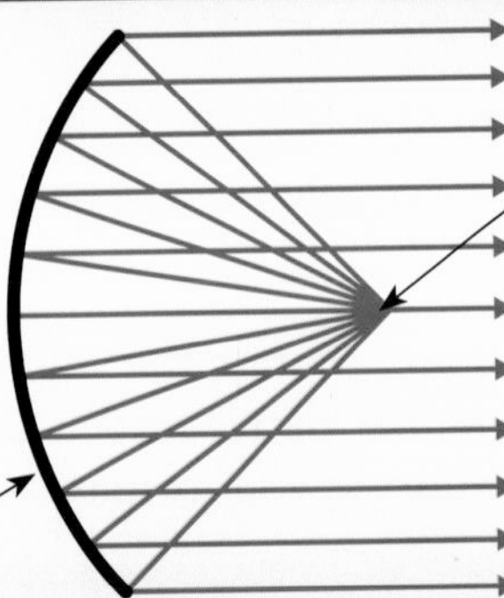

Parabolas are also used in the design of headlamps. The first headlamps were round because this was the easiest shape to make parabolic reflectors.

The Golden Gate Bridge

In a suspension bridge the weight of the bridge is supported by vertical rods that are hung from a suspension cable. This suspension cable takes up the shape of a parabola. Suspension bridges can achieve longer spans than other types of bridges – the Akashi Kaikyō Bridge in Japan is currently the world's longest suspension bridge, with a main span of 1991 metres.

Follow the instructions below to draw a scaled graph of the main span of the Golden Gate Bridge in San Francisco. The graph can be used to calculate the equation of the bridge's parabola. (The diagram below is an inaccurate sketch of what your drawing should look like.)

- Draw a set of axes from –700 to 700 on the x-axis and –100 to 200 on the y-axis. Use a scale of 1cm represents 100 m.

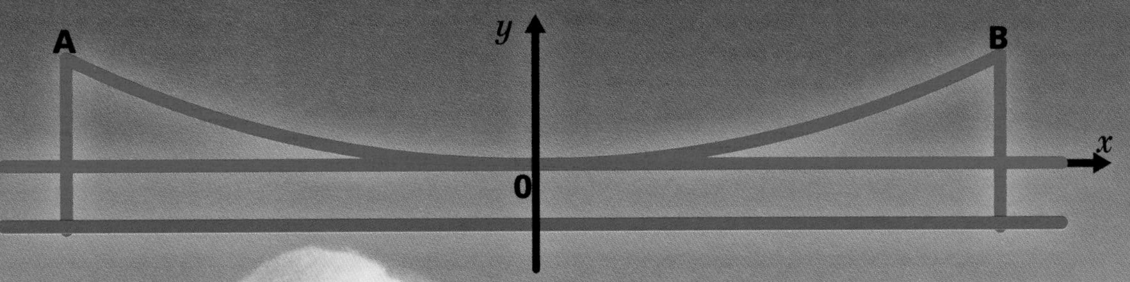

- The main span of the bridge is 1280 m, and the cable sags 143 m. Use this information to deduce and plot the coordinates of points A and B.
- Join points A, B and the origin with a smooth parabola.
- Each tower is 227 m high. Add these as vertical lines on your diagram. Draw the roadway over the x-axis.
- The bridge is approximately 80 m above the water. Write down the equation of the water line and add the line to your diagram. (The legs of the tower should go beneath the level of the water.)
- The equation of this parabola is of the form $y = ax^2$, where a is a constant. Substitute in the values of x and y from point B to find the value of a, and hence the equation of the Golden Gate Bridge parabola.

8.8 Functions and graphs in real life

- Create a function to model a real-life situation and plot its graph
- Interpret graphs from real-life situations
- Sketch a graph showing an approximate relationship between two variables

Why learn this?

A graph can help you to understand a situation.

What's the BIG idea?

- You can write a **formula** to describe a real-life situation.
  For example, the price of a holiday flat is £75 per night plus a £35 booking fee. The formula for the price P of staying n nights is $P = 35 + 75n$. **Level 6**
- You can use graphs to find out more about real-life situations.
  For example, the **gradient** of a **distance–time** graph shows the speed. **Level 6 & Level 7**
- A sketch of a graph shows the approximate relationship between two variables. **Level 6 & Level 7**

Watch out!

Take care with your units of measurement when you write a formula. For example don't work in a mixture of minutes and hours, or pounds and pence.

Practice, practice, practice!

Level 6

6b I can construct formulae to describe mathematical models of real-life situations

1 Write a formula for each of these.

a the cost C, in pounds, of k kilograms of apples at 50p per kilogram

b the cost C of using e electricity units, at 10p per unit, with a standing charge of £15

Level 7

7b I can interpret and discuss linear and non-linear sketch graphs

2 The sketch graph shows the distance run by Arran and Millie during a 30-minute period. Describe their runs.

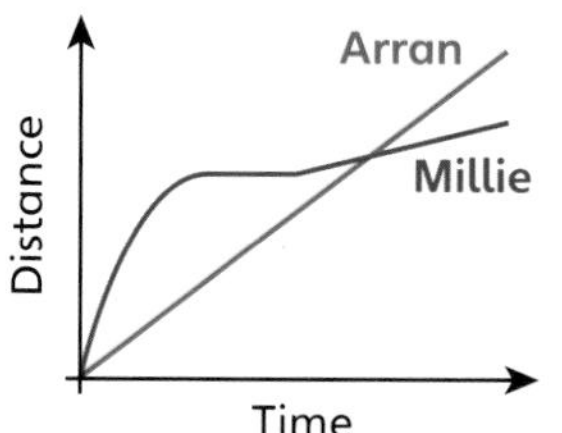

3 These graphs show how the height of sand in a container changes with time as sand is steadily poured into it.

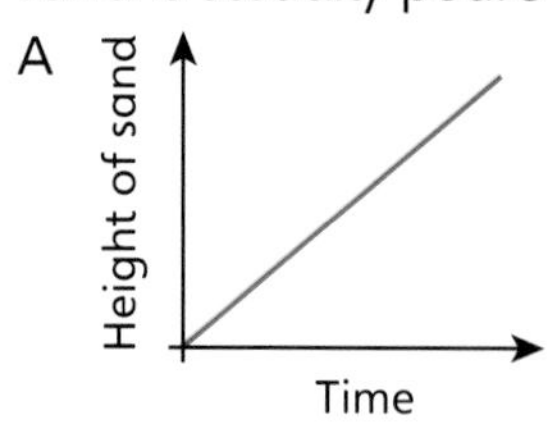

B

Height of sand

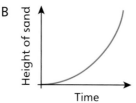

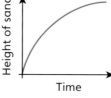

Time

C

Height of sand

Time

Match each of these containers to the correct graph.

a

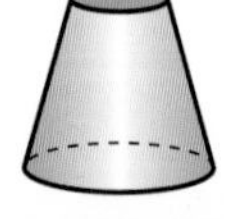

b

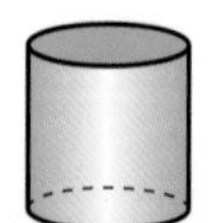

c

Did you know?

A straight line graph shows that the rate of change is steady.
A curved graph shows that the rate of change varies.

distance–time graph formula

4 For each situation in Q1, sketch a graph to show the approximate relationship between the cost C and the other variable.

5 For each situation in Q1, plot and draw an accurate graph for these values.

a for k from 0 to 10 kg

b for e from 0 to 2000 units

6 Here is part of a distance–time graph of Siân's journey from home to the airport and back to pick up a friend.

Distance (km) from Siân's house
24
20
16
12
8
4
0
0 10 20 30 40 50 60 70
Time (minutes)

a Siân got stuck in a traffic jam 8 km from home. For how long was she stuck?

b How far is the airport from Siân's house?

c Work out Siân's speed for the first 10 minutes of her journey. Give your answer in kilometres per hour (km/h).

d Siân spent 15 minutes at the airport. She then travelled home at 60 km/h. Copy and complete the graph.

7 Nathan wants to find the best mobile phone deal for various family members.

Tariff A (pay as you go): 10p per minute
Tariff B (contract): £10 per month plus 5p per minute
Tariff C (contract): £30 per month plus 1p per minute

a Work out a formula for each tariff.

b Draw graphs for each tariff on the same coordinate grid.

c Explain which tariff will be most suitable for

i his grandmother, who seldom uses her phone

ii his sister, who uses her phone for about 600 minutes a month.

d Explain the significance of the points where the line for one tariff crosses the line for another tariff.

Level 7

7b I can sketch a line graph to show the relationship between two variables

7b I can plot and draw graphs from real-life situations

7a I can interpret, plot and draw graphs from real-life situations

7a I can interpret graphs, including the intersection points of pairs of lines

Now try this!

A Fancy-dress hire

A game for two players. When a fancy-dress company hires out a costume, it charges a fixed amount for cleaning plus a daily rate. Each player chooses a cleaning charge and a daily charge so that the price of three days' hire is £30. Plot a graph to show the hire price for between 1 and 6 days. Player 1 rolls a dice, and Player 2 works out how much they would be paid for hiring out a costume for the number of days thrown. Swap roles and repeat. The winner is the player who has taken the most money after four rounds.

B Train trip

Work with a partner.
Look at this train timetable.
Use a map to find out the approximate distance for each stage of the journey. Decide how long you think the train spends at a station. Plot a distance–time graph for the journey and work out the average speed of the train for different parts of the journey. What conclusions can you make?

London	11:48
Grantham	12:54
Retford	13:16
Doncaster	13:29
Selby	13:45
Howden	13:55
Brough	14:08
Hull	14:25

gradient

8.9 Investigating patterns

- Represent problems in algebraic, geometric or graphical form
- Move from one representation to another to gain a different view of the problem

What's the BIG idea?

- Plotting a graph can help you to find a solution to a problem. **Level 7**
- One approach to investigating patterns or sequences is to start working with numbers and then to use algebra to **generalise**.

Stage 1	Stage 2	Stage 3 (if required)
Work with small numbers and look for a **pattern**	Work with algebra and look for a pattern	Generalise your algebra and look for a pattern
↓	↓	↓
When you spot a pattern write down the **rule**	When you spot a pattern write down the **formula**	Generalise your formula from Stage 2
↓	↓	↓
Test your rule	Test your formula	Test your generalisation

Level 6, Level 7 & Level 8

Why learn this?

The ability to think logically can be applied when solving everyday problems.

Practice, practice, practice!

Level 6

6a I can use the patterns of differences to find and justify the nth term

1 Look at these number sequences. They are based on patterns made from sticks.

For triangles: 3, 5, 7, 9, … For squares: 4, 7, 10, 13, …
For pentagons: 5, 9, 13, 17, … For hexagons: 6, 11, 16, 21, …

a Find the next two terms in each sequence. **stage 1**

b Find the nth term of each sequence. **stage 2**

c Find the nth term of the sequence for a shape with s sides. **stage 3**

Level 7

7c I can use algebra to investigate a pattern

2 An L shape is placed on a 10 × 10 number grid.
The L-number is the number in the bottom left-hand corner.
The L-total is the sum of all the numbers in the L.
In this case, the L-number is 34; and the L-total is 267.

63	64	65	66
53	54	55	56
43	44	45	46
33	34	35	36
23	24	25	26

a Find the L-total for the L-numbers 35, 36 and 37. **stage 1**

b Find an expression for the L-total when the L-number is n.

c Repeat parts **a** and **b** for an L shape on a 9 × 9 grid, an 8 × 8 grid, a 7 × 7 grid, and so on. **stage 2**

d Find an expression for the L-total for a $G \times G$ grid. **stage 3**

formula generalise

3 A farmer has 1000 m of fencing to completely enclose a section in the middle of a large field. The shape must be rectangular.

a Investigate what size rectangle gives the maximum area.

b Use a graph to prove your answer.

Level 7

7b I can present a reasoned argument using graphs

4 A 3 × 2 rectangle is placed on a 10 × 10 grid. To find the product difference, you multiply the numbers in the opposite corners of the rectangle and subtract them. In this case, you work out (44 × 36) − (34 × 46) = 20.

63	64	65	66	67
53	54	55	56	57
43	44	45	46	47
33	34	35	36	37
23	24	25	26	27

a Find the product difference for other 3 × 2 rectangles on a 10 × 10 grid.

b Explain algebraically why, for any 3 × 2 rectangle with the number n in its top left-hand corner, the product difference is 20.

c Find the product difference for an $l \times w$ rectangle with the number n in its top left-hand corner on a 10 × 10 grid.

d Find the product difference for an $l \times w$ rectangle with the number n in its top left-hand corner on a $G \times G$ grid.

5 A number stair is placed on a 10 × 10 grid. The stair number is the number in the bottom right-hand corner. The stair total is the sum of all the numbers. In this case, the stair number is 36 and the stair total is 252.

63	64	65	66	67
53	54	55	56	57
43	44	45	46	47
33	34	35	36	37
23	24	25	26	27

a Investigate stair totals for a 10 × 10 grid.

b Find an expression for the stair total for a $G \times G$ grid.

Tip
Remember to check your generalisations.

6 For the shapes in Q2 and Q5, find an expression for the shape total when the shape undergoes a translation of x horizontally and y vertically.

Level 8

8b I can present a reasoned argumen using algebra

8a I can use algebra to investigate an extension to a problem

Now try this!

A How high?

Work in groups of three. You need a selection of balls from different sports. Drop the balls, one at a time, from a height of 1 m and measure how high they bounce. Don't forget to measure from the same part of the ball each time. Do this three times for each ball and find the average result. Display your results graphically. What do you notice?

Tip You could use a video camera to help you make accurate measurements.

B How dense?

Find the density of each ball from Activity A and investigate whether there is a relationship between the density of a ball and the height to which it bounced. Describe your results.

Tip The formula for density (D) is $D = \frac{m}{V}$ where m is mass and V is volume.

Tip The formula for volume (V) of a sphere is $V = \frac{4}{3}\pi r^3$ where r is the radius of the sphere.

Primed for action

Humans have 23 pairs of chromosomes.

The biological significance of prime numbers was discussed in the opener to this unit. Marcus du Sautoy's comment that 'primes seem to underpin the universe and the way nature works' is further highlighted by the fact that there are connections between prime numbers and quantum physics. The more you start to look for primes, the more you will find!

1

- Pick a prime number bigger than 5
- Square it
- Add 17
- Divide by 12

a Follow the procedure set out in the box above. Does your answer have a remainder? What is it?

b Repeat the procedure with a different prime number. What is the remainder this time? **Level 6**

2 Use prime factor decomposition to find the highest common factor (HCF) of

a 34 and 102 **b** 36 and 234 **Level 6**

3 **a** Find the smallest number that has four *different* prime factors.

b Find the smallest number that has five prime factors. **Level 6**

4 **a** What is the first prime number greater than 3? Square it and subtract 1.

b What is the next prime number? Square it and subtract 1.

c Do the same for the next prime number.

d Look at your answers to parts **a**, **b** and **c**. What is the connection between these numbers? **Level 7**

5 Use prime factors and the appropriate index laws to show that

a $24 \times 64 = 2^9 \times 3$

b $\frac{72 \times 36}{12} = 2^3 \times 3^3$ **Level 7**

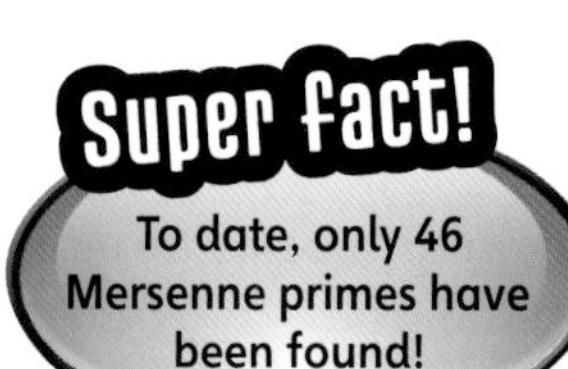

6 A Mersenne prime is a number of the form $M_n = 2^n - 1$ that is prime.

a Are these numbers Mersenne primes? 3, 5, 7, 13, 31, 63

b Based on this investigation, can you draw a conclusion about Mersenne primes? **Level 8**

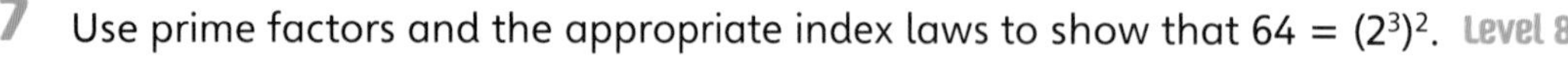

7 Use prime factors and the appropriate index laws to show that $64 = (2^3)^2$. **Level 8**

8 Use prime factors and the appropriate index laws to show that

a $\sqrt{576} = 2^3 \times 3$ **b** $\sqrt[3]{512} = 2^3$ **Level 8**

9 Use Q8 to evaluate

a $\sqrt{576}$ **b** $\sqrt[3]{512}$ **Level 8**

The BIG ideas

- A **factor** divides into a number exactly. The **highest common factor** (**HCF**) of a group of numbers is the highest number that is a factor of all the numbers.
  For example, 9 is the HCF of 27, 45 and 54. **Level 6**
- A **prime number** has exactly two factors, 1 and itself. Any other number can be written as a **product** of its prime factors. This is called **prime factor decomposition**.
  For example, the prime factor decomposition of 60 is $2 \times 2 \times 3 \times 5$, or $2^2 \times 3 \times 5$. **Level 6**
- You can work out the HCF and LCM of a group of numbers from the prime factor decomposition of each number. **Level 6 & Level 7**
- When you multiply numbers or variables written as **powers** of the same number, add the powers (or **indices**). For example, $2^4 \times 2^2 = 2^{4+2} = 2^6$ **Level 6 & Level 7**
- When you divide numbers or variables written as powers of the same number, subtract the powers (or indices). For example, $2^4 \div 2^2 = 2^{4-2} = 2^2$ **Level 6 & Level 7**
- When you raise a number or variable written as a power to another power, multiply the powers (or indices). For example, $(2^4)^2 = 2^{4 \times 2} = 2^8$ **Level 8**
- A fractional power can be written as a root. For example, $4^{\frac{1}{2}} = \sqrt{4}$ **Level 8**
- Raising a number or variable written as a negative or fractional power to another power follows the same rule as for positive integer indices. For example, $(2^{\frac{1}{4}})^2 = 2^{\frac{1}{4} \times 2} = 2^{\frac{2}{4}} = 2^{\frac{1}{2}}$ **Level 8+**

Find your level

Level 6

Q1 The graph shows the straight line with equation $y = 2x - 2$.

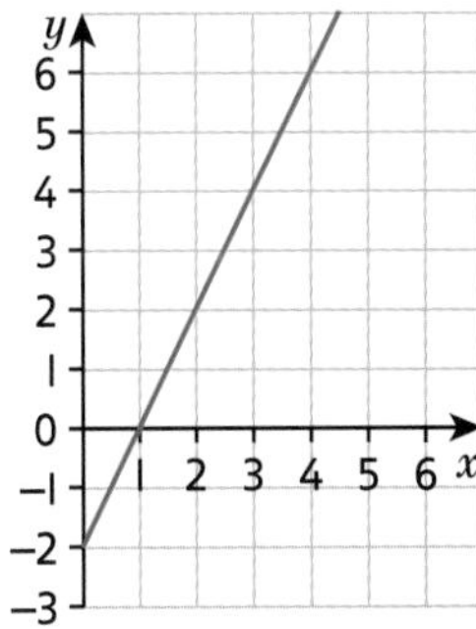

a A point on the line has an x-coordinate of 10. What is the y-coordinate?

b A point on the line has a y-coordinate of 10. What is the x-coordinate?

c Is the point (12, 26) on the line? Give a reason for your answer.

Q2 The prime factorisation of a certain number is $3^2 \times 5 \times 11$.

a What is the number?

b Write down the prime factorisation of 165.

Level 7

Q3 a $48 = 3 \times 2^a$
What is the value of a?

b $36 = 3^b \times 2^2$
What is the value of b?

c Show that $48 \times 36 = 2^6 \times 3^3$

Q4 Find the values of a and b.

$4^8 \times 4^3 = 4^a$

$\frac{4^b}{4^3} = 4^5$

Q5 Match each graph to its equation.

$y = x^2 - 3$ $y = x^3$ $y = \frac{1}{2x}$ $y = 5 - x$

a

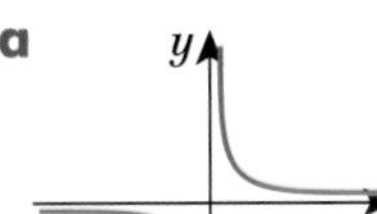

b

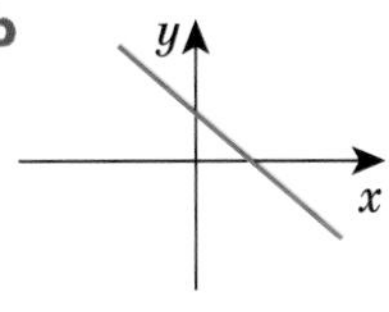

c

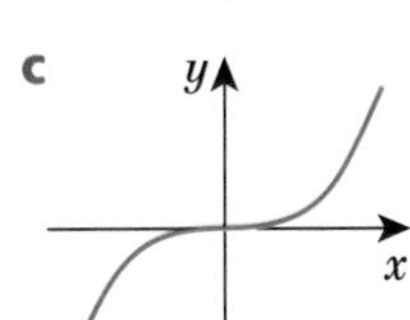

d

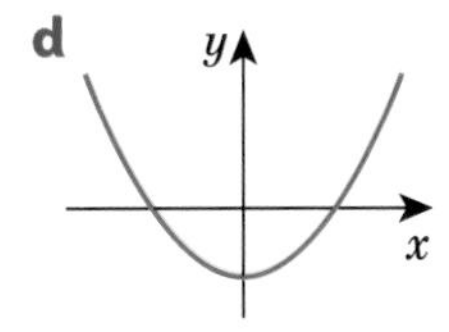

9 Strictly come chancing

This unit is about probability. Chances and probabilities aren't always what you might first suppose.

Every number starts with 1, 2, 3, 4, 5, 6, 7, 8 or 9.

You might think that in a set of numbers with two or more digits, roughly one ninth of them would start with 1, one ninth would start with 2, and so on. Experiment and experience show that this isn't so.

In the 1930s, US physicist Frank Benford studied lots of data sets, including figures from the stock exchange, the numbers on the front page of *The New York Times*, the populations of towns, the half-lives of radioactive elements and much more. He discovered that in real-life data, the probability that the first digit is 1 is about 30%, and the probability that it is 2 is around 17%. This table shows the probabilities for 1 to 9.

First digit	1	2	3	4	5	6	7	8	9
Probability (%)	30.1	17.6	12.5	9.7	7.9	6.7	5.8	5.1	4.6

40

Activities

A Circle all the numbers on the front page of a newspaper. What percentage begin with 1, 2, 3, …? Do these percentages agree with Benford's law?

B You need a 100 square.

- Look at the numbers 1 to 9. If a number is picked at random from this set, what is the probability that its first digit is 1? Write down the probabilities for the first digit being 1, 2, 3, 4, 5, 6, 7, 8 or 9.
- Now look at the numbers 1 to 20. Write down the probabilities for the first digit.
- Next do the same for the numbers 1 to 30, then 1 to 40. Continue up to 1 to 100.
- How do your findings relate to Benford's law?

Did you know?
In the US, Benford's law has been used to detect income tax fraud. If the numbers entered on a tax return more or less match Benford's probabilities, then they are probably honest. If they don't, accountants will check the details much more carefully.

Before you start this unit...

1 Which of these events are impossible, unlikely, equally likely, likely, certain?

a There will be a hotel on the Moon by 2020.
b The world population will continue to increase.
c An astronaut will broadcast from the Sun this year.
d It will rain in Britain next year.
e If you choose a time at random, it will be a.m.

2 **a** If you roll a dice 30 times, how many 6s would you expect to get?

b Would you be likely to get no 6s at all?

3 If you use a normal dice

a what is the probability of rolling an even number?
b what is the probability of rolling a 1 or a 2?

9.1 Fair play
9.2 It's exclusive!
9.3 Tree diagrams
9.4 Relative frequency
maths! Pollyopoly
Unit plenary: I've got the 'skill' factor
Plus digital resources

9.1 Fair play

- Talk about the likelihood of different outcomes occurring
- Work out if a game is fair

What's the BIG idea?

- You can compare the **probabilities** of different **outcomes** when they are expressed as fractions.
  For example, an outcome with probability $\frac{3}{7} = \frac{15}{35}$ is more likely than one with probability $\frac{2}{5} = \frac{14}{35}$. **Level 6**
- You can use probabilities to say how **likely** something is to happen. **Level 6**
- You can use probabilities to look at whether or not a game is **fair**. Does everyone have the same **chance** of winning? **Level 6**

Why learn this?

It can be helpful to decide whether a game of chance is fair before you agree to play it.

Practice, practice, practice!

Level 6

6c I can compare probabilities using fractions with the same numerator

1 There are two lucky dip barrels at the school fair. Barrel A has 16 prizes and barrel B has 20 prizes. Six of the prizes in each barrel are toy cars. Which barrel should you pick from to have the best chance of winning a toy car? Explain your answer.

2 Two pupils are handing out pencils randomly to pupils. Tom's box contains 18 pencils. Sam's box contains 14 pencils. There are five blue pencils in each box. Do you have more chance of getting a blue pencil from Tom or from Sam? Explain your answer.

6b I can compare probabilities using equivalent fractions

3 All the pupils in a music class were asked to bring in a CD, and the CDs were put into two piles. There were 16 CDs in the first pile, four of which were rap. There were 24 CDs in the second pile, five of which were rap. Which pile should you pick a CD from to have the better chance of picking rap? Explain your answer.

4 Two pupils did a survey to find out how many people wanted the centre of town to become a pedestrian zone. Sophie asked 64 people and 48 said yes. Helen asked 72 people and 60 said yes. Compare the probabilities that people picked at random from the two groups will want a pedestrian zone, writing them as simply as possible.

6b I can identify conditions for a fair game from a small set of simple options

5 Three friends are playing darts. Which of these conditions are fair, and which are no

a Everyone takes a turn at throwing three darts.
b The tallest person stands further back.
c If you get an even number you throw again.
d The first to get to an agreed total wins.
e Each round, the person with the lowest score takes a step back.

Watch out!

When you are assessing 'fairness', you need to look at whether something is mathematically fair, not morally fair.

chance fair likelihood

6 Three triangular spinners, each labelled 1–3, are spun simultaneously.

a Describe the possible outcomes and their probabilities.
b What is the probability of getting a total of less than 4?
c What is the probability of getting a total of more than 7?

7 Two friends take turns to guess how many coins will land heads when four coins are spun simultaneously.

a Describe the possible outcomes and their probabilities.
b What would *you* guess?

8 Dave has 28 dominoes placed face down on the table for a game. He says that you can choose the rule about how to play the game.

> Rule 1: You win if you turn over a domino with at least one blank.
> Rule 2: You win if both numbers on the domino are even (not zero).
> Rule 3: You win if the sum of the numbers is greater than 8.

a Which rule will give you the most chance of winning?
b Which rule will give you the least chance of winning?
c What is the probability of winning in each case?
d Make up a rule that gives you the same chance of winning as rule 2.

9 Four friends make up a card game using a normal set of playing cards that have been shuffled. All the picture cards are worth 10 and the ace is worth 1. They each select a 'rule card' and then get dealt one playing card.

Rule 1: You must have less than 5.

Rule 2: You must have a picture card.

Rule 3: You must have a red card.

Rule 4: You must have a 6.

They win if their playing card agrees with their rule card.

a Which rule gives the most chance of winning?
b Which rule gives the least chance of winning?
c What is the probability of winning in each case?
d Make up two 'rule cards' that both offer the same chance of winning.

Level 6

6a I can describe the probability of a complex event

6a I can identify conditions for a fair game from a set of options

Tip
You will need a set of dominoes for this question.

Now try this!

A Card shuffle

Work in pairs. Using a sheet of A4 paper, make a set of 12 cards, four marked A, four B and four C.

a Shuffle and turn over the top card. Record the result. Repeat three times and then see if you can predict the next result based on the previous ones.
b Calculate the probabilities of turning over an A card, a B card and a C card. Now shuffle and turn over the top card. Record the result. Repeat 60 times. Do your results agree with your calculated probabilities.

B Make it unfair

Work in pairs. Choose a couple of well-known games, such as snap or ludo. Invent new rules that will make them unfair.

Did you know?
Dice can be weighted to alter the probabilities of them landing on particular numbers.

likely outcome probability unlikely

9.2 It's exclusive!

- Identify mutually exclusive outcomes of an experiment
- Write the probability of event n as P(n)
- Know that the sum of the probabilities of all mutually exclusive outcomes is 1 and use this to calculate probabilities

Why learn this?

You can work out the chances of winning the jackpot, winning a smaller amount or losing altogether.

What's the BIG idea?

- **Mutually exclusive** events are things that cannot happen together.
  For example, the two outcomes for tossing a coin are mutually exclusive – either heads or tails. **Level 6**
- Probability of event A or event B happening P(A or B) = P(A) + P(B). **Level 6**
- The sum of the probabilities of all mutually exclusive outcomes is 1.
  For example, the **probability** of getting a head when you toss a coin is $\frac{1}{2}$ and the probability of a tail is $\frac{1}{2}$, so the probability of a head or a tail is $\frac{1}{2} + \frac{1}{2} = 1$. You will definitely get a head or a tail. **Level 6**
- P(A) is mathematical shorthand for 'the probability of event A occurring'. **Level 6**
- The probability of event A *not* occurring is 1 – P(A). **Level 6**
- Probability of a combined event occurring = 1 – the sum of the probabilities for the other combinations. **Level 6 & Level 7**

Practice, practice, practice!

Level 6

6c I can identify mutually exclusive events

1 Which of these events are mutually exclusive, and which are not?

- **a** landing on a black or a white square in a game of chess
- **b** liking pizza, pasta or curry
- **c** being in class 9 or class 8
- **d** liking playing football, tennis or basketball
- **e** picking a hard centre or a soft centre from a box of sweets

6b I can list all the mutually exclusive outcomes for an event and check that the total of their probabilities is 1

2 A simple spinner can land on a cherry, a plum or a banana.

- **a** What are the possible outcomes for two spins?
- **b** What is the probability of each outcome if it doesn't matter which order the fruits are in?
- **c** How do you know that you haven't missed out a possible outcome?

Learn this

The sum of the probabilities of all the mutually exclusive outcomes is 1.

3 A bag contains four red, four blue and four green counters. Erin picks one counter, records its colour and puts it back. Then she repeats the process.

- **a** What are the possible outcomes for picking two counters in this way?
- **b** What is the probability of each outcome if the order of colours doesn't matter?
- **c** How do you know that you haven't missed out a possible outcome?

event mutually exclusive

4 A pupil uses lettered tiles to write her full name. She puts the letters in a bag. A friend picks out a letter at random. The probability of picking an A, P(A), is $\frac{3}{11}$, P(B) = $\frac{1}{11}$, P(E) = $\frac{2}{11}$ and P(N) = $\frac{3}{11}$.

- **a** What is the probability of picking **i** an A or a B **ii** an E or an N?
- **b** Are there any other letters in the bag? How can you tell?

5 A game involves a simple 'dartboard' with five equal sectors numbered 1 to 5. Players throw two darts each.
If both darts hit the board, what is the probability of getting

- **a** two identical numbers
- **b** two different numbers
- **c** a total of more than 8
- **d** a total of 8 or less?

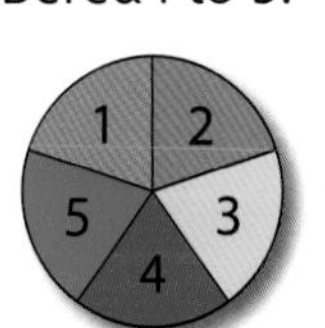

Level 6

6b I can list all the mutually exclusive outcomes for an event and check that the total of their probabilities is 1

6a I can calculate the probability of the final event of a set of mutually exclusive events

6 A bag of sweets contains strawberry, lemon and lime flavours. There is a $\frac{2}{5}$ probability of getting strawberry, and a $\frac{1}{3}$ probability of getting lime.

- **a** What is the probability of getting a lemon sweet?
- **b** If there are 12 strawberry sweets, how many sweets are there in the bag?

7 A drawer contains pairs of socks in four colours – black, white, red and blue. Jim picks a pair at random. The probabilities of picking one of three of the colours are P(black) = $\frac{1}{3}$, P(white) = $\frac{1}{4}$, P(red) = $\frac{1}{3}$.

- **a** What is the probability Jim picks a blue pair?
- **b** What is the probability Jim picks a black or a white pair?

8 There are two pyramid dice, each with four triangular faces. Each dice is marked with a different letter on each face: A, B, C, D. The two dice are rolled.

- **a** How many possible outcomes are there?
- **b** What is the probability of getting at least one vowel?
- **c** What is the probability of getting two consonants?
- **d** What is the probability of getting two vowels?
- **e** Are the outcomes in parts **b** and **c** mutually exclusive?

9 Repeat Q8 for three pyramid dice.

10 A box contains 120 crayons. The probability of picking: a blue crayon is $\frac{1}{3}$, a red is $\frac{1}{10}$, a green is $\frac{3}{20}$, a pink is $\frac{1}{12}$, and a yellow is $\frac{1}{6}$.

- **a** What is the probability of picking a crayon that is not one of these colours?
- **b** How many 'other colour' crayons are there in the box?

Level 7

7c I can use the fact that the probabilities of mutually exclusive outcomes total 1 to solve problems

Now try this!

A Shuffling cups 1

Take three paper cups and turn them upside down. Put a coin under one. Player 1 shuffles the cups while Player 2 looks away. After the cups have been shuffled, Player 2 picks up a cup. Record whether or not the coin is under the cup. Repeat a total of 12 times. Are Player 2's results what you would expect?

B Shuffling cups 2

Do Activity A. Repeat it but this time Player 2 watches the shuffling. Player 1 shuffles as quickly as possible. Is Player 2 more accurate at finding the coin this time? What factors affect Player 2's chances?

outcome probability

9.3 Tree diagrams

- Understand where two or more events are independent
- Draw and use tree diagrams

Why learn this?

Tree diagrams are a good visual representation of possible outcomes. Although family trees are used differently from probability trees, they both show all the results from a single starting point.

What's the BIG idea?

- Two events are **independent** if the outcome of one has no effect on the outcome of the other.
  For example, if you toss two coins, getting heads with the first coin does not affect the probability of getting heads with the second. Level 8
- **Tree diagrams** help you to show all possible **outcomes** of a series of events and to calculate their probabilities.
  For example, this tree diagram represents two coins being tossed.

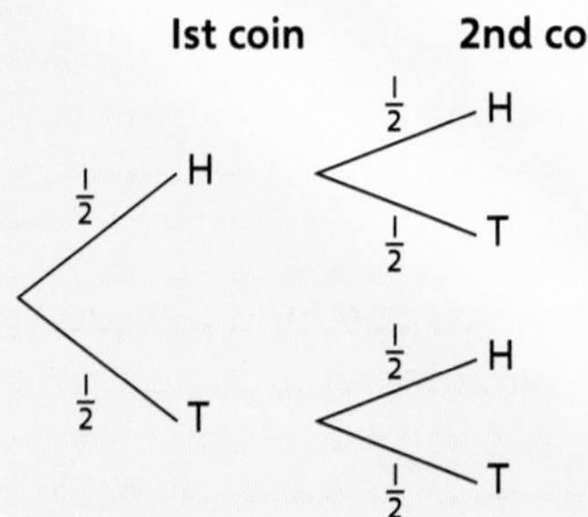

  Each branch represents a possible outcome. The probability of the outcome is written beside its branch Level 8
- For independent events, to find the probability of one outcome followed by another, multiply their probabilities.
  The probability of two heads is $P(H) \times P(H) = \frac{1}{2} \times \frac{1}{2} = \frac{1}{4}$ Level 8

Practice, practice, practice!

1 Imagine spinning this spinner twice.

a Draw a tree diagram to represent the possible outcomes.

b What is the probability of it landing on red both times?

c What is the probability of it landing on red once and on yellow once?

Learn this

The sum of the probabilities on any set of branches should always be 1.

Level 8

8C I can draw and use tree diagrams to represent outcomes of two independent events and calculate probabilities

2 There are eight black counters and eight white counters in a bag. One counter is picked at random and then put back. This is repeated once.

a Draw a tree diagram to represent the possible outcomes.

b What is the probability of getting two black counters?

c What is the probability of getting one black and one white counter?

independent

3 Box A contains five red counters and three blue counters.
Box B contains six red counters and four yellow counters.
One counter is selected at random from each box.

a Copy and complete this tree diagram.

Box A | Box B

$\frac{5}{8}$ red — $\frac{6}{10}$ red; □ yellow

□ blue — □ red; □ yellow

b What is the probability of getting two red counters?

c What is the probability of getting a blue then a red counter?

d What is the probability of getting no red counters at all?

4 At a school fair there are two bags of tickets. The first bag has 20 green tickets and 10 red tickets. The second bag has 15 green, 20 red and 10 blue tickets.

a Copy and complete this tree diagram.

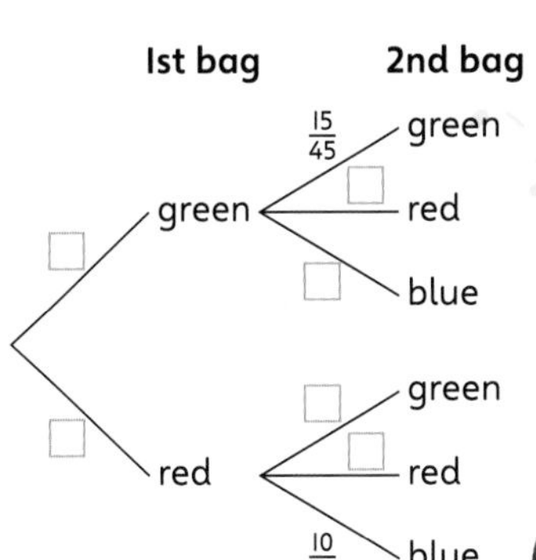

If you pick two tickets of the same colour, you win a prize. If you pick a red ticket and a blue ticket, you get another go.

b What is the probability of winning a prize?

c What is the probability of getting another go?

5 In a survey of 80 pupils, 48 had school dinners and the rest had packed lunches.
Of the people who had school dinners, six were vegetarian, nine didn't eat lamb and the rest ate any meat.
Of the people who had packed lunches, 12 were vegetarian, 10 didn't eat lamb and the rest ate any meat.

a Draw a tree diagram to represent this.

b What is the probability that someone who has packed lunch will eat any meat?

c What is the probability that someone who has school dinner is a vegetarian?

d What is the probability that a pupil picked at random has school dinners and is a vegetarian?

6 Repeat Q5 using real data from your class.

Level 8

8C I can draw and use tree diagrams to represent outcomes of two independent events and calculate probabilities

Tip

Remember that in a tree diagram you multiply horizontally and add vertically.

Now try this!

A Which hand?

Work in pairs or small groups. Carry out a survey of the class to find out who is left-handed and who is right-handed, and who plays a musical instrument and whetherit is a stringed, wind or percussion instrument. If pupils play two instruments, record their favourite instrument. Draw a tree diagram to represent the information. Work out probabilities, for example that a person picked at random is left-handed and plays a stringed instrument.

B Tree diagram posters

In small groups carry out your own survey and then represent the information as a tree diagram on A3 sized paper to show to the class.

outcome tree diagram

9.4 Relative frequency

- Use probabilities from experimental data to predict outcomes
- Understand relative frequency
- Use relative frequency to compare experiments
- Plot and use relative frequency diagrams

What's the BIG idea?

- You can use the results from a small experiment to work out how many successes you would expect in a larger one. If you observe 50 cars and see six reds ones, then you could expect to see 24 red cars if you observe 200. **Level 6**
- You need to use **relative frequency** to **predict** the frequencies of outcomes that are not equal. **Level 6**
- Relative frequency = $\frac{\text{number of times the outcome happens}}{\text{total number of trials}}$

  For example, if you observe 100 people and find that 18 are wearing coats, the relative frequency is $\frac{18}{100}$. **Level 6**
- The more times you repeat an experiment, the more accurate the result. The **probability** tends to a **limit**.

  For example, if you roll a dice six times, you are unlikely to get one of each number, but if you roll it 600 times you'll get close to 100 of each number. **Level 8**

Why learn this?

Surveys allow you to estimate results for a large group of people by finding the relative frequency with a smaller group.

Watch out!

Make sure that the conditions are the same when you compare experimental results. The relative frequency of car colours would be affected if you collected data near a black cab taxi rank.

Practice, practice, practice!

Level 6

6c I can estimate probabilities from experimental data and use them to predict results

1 A supermarket survey of 100 shoppers found that 72 of them bought at least £5 worth of fruit and vegetables.
In an average week, the supermarket has 5000 shoppers.

- **a** How many shoppers each week are likely to buy at least £5 worth of fruit and vegetables?
- **b** What are the supermarket's minimum likely takings from fruit and vegetables each week?

6a I can use probabilities of two-step outcomes to predict how many successes to expect

2 Five cards, numbered from 1 to 5, are laid face down on a table. One card is turned over, the result is recorded, and then it is turned face down again and mixed with the others. The same process is repeated and the two results are added to give a total score. This is repeated to give 100 totals.
How many times in the 100 trials would you expect to get each of these totals?

a 10 **b** 7 **c** 3

6a I know how to calculate the relative frequency of an outcome

3 What is the relative frequency of each of these outcomes?

- **a** In a survey of 200 people, 45 did not own an MP3 player.
- **b** 48 out of 60 pupils had school lunches.
- **c** 65 out of 90 cars surveyed used unleaded petrol.

estimate limit predict

4 Use the relative frequencies in Q4 to estimate these.
- a In a group of 5000 people, how many are likely not to own an MP3 player?
- b In a school of 900 pupils, how many are likely to have school lunches?
- c In a petrol station that serves 600 cars per day, how many cars are likely to need unleaded petrol?

Level 6

6a I know how to use the relative frequency of an outcome to make estimates

5 A canteen counted how many free biscuits 30 staff had with their coffee or tea. They did this for two days and averaged the results. Then they recorded the results in this frequency diagram.

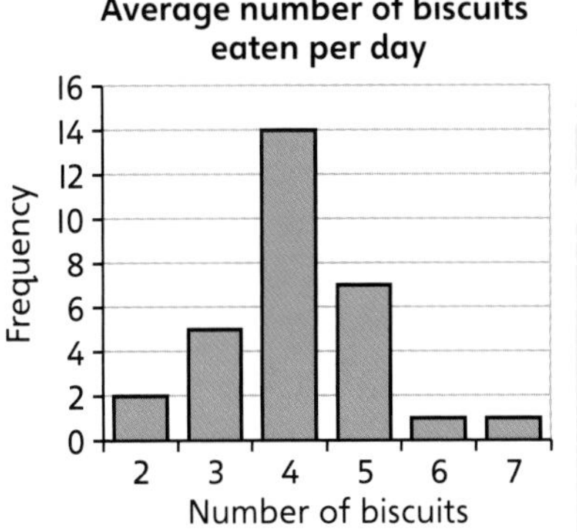

- a What is the most likely number of biscuits that someone will eat?
- b What is the estimated probability that that someone will eat a total of three biscuits?
- c What is the relative frequency of a member of staff eating an average of more than two biscuits per day?
- d How could you make a more accurate estimate?
- e If the experiment included 300 staff, what would you expect the frequencies to be? Justify your answers.

Level 7

7b I understand that relative frequency is an estimate of probability and I can use it to predict outcomes of an experiment

Tip
Remember that relative frequency can only be used as an estimate.

6 Use a square of cardboard to represent a slice of toast. On one side, write 'butter'.

Relative frequency of 'butter' side up (0, 0.2, 0.4, 0.6, 0.8, 1) against Number of throws (0, 10, 20, 30, 40, 50, 60, 70, 80, 90, 100)

- a Throw the 'toast' in the air 10 times and record how many times it lands 'butter' side up.
- b Estimate the probability from these trials.
- c Copy the frequency diagram and plot your result.
- d Repeat the experiment another 90 times. Combine your results, and calculate and record the probability (relative frequency) every 10 throws.
- e What do you notice about the results? Can you explain this feature?

7 Stick a small piece of Plasticine on one side of the 'toast', then repeat Q6. Discuss any differences between your results and other pupils' results.

Level 8

8c I can plot and use relative frequency diagrams, and recognise that experimental probability tends to a limit

Now try this!

A Spinner segments

Work in pairs. Make a circular spinner, divide it into three unequal sectors and colour each sector a different colour. Record for 10 spins and work out the relative frequency of each colour. Record another 10 spins, combine the results with the first 10 spins and compare the relative frequencies with the first results. Predict the results from 100 spins.

B Many spins

Record the results from Activity A on a relative frequency table. Continue the experiment to 100 spins, calculating and recording the relative frequencies after every 10 spins. How close is the relative frequency, after 100 spins, to your prediction in Activity A?

probability **relative frequency**

maths!

Pollyopoly

A board game of chance

The aim of the game is to buy animal enclosures and animals at a wildlife park and charge visitors to see them. Players roll two dice at the same time and add the two scores together to find how many spaces they can move around the board.

When working out probabilities of events in games like Pollyopoly, you need to consider the roll of the two dice as well as the position of the items on the board.

Imagine you are starting on the Entrance space.

- What is the probability of landing on the Screech! space in one roll of two dice?
- What is the probability of landing on a purple enclosure in one roll of two dice?
- What is the probability of landing on a brown enclosure followed by a purple enclosure in two rolls of two dice?
- Which is the most likely brown and purple combination?
- How could you carry out an experiment to test whether this is the most likely brown and purple combination?

Holly is playing Pollyopoly with three of her friends. She starts on the Entrance space. She hopes to land on a purple enclosure followed by a blue enclosure on her first two goes.

- Write out all the possible combinations of a purple enclosure then a blue one.
- Is this an exhaustive list of all the possible outcomes?

I've got the 'skill' factor

Card games are often thought to be games of pure luck – but is this really the case?
Poker requires the skills of psychology and mathematics.
Poker players need to calculate the probability of winning the hand with the cards they hold.

1 **a** In a card game, the ace is worth 1 point, the picture cards are all worth 10 points, and the other cards are worth the number written on them. Player 1 wins if the first card turned over is worth an odd number of points. Player 2 wins if it is worth an even number of points. Is the game fair?

b In another game, Player 1 wins if the first card turned over is black. Player 2 wins if it is red. If your aim was to win, would you take part in this game? Explain your answer. **Level 6**

MAKE MATHS FUNCTIONAL!

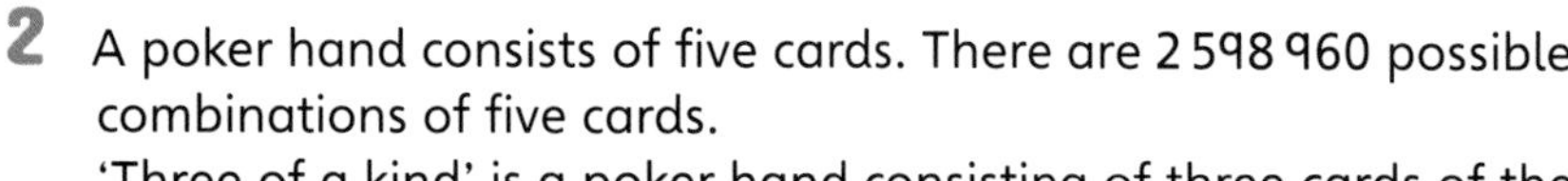

2 A poker hand consists of five cards. There are 2 598 960 possible combinations of five cards.
'Three of a kind' is a poker hand consisting of three cards of the same value (for example, three 8s) and two unmatched cards.
There are 54 912 possible three-of-a-kind hands.

a What is the probability of getting a three-of-a-kind hand?

b What is the probability of not getting a three-of-a-kind hand? **Level 6**

3 In a raffle, there is a $\frac{7}{8}$ chance of not winning anything and a $\frac{1}{12}$ chance of winning a runner-up prize. What is the probability of winning one of the main prizes? **Level 7**

4 Look back at the card game in Q1a.
Emma selects one card at random and then replaces it in the pack.
She shuffles the pack and then selects another card.
What is the probability that both cards are worth 10 points? **Level 8**

5 A bag contains equal numbers of green, red and blue tokens.
Anil selects one token at random, records the colour, then replaces it in the bag.
Here are the results set out in groups of 10.

b b g r r b r g r r   r b g g g g b r b b   r b b b g g r b b r   g b b r b g b b r b

a Calculate the relative frequencies of green, red and blue tokens
  i after 10 selections **ii** after 40 selections.

b Copy the frequency diagram and plot the results for the relative frequency of red.

c What would you expect to happen to the relative frequency of red if the selection of tokens continues many more times? **Level 8**

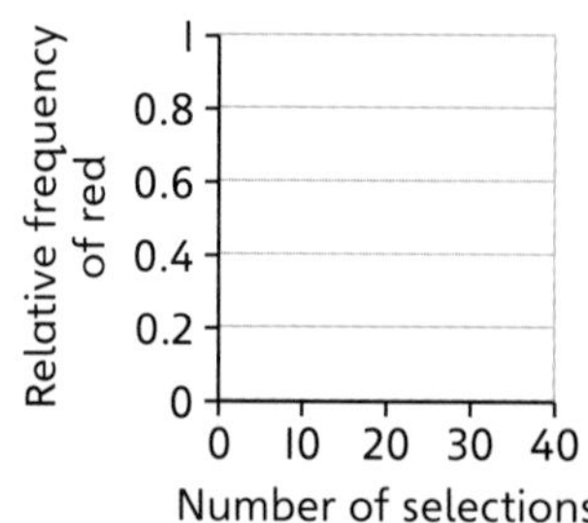

The BIG ideas

- You can use probabilities to look at whether or not, a game is **fair**. Does everyone have the same **chance** of winning? Level 6
- **Mutually exclusive** outcomes are things that cannot happen together. Level 6
- The sum of the probabilities of all mutually exclusive outcomes is 1. Level 6
- P(n) is mathematical shorthand for 'the probability of n occurring'. Level 6
- The probability of outcome n *not* occurring is 1 – P(n). Level 6
- You can use the results from a small experiment to work out how many successes you would expect in a larger one. Level 6
- You need to use **relative frequency** to **predict** the frequencies of outcomes that are not equal. Level 6
- Relative frequency = $\frac{\text{number of times an event happens}}{\text{total number of trials}}$. Level 6
- **Tree diagrams** help you to show all possible **outcomes** of a series of events and to calculate their probabilities. Each branch represents a possible outcome. The probability of the outcome is written beside its branch Level 8

Find your level

Level 6

Q1 Two classes are holding raffles.
Class A has 200 tickets and 15 prizes.
Class B has 150 tickets and 12 prizes.

a What is the probability of winning a prize
  i at class A's raffle
  ii at class B's raffle?

b Which raffle offers the greater chance of winning a prize?

Level 7

Q2 A bag contains three types of sweets: toffees, mints and chocolates.
I am going to take a sweet at random.

a Copy and complete the table below.

Type of sweet	Number of sweets	Probability
toffees	12	
mints		$\frac{1}{7}$
chocolates	12	

b Before I take a sweet out of the bag, I put in 2 extra mints.
What does this do to the probability that I will select a toffee?

Level 8

Q3 A bag contains nine coloured counters.
Each counter is blue, green or red.
Dirk takes a counter from a bag and records its colour, before replacing it in the bag. He does this 100 times.

Here are his results.
blue 45   red 21   green 34
How many counters of each colour do you think there are?

Q4 I have three fair dice, each numbered 1 to 6.
I am going to roll all three dice.
What is the probability that all three dice will show the same number?

Q5 Elin travels through two sets of traffic lights on her way to work. There is a 0.6 probability of a green light at the first set and a 0.5 probability of a green light at the second set.

a Draw a tree diagram to show all the possible outcomes.

b What is the probability of passing through two green lights?

c What is the probability of meeting two red lights?

10 Shape shifter

This unit is about the mathematics of similar and congruent shapes.

Patterns with a large number of similar shapes can trick the brain into thinking that there is motion going on. The illusion shown here consists of many similar copies of one shape. The colours, curves and relative sizes of the shapes are chosen to maximise the illusory effect, which is known as the 'peripheral drift illusion'.

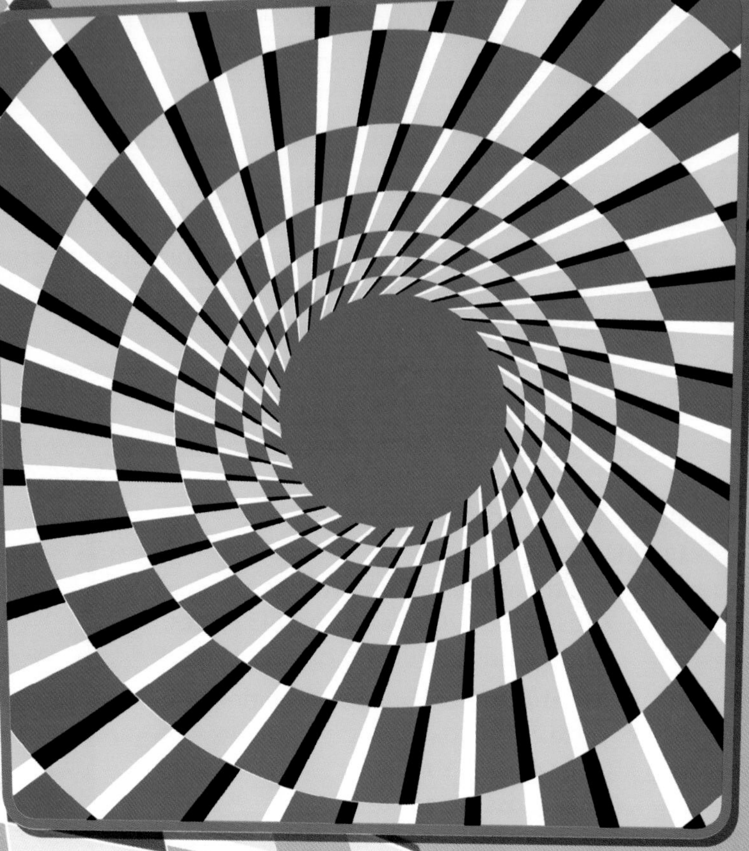

Activities

A Look at this diagram. Do the two horizontal lines look the same length? Now measure the horizontal lines. What do you find?

B Draw three circles with radius 4 cm so that all three intersect at one point. Follow the steps below to draw a fourth circle that goes through the three points where two of the original circles intersect.

- Join up the points where two of the circles intersect, to make a triangle.
- Construct the perpendicular bisector of each side.
- Find the point where the bisectors intersect. This will be the centre of your new circle.
- Draw the new circle and measure its radius.

You should find that the new circle has the same radius as the original circles, making it congruent to them. This fact is known as Johnson's theorem.

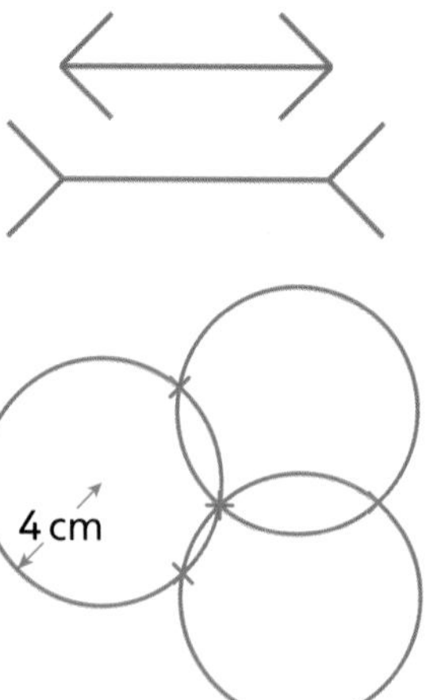

+ Point where all three circles intersect

× Points where two of the circles intersect

Before you start this unit...

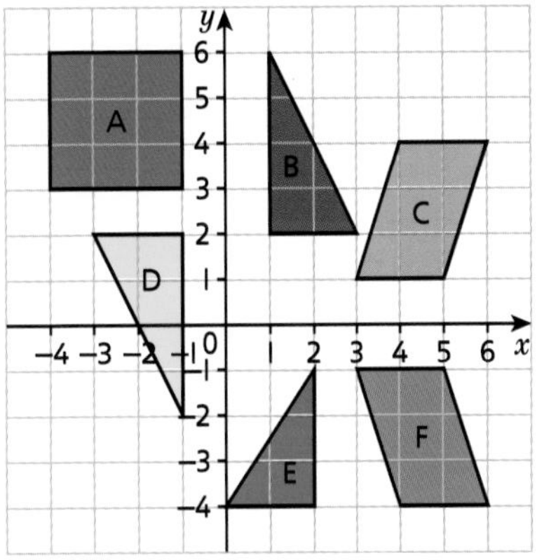

1 Which sets of shapes are congruent in the diagram above?

Level Up Maths 5-7 page 152

2 Describe fully a single transformation to take
a shape B to shape D
b shape C to shape F.

Level Up Maths 5-7 page 154

3 Write down the number of lines of symmetry and the order of rotational symmetry of shapes A, C and E.

Level Up Maths 5-7 page 152

Patterns with a large number of similar shapes can also distort the way that we perceive things. Below you can see a photograph of Port 1010 in Melbourne, Australia. This building has been tessellated with regular black and white surfaces. The orange lines appear slanted, but are actually parallel to one another.

10.1 Testing for congruence
10.2 Transforming shapes
10.3 UFO hunt
10.4 Similarity
10.5 Introducing trigonometry
10.6 Using trigonometry 1
Unit plenary: The shape of illusions

Plus digital resources

10.1 Testing for congruence

- Know what 'congruent' means and be able to identify congruent shapes
- Know the difference between a practical demonstration and a proof
- Know what information you need about a triangle to draw it exactly

Why learn this?

Architects use congruent and similar shapes to help them design safe and attractive structures.

What's the BIG idea?

- Shapes are **congruent** if they are exactly the same shape and size. As congruent shapes have the same lengths and angles, you can use congruence to help solve shape problems. Level 6 & Level 7
- A practical demonstration tests one particular case of a mathematical idea. A **proof** establishes whether the idea works for *all* values. Level 7
- **Triangles** drawn given the information SSS, SAS, ASA or RHS are unique, and therefore congruent, but triangles drawn using the information SSA or AAA are not necessarily congruent. Level 8

Practice, practice, practice!

Level 6

6c I can use side and area information to identify congruent shapes

1 Which if any of the shapes in each set are congruent?

a a square with side length 7 cm
square with area 81 cm^2
square with area 49 cm^2

b rectangle *ABCD* with *AB* = 3 cm and area 18 cm^2
rectangle *EFGH* with *GH* = 9 m and *FG* = 2 m
rectangle *JKLM* with *JK* = 4.5 mm and *KL* = 4 mm

c rectangle *NOPQ* with *NO* = 6 cm and area 24 cm^2
rectangle *RSTU* with *TU* = 4 cm and area 24 cm^2
rectangle *VWXY* with *VW* = 4 cm

6b I can use congruence to solve simple problems in quadrilaterals

2 A computer company designs logos based on congruent rhombuses.

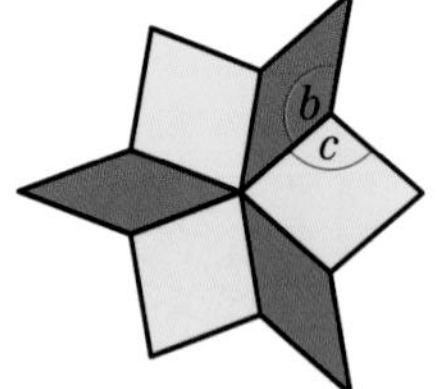

a The first logo is made from eight congruent rhombuses.
Find the size of angle a.

b The second logo is made from three congruent purple rhombuses, and three congruent yellow rhombuses. The acute angle in the yellow rhombus is twice the size of the acute angle in the purple rhombus.
Find the size of angles b and c.

Watch out!

Two shapes are congruent if you could cut one of them out and place it exactly on top of the other. It doesn't matter if you need to reflect or rotate it.

congruent proof

3 In each of these sets of triangles, two triangles are congruent and one is not. By constructing the triangles, or otherwise, find the odd triangle out in each set.

a Triangle ABC, with $AB = 5$ cm, $\angle ABC = 90°$ and $BC = 12$ cm
Triangle DEF, with $DE = 12$ cm, $\angle DEF = 60°$ and $DF = 14.5$ cm
Triangle GHI, with $GI = 12$ cm, $\angle HGI = 23°$ and $GH = 13$ cm

b Triangle JKL, with $JK = 5$ cm, $JL = 10$ cm and $\angle KJL = 70°$
Triangle MNO, with $MN = 9.5$ cm, $\angle MNO = 80°$ and $\angle NMO = 30°$
Triangle PQR, with $PR = 5$ cm, $\angle PRQ = 80°$ and $\angle RPQ = 55°$

4 **a** Find four different triangles that have whole-number side lengths, at least two equal sides and a perimeter of 18 cm.

b By putting your answers to part **a** in order, or otherwise, prove that there are not any other triangles that fit all of these conditions.

5 There is a mathematical theorem which states that any triangle formed by drawing lines from any point on the circumference of a circle to the two ends of a diameter is right angled.

a Draw a circle and check this theorem for yourself. Try three different triangles. Have you created a practical demonstration or a proof?

b Copy this diagram and follow the steps.

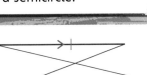

i Let B be any point on the circumference of the circle.
Label the centre of the circle.
Explain why OAB is an isosceles triangle.
Label the two equal angles in the triangle x.

ii Explain why OBC is also an isosceles triangle.
Label the two equal angles in the triangle y.

iii By looking at the triangle ABC, show that $2x + 2y = 180°$.

iv Hence prove that $x + y = 90°$ for any triangle in a semicircle.

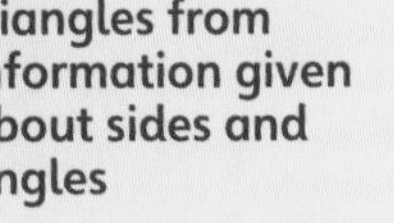

Level 7

7b I can identify congruent triangles from information given about sides and angles

7a I can recognise the difference between a practical demonstration and a proof

7a I can recognise the difference between a practical demonstration and a proof

6 Two parallel line segments have equal lengths. Their opposite ends are joined with straight lines.
By using alternate angles, or otherwise, prove that the triangles created are congruent.

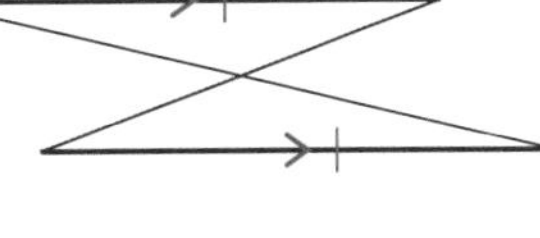

7 Prove that the two triangles created by the perpendicular bisector of the base line of an isosceles triangle are congruent.

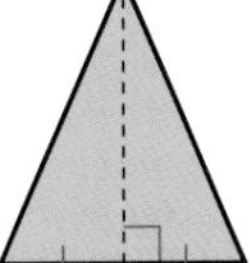

Level 8

8a I can use the side and angle information to deduce properties of triangles and whether they are congruent or not

Now try this!

A Evening out

Proofs deal with general cases, not specific ones, so we often use algebra to create a proof. By calling the first number n, create an algebraic proof that the sum of any two consecutive integers is odd.

B 3-D congruency

What do you need to measure to check whether

a two cubes are congruent

b two cuboids are congruent?

quadrilateral triangle

10.2 Transforming shapes

- Transform 2-D shapes by combinations of rotations, reflections and translations
- Know that translations, rotations and reflections map objects onto congruent images

What's the BIG idea?

- You can transform 2-D shapes using combinations of **rotations**, **reflections** and **translations**. It is important to describe these transformations fully. Level 6 & Level 7
- Translations, rotations and reflections keep the lengths and angles of shapes the same.
  The **objects** and **images** of these transformations are **congruent**. Level 6 & Level 7

Why learn this?
All movements can be described using a combination of transformations.

Practice, practice, practice!

Level 6

6C I can use and describe more complex combinations of transformations

1 a Describe fully a transformation that will map
   i shape A onto shape B
   ii shape A onto shape C
   iii shape A onto shape D.

b Are shapes A, B, C and D congruent?
Give reasons for your answer.

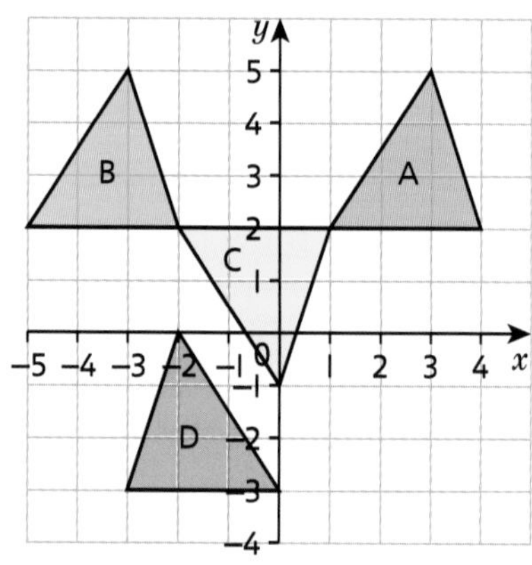

6C I can explain what effect translations, rotations and reflections have on lengths and angles

2 a Draw x- and y-axes from −6 to +6. Draw the shape with vertices at (−6, 4), (−4, 0), (−2, 4) and (−4, 6), label it A. What is the name of the shape you have drawn?

b Rotate shape A 270° antic lockwise about (−4, 0). Label your new image B.

c Reflect shape A in the line $y = 1$. Label your new image C.

d Translate shape A by 7 units right and 3 units down. Label your new image D.

e How have these transformations changed the lengths and angles of shape A?

6C I can describe fully complex combinations of transformations

3 a Draw x- and y-axes from −8 to +8. Draw the shape with vertices at (−6, −1), (−4, −1), (−1, −5) and (−5, −5) and label it A.
What is the name of the shape you have drawn?

b Rotate shape A by 180° about the point (−4, −1) and then reflect it in the y-axis. Label your new image B.

c Translate shape A by 4 squares up and reflect it in the line $x = -4$. Label your new image C.

d What single transformation will map shape C onto shape B?

centre of rotation | congruent | image | object

4 a Describe fully a combination of a reflection and a translation that will map shape A onto shape B.

b Fully describe a single transformation that will map shape A onto shape D.

c Explain why no combination of translations, reflections or rotations can map shape A onto shape C.

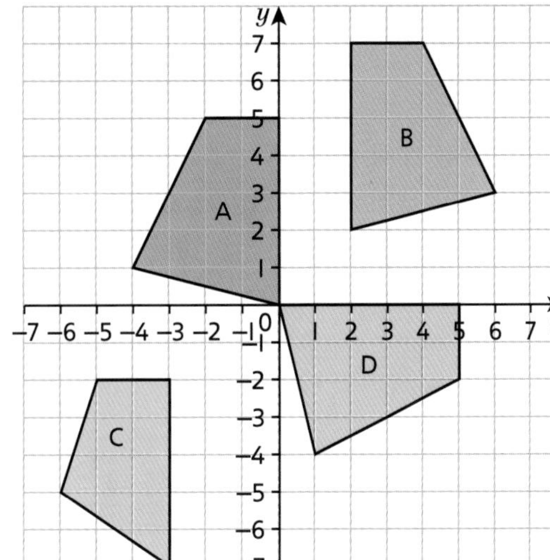

Level 7

7c I can understand the effects of rotation, translation and reflection and use these to solve problems

Tip

If a question asks you to describe a transformation 'fully', there's usually more than one thing you need to write down.

5 Shape X is a right-angled isosceles triangle that undergoes a series of three transformations. The coordinates of its vertices are given in the table.

a Copy and complete the table without drawing a diagram.

Shape X	Reflect in the y-axis	Translate 1 unit left and 4 units down	
(−6, 1)	(6, 1)	(5, −3)	(−3, −3)
(−2, 1)	(2, 1)	(1, −3)	(1, −3)
(−2, 5)			(1, −7)

b What shape is the final image? How do you know?

7a I can understand how translations, rotations and reflections map objects onto congruent images

Learn this

The image of a rotation, reflection or translation is congruent to the object, so the lengths and angles all stay the same. This can help you find any missing vertices of a shape.

6 a Use either ICT or paper. Draw an octagon with vertices at (0, 3), (2, 2), (3, 0), (2, −2), (0, −3), (−2, −2), (−3, 0) and (−2, 2).

b Translate this octagon

i 5 units up and 2 units right ii 5 units right and 2 units down.

Continue this tessellation.

c These octagons alone cannot tessellate.
What shape is created in the gaps between them?

d The original octagon shares edges with four other octagons.
Describe two translations that would map the original octagon onto each of the other two octagons not drawn in part **b**.

Now try this!

A The $y = x$ factor

Draw a shape on a set of axes and reflect it in the line $y = x$.
Compare the coordinates of the image and the object.
Experiment with different shapes? What happens to the coordinates?
What are the coordinates of the point (a, b), after reflection in the line $y = x$?

B Tessellations

Can every single-shape tessellation be generated using only translations of one original shape? Explain your answer.

reflection rotation translation

World's Greatest Maths!

10.3 UFO HUNT

You have joined a team of crack UFO hunters. Can you use your map-reading and scale-drawing skills to prove that the government has been hiding evidence of extra-terrestrial life?

RAF Bentwaters

Your first assignment is to investigate a disused air force base in Suffolk.

Checklist

The scale of this map is 1 : 50 000.
This means that 1 cm on the map represents 50 000 cm on the ground.

1 Convert 50 000 cm into
 a metres b kilometres.

2 What real distance would be represented by 3.5 cm on the map?

3 What length on the map would represent a real distance of 700 m?

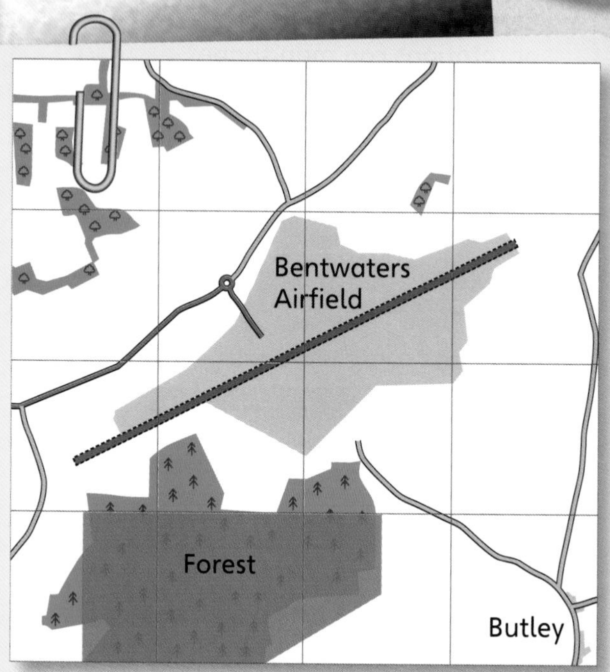

Main runway

UFO take-offs and landings have been reported at Bentwaters – but is the runway long enough?

4 The main runway has been marked on your map in red. Use a ruler to measure its length to the nearest millimetre.

5 What is the real length of the runway?

Lights in the forest

The US military reported sightings of extra-terrestrials in the forest around Bentwaters. You have highlighted a search area in purple on your map.

6 What area of land is represented by the entire map?

7 Calculate the real size of the search area.

8 You can scan an area of 0.5 km^2 every 6 hours. How long will it take you to scan the whole search area?

UFO FACT

In December 1980, military personnel at Bentwaters saw bright lights and a pyramid-shaped object hovering over the base.

TOP SECRET

Secret plans

You have discovered these UFO blueprints hidden in the forest.

1 Write the scale of the blueprints as a ratio.

2 What is the real wingspan of the UFO?

3 What is the real area of the radar dome as seen from above?

4 Find the real distance from the front wheel to one of the rear wheels.

5 Work out the real distance from point A to point B.

In Q5 you need to think in 3-D.

6 Construct a scale drawing showing what the side elevation of this UFO might look like. Make sure your drawing corresponds with the blueprints.

CLASSIFIED

Roswell

The blueprints have led you to the UFO capital of the world!

UFO FACT

In 1947, the US military reported finding a crashed 'flying disc' near Roswell in New Mexico.

1. The distance from Los Angeles to Denver is 1350 km. Use this fact to calculate the scale of this map. Write your answer as 1 cm = ___ and as a ratio.
2. You have flown into San Francisco. Calculate the distance you need to travel to reach Roswell in New Mexico.
3. You believe that the UFO took off from Area 51 in Nevada. According to the blueprints, it has a maximum range of 1000 km. Could it have reached Roswell?
4. Which of these cities would have been in range for the UFO?
   a Amarillo **b** San Francisco
   c Dodge City **d** Twin Falls
5. Use this map to estimate the area of each of these states.
   a Utah **b** Colorado
   c Nevada

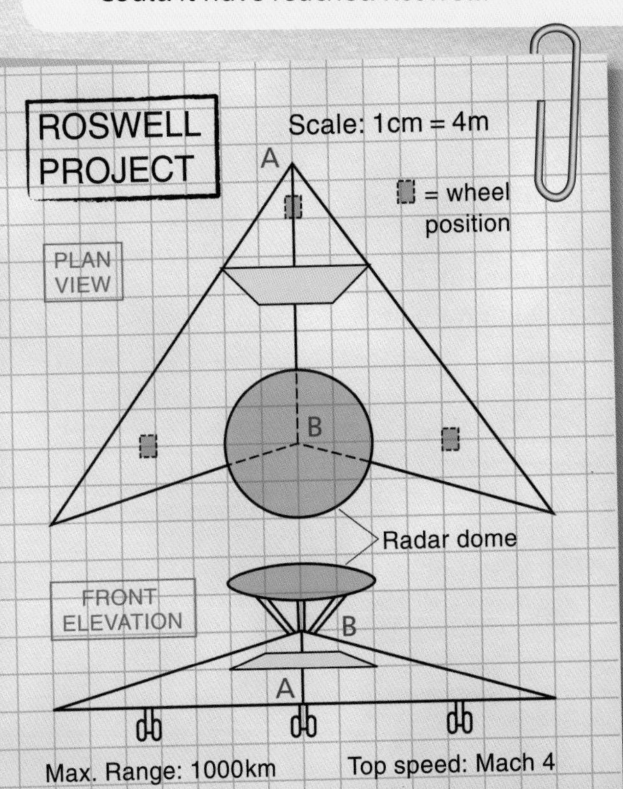

Weather balloons

The US government have released an official report stating that any UFO sightings were really just weather balloons.

This diagram shows a typical weather balloon. It is a sphere 1.8 m in diameter, with a 0.5 m long cable hanging from it. Attached to the cable is an instrument package, which is a cube with a side length of 0.7 m.

Choose an appropriate scale, then construct an accurate scale drawing of the weather balloon. Label your drawing.

10.4 Similarity

- Find a point that divides a line segment in a given ratio
- Recognise and use similarity to solve problems

What's the BIG idea?

→ You can find the coordinates of the **mid-point** of the line segment AB by finding the mean of the x-coordinates and the mean of the y-coordinates of A and B.

For example,
x-coordinate: $\frac{2 + 10}{2} = 6$

y-coordinate: $\frac{1 + 5}{2} = 3$

So the coordinates of mid-point M are (6, 3). **Level 6**

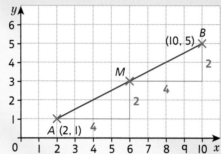

→ Shapes are **similar** if corresponding sides are in the same ratio and corresponding angles are equal. **Level 7**

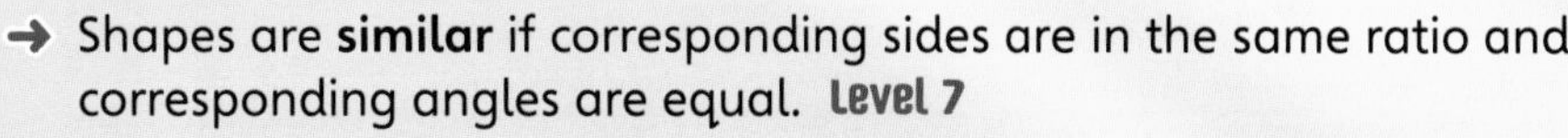

→ You can use the properties of similar triangles to find a point that divides a line in a given ratio. **Level 7**

→ **Enlargements** create images that are similar to their objects. **Level 7 & Level 8**

Why learn this?

Scale models are built for tv and film special effects using shapes similar to the real-life objects.

Learn this

Congruent shapes are the same shape and the same size. Similar shapes are the same shape but can be different sizes.

Practice, practice, practice!

Level 6

6b I can find the mid-point of a horizontal, vertical or diagonal line AB, using coordinates

1 M is the mid-point of the line AB. Find the coordinates of M for these lines.

a $A(9, 8)$, $B(9, 2)$ b $A(1, -6)$, $B(7, -6)$ c $A(5, -3)$, $B(-3, -3)$

d $A(4, 7)$, $B(-4, 5)$ e $A(0, -6)$, $B(12, -2)$ f $A(-5, -2)$, $B(3, 4)$

Level 7

7b I can find the ratio of lengths in similar triangles

2 Draw an x-axis from 0 to +18 and a y-axis from 0 to +14.
Draw triangle XYZ, where X, Y and Z have coordinates (0, 0), (16, 12) and (16, 0), respectively.

a Enlarge triangle XYZ by scale factor $\frac{1}{4}$ with centre of enlargement (0, 0). On the line XY label the new point M. On the line XZ label the new point N.

b Is triangle XMN similar to triangle XYZ? Explain your answer.

c Write down in its simplest form the ratio of these lengths.
i $XN : XZ$ ii $XM : XY$ iii $XN : NZ$ iv $XM : MY$

7b I can identify similar triangles

3 Which of these triangles are similar?

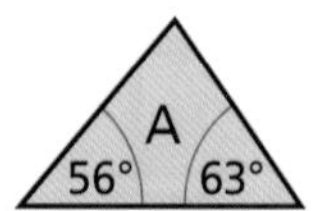

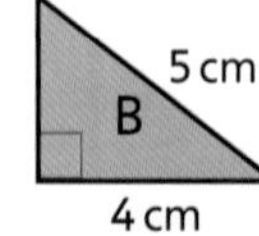

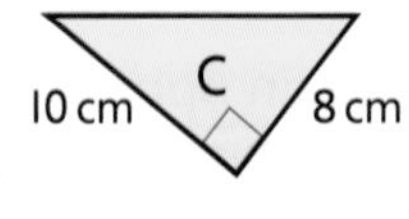

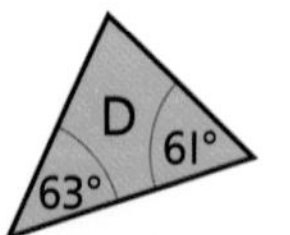

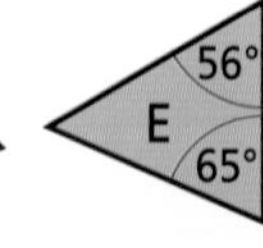

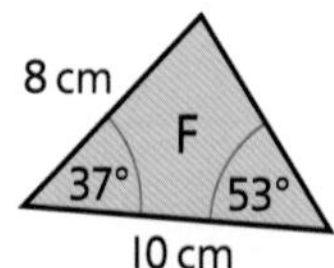

centre of enlargement congruent enlargement

4 The line segment AC is split in the ratio 1 : 2.
Work out the y-coordinate of B.

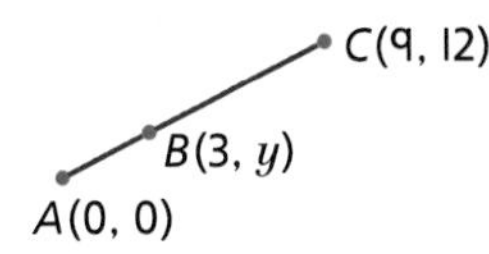

5 A is the point (2, 5) and B is the point (10, 13).
Find the coordinates of the point N that divides the line AB in the ratio 1:3.

6 The point Y divides the line XZ in the ratio 2:3.
Work out the coordinates of Y.

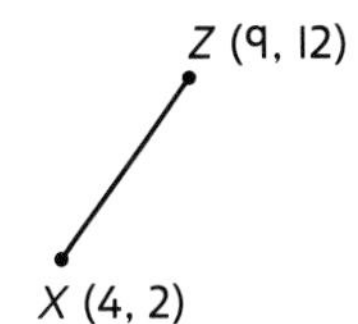

7 Triangles ABC and PQR are similar.
Find the values of

a length PQ

b $\angle PQR$

c length AC.

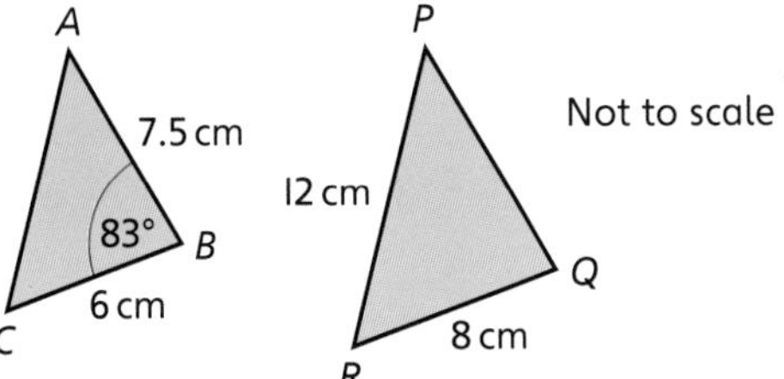

8 Are the two triangles in this diagram similar?
Give reasons for your answers.

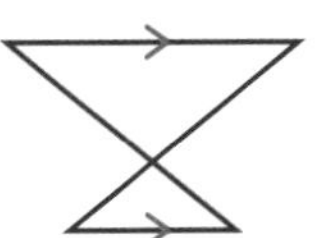

9 Find the length of

a AD

b CE

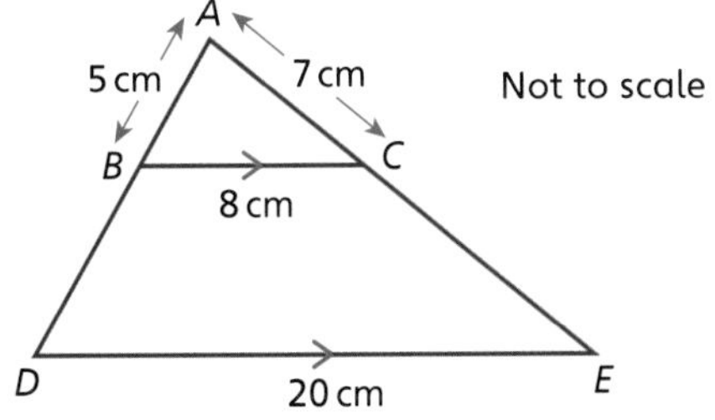

Level 7

7b I can find the coordinates of the point that divides a line in a given ratio, using the properties of similar triangles

7a I can use similarity to solve 2-D shape problems

Level 8

8c I can recognise similarity in enlargements

Now try this!

A True or false?

Decide whether these statements are true or false.
For each statement write a sentence or sketch a diagram to explain your choice.

a Every square is similar to every other square.

b Every rhombus is similar to every other rhombus.

c Every enlargement of a 2-D shape is similar to the original shape.

d Every pentagon is similar to every other pentagon.

e Every circle is similar to every other circle.

B 3-D mid-points

M (−3, y, z) is the mid-point of the line segment AB, where A has coordinates (0, 6, −2) and B has coordinates (x, 10, −8). Find the values of x, y and z.

mid-point similar

10.5 Introducing trigonometry

- Understand the trigonometric ratios of sine, cosine and tangent
- Use sine, cosine and tangent to find the lengths of sides in right-angled triangles

Why learn this?
Architects use trigonometry to find lengths and angles to draw building plans.

What's the BIG idea?

- When you enlarge a right-angled triangle, both the angles and the ratios of the lengths stay the same. We can link these two things together using the trigonometric ratios **sine**, **cosine** and **tangent**. Level 7 & Level 8
- In any right-angled triangle,

$$\sin\theta = \frac{\textbf{opposite}}{\textbf{hypotenuse}} \quad \cos\theta = \frac{\textbf{adjacent}}{\text{hypotenuse}} \quad \tan\theta = \frac{\text{opposite}}{\text{adjacent}}$$

You can rearrange these formulae to find the lengths of sides in right-angled triangles. Level 8

Learn this

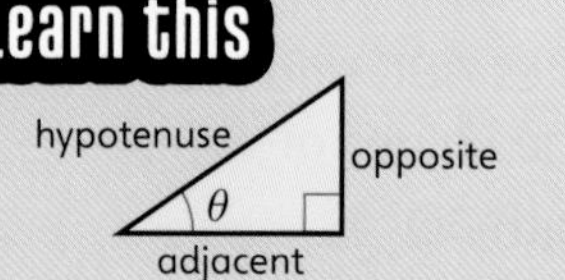

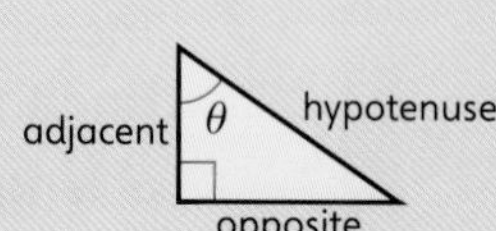

In a right-angled triangle the longest side is called the hypotenuse. The side opposite the angle of interest is called the opposite and the side next to it is called the adjacent.

Practice, practice, practice!

1 Using a calculator, write down the value of each ratio to 3 d.p.

a sin 45° **b** cos 60° **c** tan 30° **d** sin 60° **e** tan 15°
f tan 57° **g** cos 14° **h** sin 79° **i** cos 0° **j** sin 36°

2 Calculate angle x. *sin x = 0.5*
x = sin⁻¹ 0.5 = 30°

a $\sin x = 0.6427$ **b** $\cos x = 0.5$ **c** $\tan x = 1$
d $\cos x = 0.9063$ **e** $\tan x = 3.732$ **f** $\sin x = 0.7880$

Watch out!
Check that your calculator is in 'degree' mode before you start.

Level 7

7b I understand the link between similarity and trigonometry in right-angled triangles

7a I can use the $\sin^{-1}$, $\cos^{-1}$ and $\tan^{-1}$ keys on a calculator

3 Calculate the missing angles. Give your answers to 1 d.p.

a

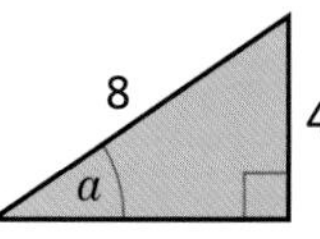

b

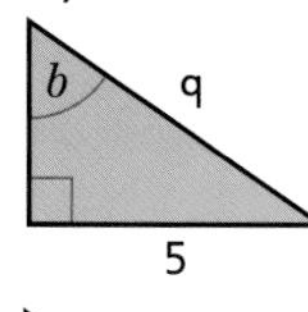

c

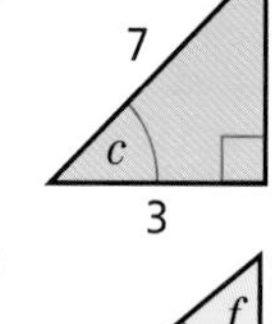

d

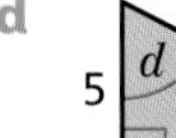
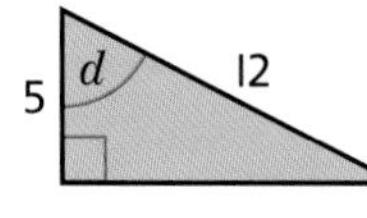

e

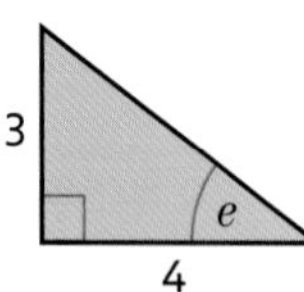

f (f, 6, 7)

4 For each of these triangles, use the sine ratio to find the length of the side opposite the given angle. Give your answers to 2 d.p.

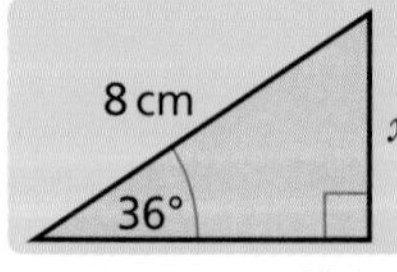

$\sin 36° = \frac{\text{opp}}{\text{hyp}} = \frac{x}{8}$

so $x = 8 \times \sin 36°$ so $x = 4.70$ cm

a

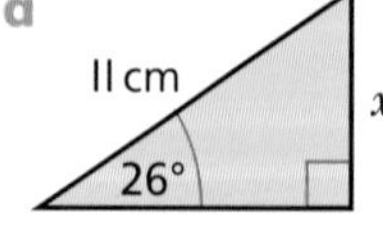

b (9.5 cm, 54°, x)

c

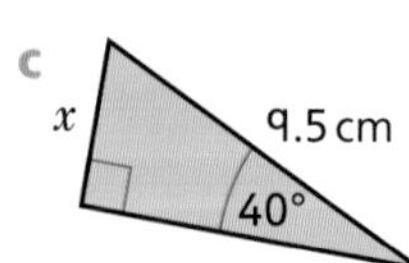

Level 8

8a I can use trigonometry to find the size of an angle in a right-angled triangle

8c I can use trigonometry to find the lengths of unknown sides in a right-angled triangle, using straightforward algebraic manipulation

adjacent cosine (cos) hypotenuse

Level 8

8c I can use trigonometry to find the lengths of unknown sides in a right-angled triangle, using straightforward algebraic manipulation

5 For each of these triangles, use the cosine ratio to find the length of the side adjacent to the given angle. Give your answers to 2 d.p.

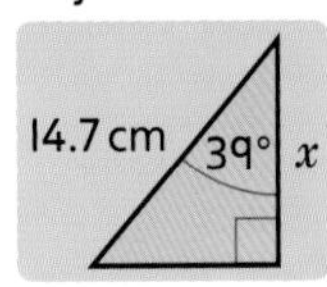

a

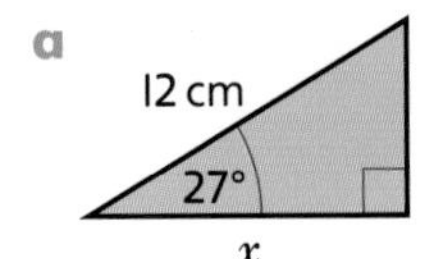

b

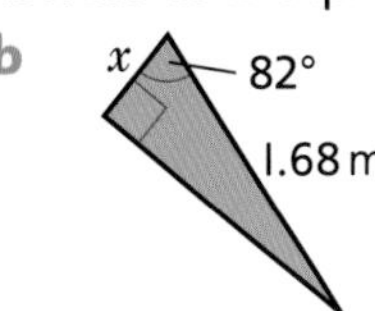

c

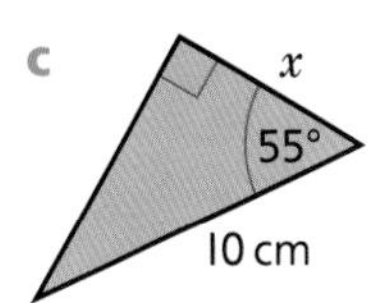

$\cos 39° = \frac{x}{14.7}$

so $x = 14.7 \times \cos 39°$ so $x = 11.42$ cm

6 For each of these triangles, use the tangent ratio to find the length of the side opposite to the given angle. Give your answers to 2 d.p.

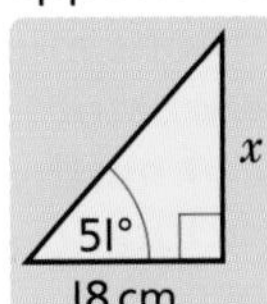

a

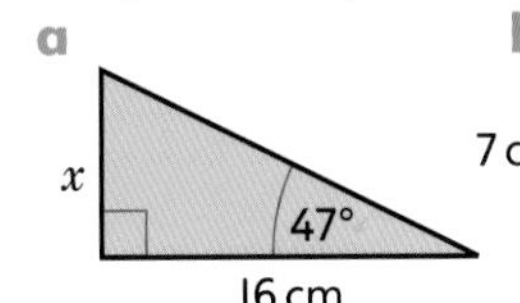

b

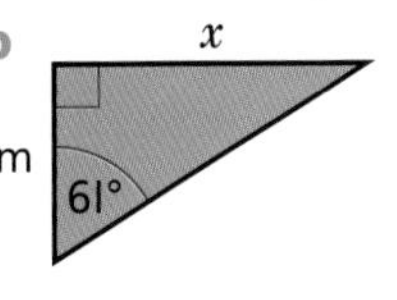

c

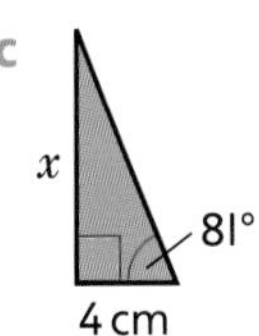

$\tan 51° = \frac{x}{18}$

so $x = 18 \times \tan 51°$ so $x = 22.23$ cm

7 For each of these triangles, use the appropriate trigonometric ratio to find the length of the side labelled x. Give your answers to 2 d.p.

a

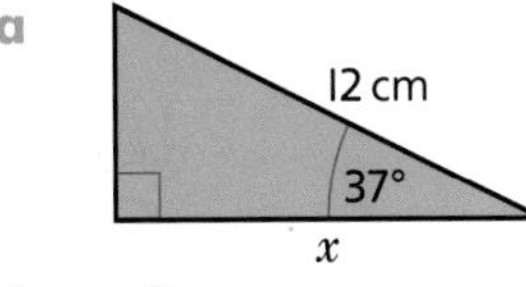

b

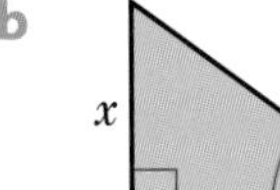

c

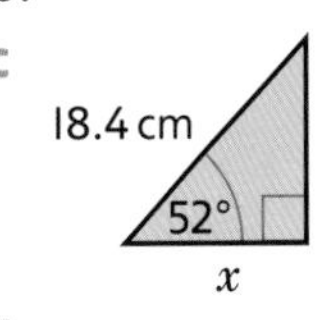

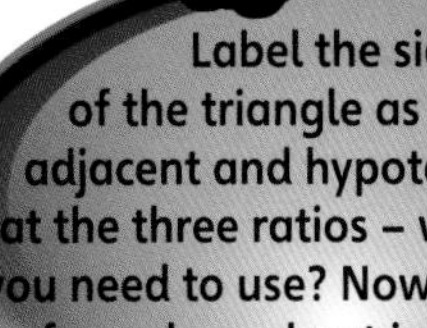

d

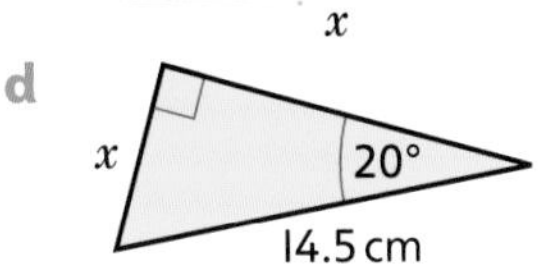

e

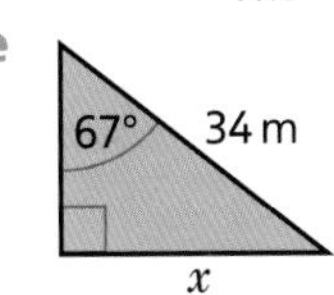

f

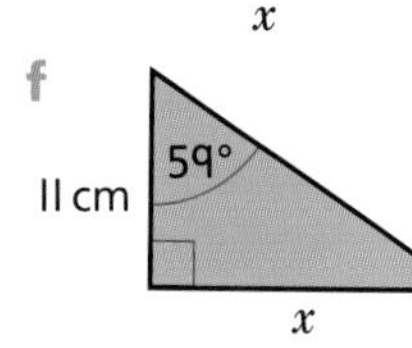

8b I can use trigonometry to solve problems

8 The two arms of a symmetrical hand-held catapult are 12 cm long. A length of elastic is stretched between the two arms as shown in the diagram.

a Calculate the length of the elastic.

b The middle of the elastic is now pulled straight back 3 cm by a finger. Calculate the new length of the stretched elastic.

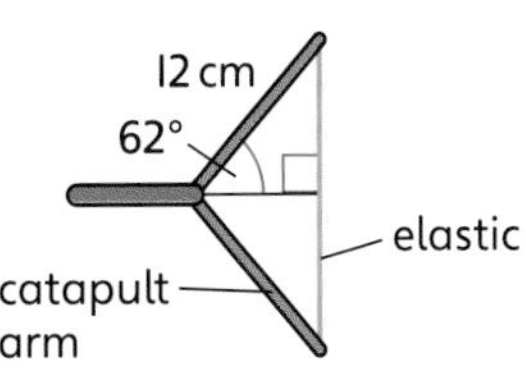

Hint: Draw a diagram.

Now try this!

A The root of trigonometry

Many key values of the sine, cosine and tangent ratios can be expressed exactly using square roots. For example, you can use a right-angled isosceles triangle with equal sides 1 unit in length to show that $\sin 45° = \frac{1}{\sqrt{2}}$.

hypotenuse = $\sqrt{2}$ (by Pythagoras)

So $\sin 45° = \frac{1}{\sqrt{2}}$

1, 45°, 45°, 1

Use an equilateral triangle with sides 2 units long to find out exact representations for sin 60°, cos 30° and tan 60°.

B Tan 90°

If you ask a calculator to work out tan 90° it will give you an error message.

Why can't you calculate tan 90°?

Tip

Try drawing a triangle that you would calculate tan 89° from, and then imagine extending it to tan 90°.

opposite sine (sin) tangent (tan)

10.6 Using trigonometry 1

- Understand the trigonometric ratios of sine, cosine and tangent
- Use sine, cosine and tangent to find the lengths of sides in right-angled triangles and to solve problems

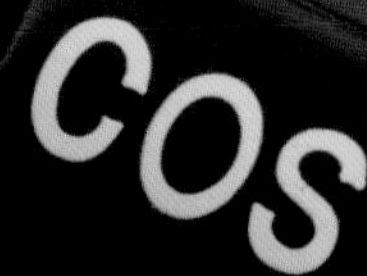

Why learn this?

The trigonometric functions are powerful tools in many areas of maths and science – including acoustics, engineering and astronomy.

What's the BIG idea?

→ In any right-angled triangle,

$$\sin\theta = \frac{\text{opposite}}{\text{hypotenuse}} \qquad \cos\theta = \frac{\text{adjacent}}{\text{hypotenuse}} \qquad \tan\theta = \frac{\text{opposite}}{\text{adjacent}}$$

You can rearrange these formulae to find the lengths of sides in right-angled triangles. Level 8

Practice, practice, practice!

1 For each of these triangles, use the sine ratio to find the length of the hypotenuse. Give your answers to 2 d.p.

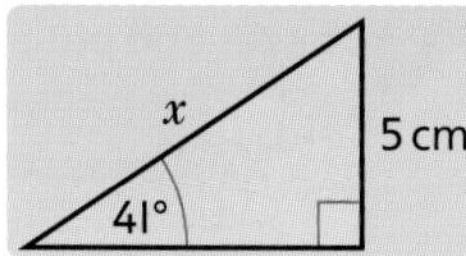

a

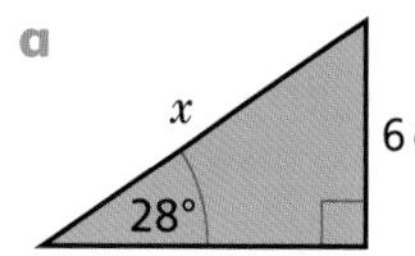

b

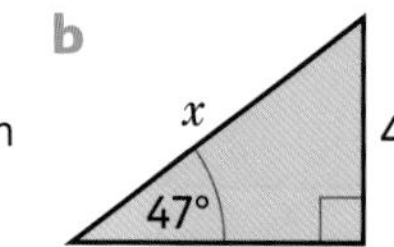

c

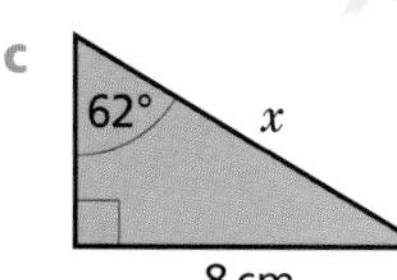

$\sin 41° = \frac{\text{opp}}{\text{hyp}} = \frac{5}{x}$

so $x \times \sin 41° = 5$

so $x = \frac{5}{\sin 41°}$

so $x = 7.62$ cm

Watch out!

Rememeber to check that your calculator is in 'degree' mode.

2 For each of these triangles, use the cosine ratio to find the length of the hypotenuse. Give your answers to 2 d.p.

a

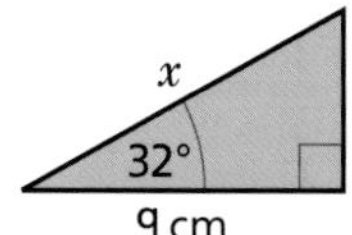

b

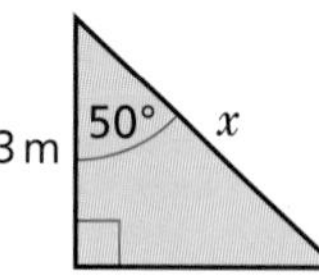

c

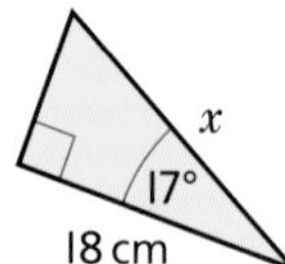

3 For each of these triangles, use the tangent ratio to find the length of the side adjacent to the given angle. Give your answers to 2 d.p.

a

75°

x

11 cm

b

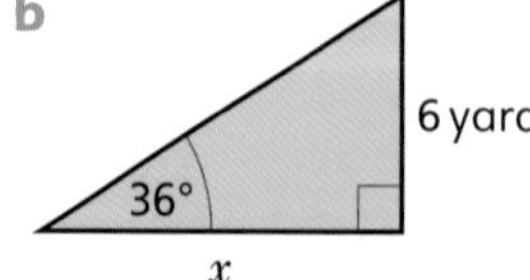

c

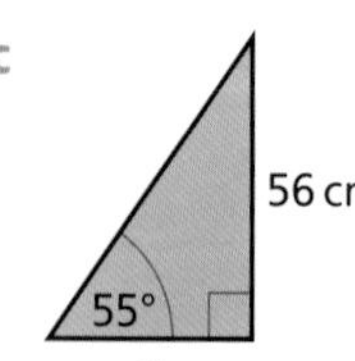

Level 8

8b I can use trigonometry to find the lengths of unknown sides in a right-angled triangle using more complex algebraic manipulation

adjacent cosine (cos) hypotenuse

4 For each of these triangles, use the appropriate trigonometric ratio to find the length of the side labelled x. Give your answers to 2 d.p.

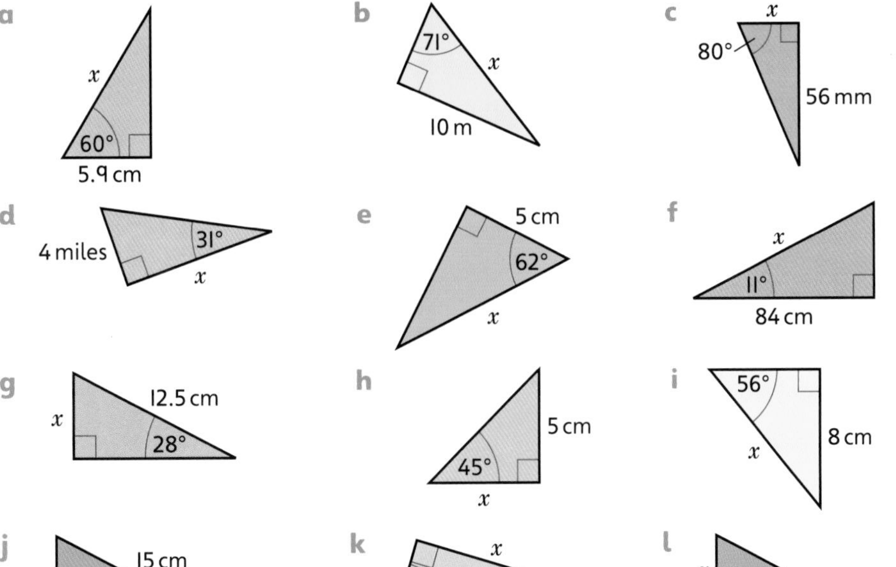

Level 8

8b I can use trigonometry to find the lengths of unknown sides in a right-angled triangle using more complex algebraic manipulation

5 When a surveyor stands 10 m away from the base of the tree, the angle between the ground and the top of the tree is 68°. Calculate the height of the tree.

68°
10 m

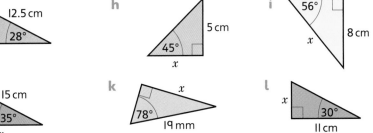

Watch out! Have you rearranged the formula correctly and have you used the correct ratio?

8a I can use trigonometry to solve problems

6 Astronomers sometimes use trigonometry to estimate the distance to faraway stars. The distance from the Earth to the Sun is 1 AU (astronomical unit) and the star IB96X appears to move 2° against the background galaxies between summer and winter. Calculate the distance from the Sun to the star IB96X.

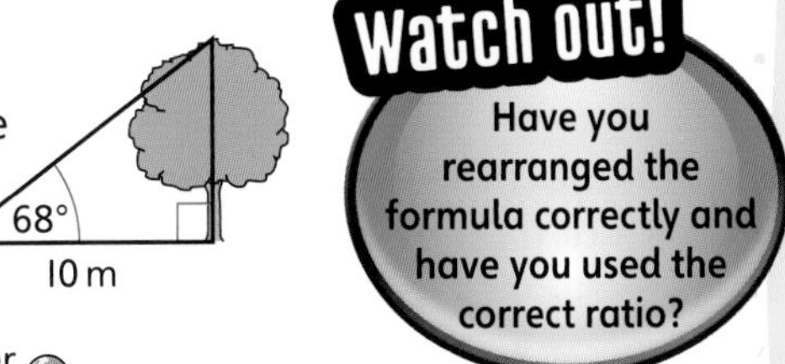

Tip Split the isosceles triangle into two congruent right-angled triangles.

Now try this!

A Trigonometry is sound

Sound waves can be modelled using sine and cosine graphs. A simple example is given by the graph of $y = \cos x$. Draw a set of axes from −1 to +1 in the y-direction and 0 to 360° in the x-direction. Use your calculator to find the values of cos 0°, cos 10°, cos 20° etc. up to cos 360° and plot them on the graph. Join the points with a smooth curve.

B Making tan

Choose an angle θ, and use a calculator to find and write down the values of $\sin \theta$, $\cos \theta$ and $\tan \theta$. Divide the first result by the second result to perform a practical demonstration that $\frac{\sin \theta}{\cos \theta} = \tan \theta$. Try it for several angles.
Now find a proof that this formula is true for all angles, by using the word formulae for the ratios.

opposite sine (sin) tangent (tan)

Unit 10 plenary

The shape of illusions

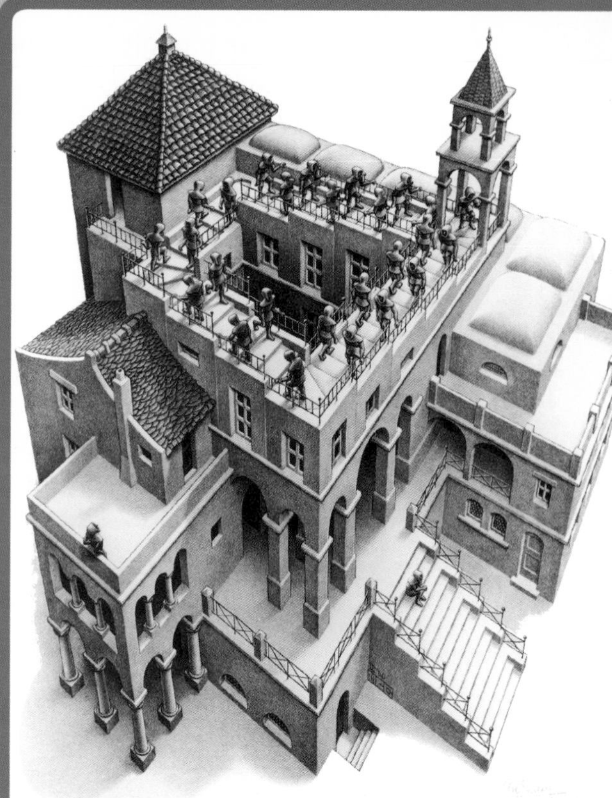

This picture is another type of optical illusion – which way is 'up' on the staircase on the roof?

The opener to this unit showed some examples of optical illusions. Another type of optical illusion is the autostereogram (for example, Magic Eye® images). These use repeated patterns and images in 2-D to give an illusion of a 3-D image.

1 These bears are a close-up of one level of an autostereogram. Describe fully the transformation that maps Bear A onto Bear B. **Level 6**

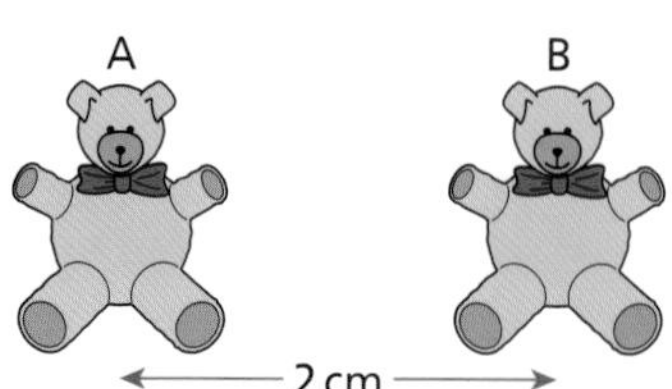

2 The bears are exact replicas of each other. What mathematical term describes this relationship? **Level 6**

3 An autostereogram tricks the eyes into focusing on an imaginary object at a point behind the paper.

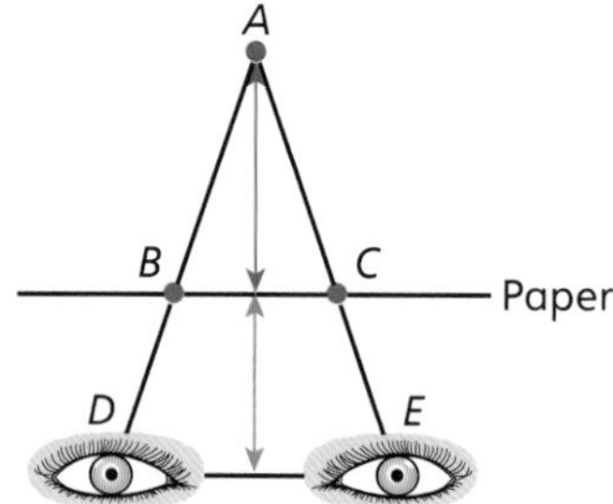

In this diagram the paper is being held parallel to the plane of the face. The focus point (*A*) is halfway between the images (*B* and *C*) and halfway between the eyes (*D* and *E*). Briefly, justify why triangle *ABC* and triangle *ADE* are similar. **Level 7**

4 In the diagram in Q3, the distance between the two images, *BC*, is 2 cm. The distance between the eyes, *DE*, is 7 cm. Find the scale factor and the centre of enlargement that would map triangle *ABC* onto triangle *ADE*. **Level 7**

5 In the diagram in Q3, angle *BAC* is 6°.

- **a** Find the perpendicular distance of the focus point from the paper, as shown on the diagram by the red arrow.
- **b** Find the perpendicular distance from the eyes to the paper, as shown on the diagram by the green arrow. **Level 8**

6 In another autostereogram, the distance between the two images, *BC* is l cm. The perpendicular distance of the focus point from the paper is 8 cm. Find angle *BAC*. **Level 8**

The BIG ideas

- Shapes are **congruent** if they are exactly the same shape and size. As congruent shapes have the same lengths and angles, you can use congruence to help solve shape problems. Level 6 & Level 7
- You can transform 2-D shapes using combinations of **rotations**, **reflections** and **translations**. It is important to describe these transformations fully. Level 6 & Level 7
- Translations, rotations and reflections keep the lengths and angles of shapes the same. The **objects** and **images** of these transformations are congruent. Level 6 & Level 7
- Shapes are **similar** if the corresponding sides are in the same ratio and corresponding angles are equal. Level 7
- An enlargement can be described by a **scale factor** and a **centre of enlargement**. Level 7
- **Enlargements** create images that are similar to their objects. Level 7 & Level 8
- When you enlarge a right-angled triangle, both the angles and the ratios of the lengths stay the same. We can link these two things together using the trigonometric ratios **sine**, **cosine** and **tangent**. Level 7 & Level 8
- In any right-angled triangle

$\sin\theta = \dfrac{\textbf{opposite}}{\textbf{hypotenuse}}$ $\quad \cos\theta = \dfrac{\textbf{adjacent}}{\text{hypotenuse}}$ $\quad \tan\theta = \dfrac{\text{opposite}}{\text{adjacent}}$

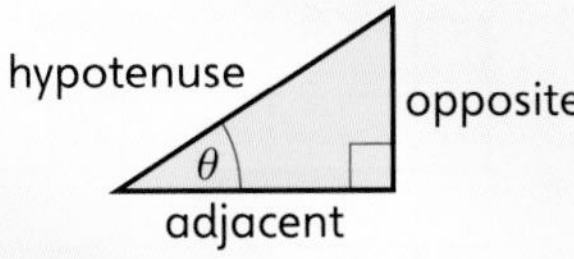

You can rearrange these formulae to find the lengths of sides in right-angled triangles. Level 8

Find your level

Level 6

Q1 This shape is made from four congruent kites. Calculate the size of angle y.

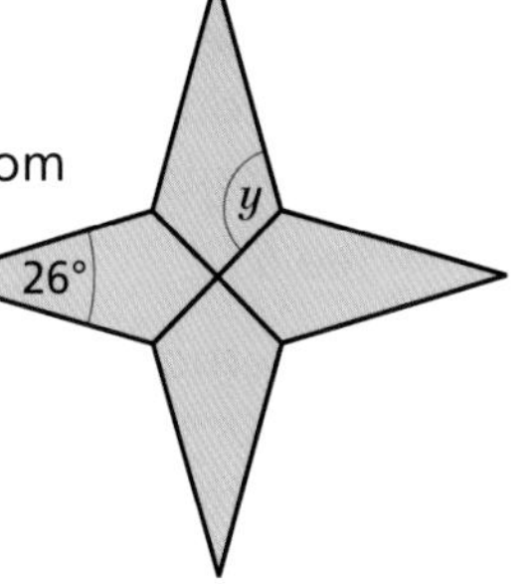

Q2 Copy this grid onto squared paper and rotate the triangle 90° anticlockwise using the dot as the centre of rotation.

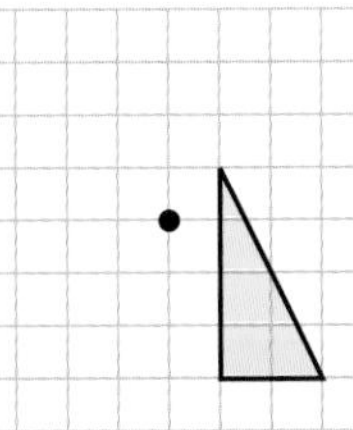

Level 7

Q3 These two rectangles are similar.

8 cm
3 cm A
4.5 cm B

a Work out the scale factor of the enlargement from rectangle A to rectangle B.
b What is the area of rectangle B?

Q4 a What is the mid-point of the line segment made by joining the points with coordinates (3, 4) and (–1, 10)?
b The mid-point of the line segment made by joining the points (–3, 5) and (x, y) is (3, 0). Find the values of x and y.

Level 8

Q5 a Calculate the height of this triangle.
b Work out the area of this triangle.

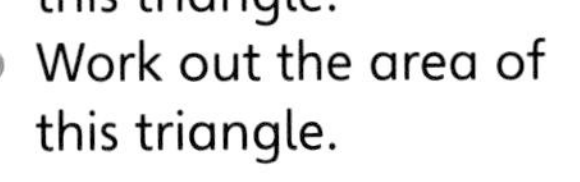

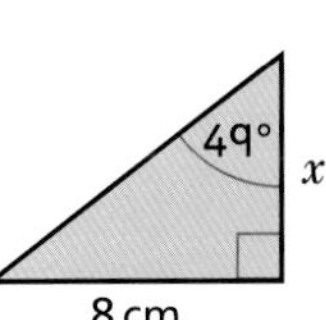

Q6 This circle has radius 10 cm.
Find the length of the chord AB.

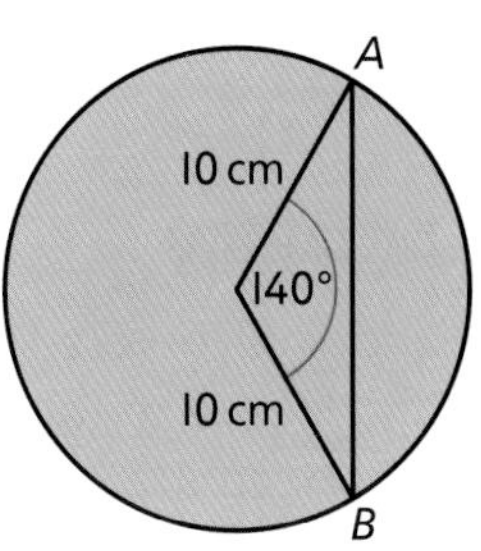

Q7 Is it possible to have a triangle with the angles and lengths shown?

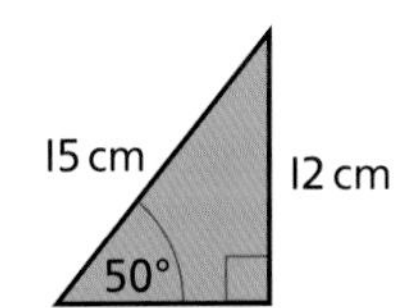

Revision 2

Quick Quiz

Q1 Work out the volume of this cuboid.

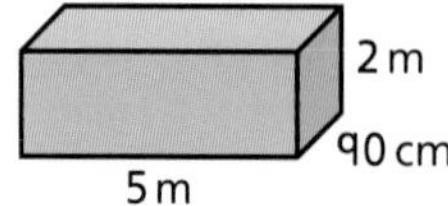

→ *See 6.6*

Q2 Use a mental or written method to work out the area of this rectangle.

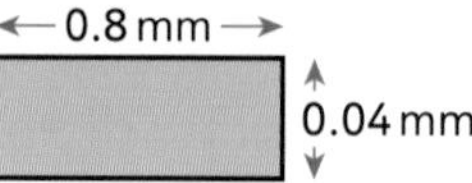

→ *See 7.6*

Q3

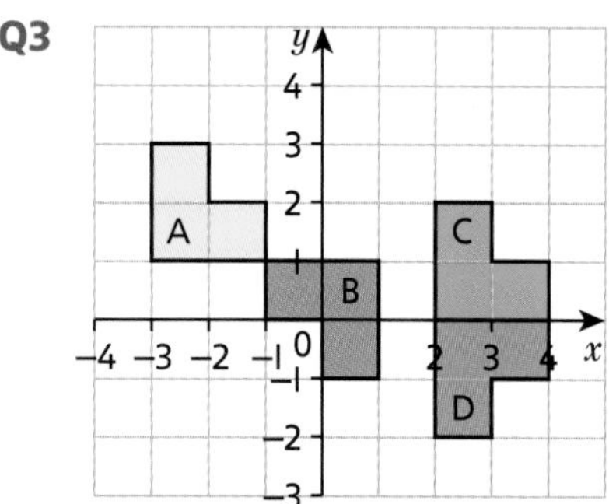

Describe fully a transformation that will map

a shape A onto shape B

b shape A onto shape C

c shape C onto shape D.

→ *See 10.2*

Q4 Work out the coordinates of the midpoint of the line joining A(2, 4) to B(8, −10).

→ *See 10.4*

Q5 The circumference of a circle is 56 cm. What is the radius of the circle? Give your answer to 1 d.p.

r

→ *See 6.3*

Q6 **a** What is the reciprocal of $\frac{3}{4}$?

b What is $\frac{3}{4}$ multiplied by its reciprocal?

→ *See 7.3*

Q7 David says

Is David correct? Explain your answer.

→ *See 8.2*

Q8 A bag of cookies contains choc-chip cookies, sultana cookies and vanilla cookies. The probability of getting a choc-chip cookie is $\frac{1}{3}$ and the probability of getting a vanilla cookie is $\frac{3}{5}$.
What is the probability of getting a sultana cookie?

→ *See 9.2*

Q9 The length and the width of this rectangle are given to the nearest cm.

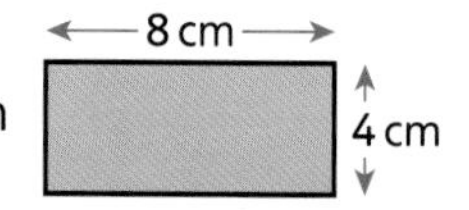

a Write down the range of the smallest and largest possible values for the length and the width.

b Work out the range for the area of the rectangle.

→ *See 6.1*

Q10 A fingernail grows at an average rate of 6.3×10^{-8} cm per second.
On average, how much will a fingernail grow in one year?

→ *See 7.5*

Q11 Work out the value of $27^{-\frac{2}{3}}$.

→ *See 8.4*

Activity

Imagine you are the manager of a leisure centre. Your leisure centre is undergoing some refurbishment.

LEVEL 6

Q1 A circular hot-tub of radius 1.5 m is being installed. The manufacturers recommend that you allow 1.1 m of the circumference per person.

a Calculate the circumference of the hot-tub.

b What is the maximum number of people you should allow in the hot-tub at one time?

Q2 The diving pool has the dimensions shown with a depth of 3.6 m. The pool is filled to the top with water.
What is the volume of water needed to fill the pool?

6 m

10.5 m

Q3 The Olympic-size swimming pool holds 2500 m³ of water. How much is this in litres?

LEVEL 7

Q4 When competitions are being held in the pool, the temperature of the water must be 79°F.
You know that 0°C = 32°F and 30°C = 86°F.

a On graph paper, use the information above to draw a conversion graph for °C to °F.

b Use your graph to work out the pool competition temperature in °C.

Q5 This is the swimming price list.

Adult: £4.20 Junior: £3.20 OAP: £3.80

The leisure centre reopened on the first day of August. The number of people using the pool that day was:

85 Adults, 150 Juniors, 15 OAPs

a What is the relative frequency of an adult, a junior and an OAP using the pool?

b For August it is predicted that a total of 8000 people will use the pool. Use the relative frequencies to work out how much money you can expect the leisure centre to take in August.

LEVEL 8

Q6 Part of the water slide into the swimming pool is held in position by two metal struts. Calculate the length, x, of the longer metal strut.

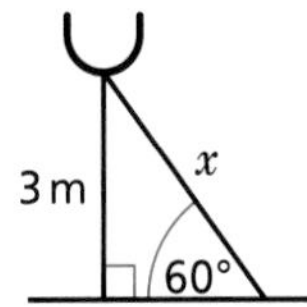

Find your level

LEVEL 6

Q1

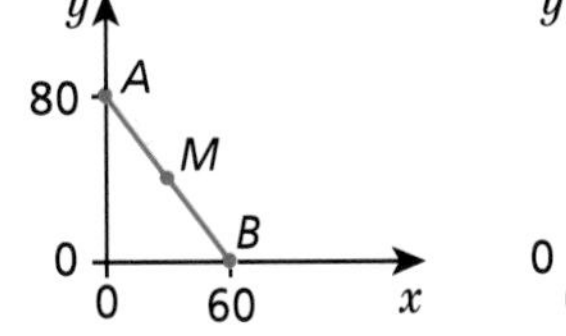

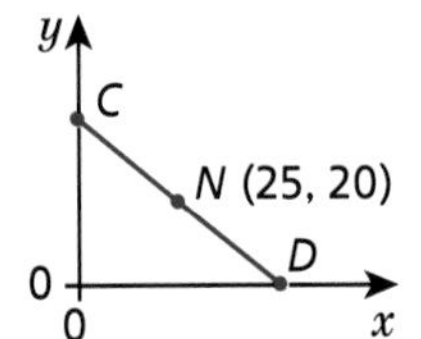

a M is the midpoint of the line AB. What are the coordinates of the point M?

b N is the midpoint of the line CD. What are the coordinates of the points C and D?

Q2 Carlos is working out the perimeter of a circle with radius 5 cm.
He writes:

Perimeter = π × 25

Explain why Carlos is wrong.

5 cm

LEVEL 7

Q3 Work out the values of x and y.

$4^9 \times 4^3 = 4^x$ $\qquad \frac{4^9}{4^3} = 4^y$

Q4 Bethan and Bryn buy marrow seeds and grow them in identical conditions.

Bethan's results:

Number of packets	Number of seeds in each packet	Total number of seeds that germinate
4	10	7, 8, 8, 6

Bryn's results:

Number of packets	Number of seeds in each packet	Total number of seeds that germinate
12	10	92

a Using Bethan's and Bryn's results, calculate two different estimates of the probability that a marrow seed will germinate.

b Whose results are likely to give the better estimate of the probability?
Explain your answer.

LEVEL 8

Q5 a In 2007, Russia produced 654 000 000 000 m³ of natural gas.
Write this number in standard form.

b In 2007, Russia produced approximately $\frac{1}{5}$ of the total world production of natural gas.
Work out the approximate total world production of natural gas.
Write your answer in standard form.

Q6 The diagram shows two shapes that are mathematically similar.

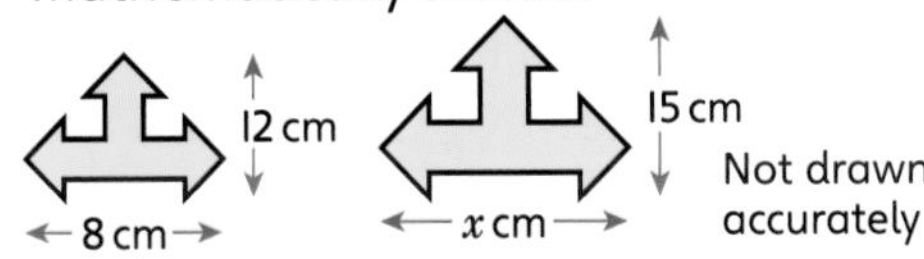

a What is the value of x?

b Sandra wants to draw a shape that is mathematically similar to these in the diagram, but a different size. Write the length and width of a shape that Sandra could draw.

11 Magic formula

This unit is about writing the same mathematics in equivalent ways. Then you can write maths in the most useful way for your purpose and do clever tricks with arithmetic.

Area and algebra have been linked together for thousands of years. This clay tablet is from the Old Babylonian period and was made around 1800–1600 BC. It shows a set of problems involving shapes, with their solutions. Some of the problems involve squares and their diagonals. Showing that two lengths or areas are equivalent is a form of proof.

Activities

A Find this diagram on the clay tablet. The Babylonians used this to show that the diagonal length of a square is the same as the side length of a square with double the area.

Draw a diagram like this on squared paper. The vertices of the inner square are at the mid-points of the sides of the outer square. Cut out the inner square. Arrange the four remaining triangles to make a square. What do you notice?

B
- Work out 14^2. Now work out 13×15.
- Work out 15^2. Now work out 14×16.
- What do you notice?
- Work out 20^2. Now predict 19×21. Check your prediction. You will find out how this arithmetic trick works in 11.2.

Did you know?
The remains of the ancient city of Babylon (pictured here) are located in modern-day Iraq.

Before you start this unit...

1 Work out
a 15^2 b $15^2 - 7^2$
c $(15 - 7)^2$

2 Work out
a $(5 + 3)(5 - 3)$
b $(13 - 5)(13 + 5)$

3 a Which one of these shapes has a different perimeter?

page 22

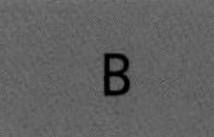

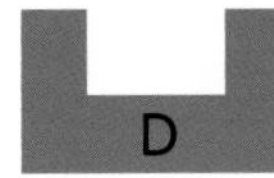

b Use colour or letters to explain why three of the diagrams have the same perimeter.
c Explain a trick for finding the perimeter of an L shape. What other shapes will this work for, and why?

11.1 Simplifying algebraic expressions
11.2 Working with double brackets
11.3 Factorising quadratics
11.4 Using and writing formulae
11.5 Working with formulae
11.6 Inequalities
maths! Algebra from the edge
Unit plenary: Accessorised with algebra

Plus digital resources

Did you know?
The Babylonians investigated the relationship between the side length of a square and the length of its diagonal, long before Pythagoras' theorem was stated.

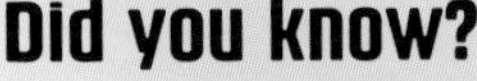

11.1 Simplifying algebraic expressions

- Simplify expressions by multiplying out single brackets and collecting like terms
- Factorise expressions involving a single bracket

Why learn this?

Working out quantities and costs for large building projects means working with many variables. You need to be able to understand different ways of writing expressions so that you can write an expression in its simplest way.

What's the BIG idea?

→ To **expand** (multiply out) a bracket, remember to multiply every term inside the bracket by every term outside the bracket.
For example, $4(p + 7) = 4p + 28$ **Level 6**

→ To simplify an expression with brackets, expand the brackets, then collect **like terms**.
For example,
$3(x - 2y + 5) - 5(3x - y) = 3x - 6y + 15 - 15x + 5y$
$= 15 - 12x - y$
Remember to look at the sign in front of each term. **Level 6 & Level 7**

→ **Factorising** and expanding are inverse (opposite) processes.
- To fully factorise an expression, take out the **highest common factor** and put it outside the bracket. When an expression is fully factorised, the terms inside the bracket have no common factor other than 1.
- To find algebraic factors, find the highest power of each letter that occurs in all the terms.
  Example 1: Fully factorise $c^3 + c^7$.
  $c^3 + c^7 = c \times c \times c + c \times c \times c \times c \times c \times c \times c$
  The highest common factor is c^3, so $c^3(1 + c^4)$ is the fully factorised form.
  Example 2: Fully factorise $8c^3y^5 + 12c^7y^2z$.
  - Find the highest numeric factor common to both terms – this is 4.
  - Find the letters that appear in both terms – these are c and y.
  - Find the highest power of each letter that occurs in both terms – these are c^3 and y^2.
  - The fully factorised form is $4c^3y^2(2y^3 + 3c^4z)$. **Level 7**

Practice, practice, practice!

1 Multiply out each bracket and collect like terms.

a $5(3x + 5y) + 5x$
b $5(3x - 7) - 8$
c $6(6a - 13b) + 7a(5 + 2a) - 7b$
d $9(11 - 4c) + 5(13 - c + d) - d$

2 Copy and complete the table.

Unfactorised	Fully factorised
$18x + 12$	$6(3x + 2)$
	$12(3x - 2)$
$25x + 35$	
$26x - 39$	

Tip
Remember that expanding and factorising are opposites. You can check that you have factorised correctly by expanding. You can also check that you have factorised fully by checking that the terms inside the bracket have no common factor other than 1.

Level 6

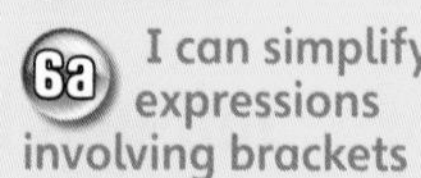

6a I can simplify expressions involving brackets

6a I can factorise an expression by taking out the common number factor

expanding factorising highest common factor

3 Factorise these expressions completely.

a $12xy + 6y^2 + 9yz$ b $15pqr + 6prs + 9prt$ c $xy^2 + 5x^2y + yz^2$

4 Expand each bracket and simplify each expression where possible.

a $-5(x + 2 - 3)$ **Hint:** $-5 \times x = -5x$

b $-7(2 - x + 3y)$

c $4(2 + 5s - 8t) - 3s(2 + t)$

d $7(10 - 3p) - 5(2p - 3)$

e $6(3x - 4) - 2(8 + 4x)$

f $3(2x + y) - 4(3y - x)$

g $p(7 - q) - q(p + 5)$

h $3a(1 - b) - a(5 - 3b)$

Watch out!

If you have a minus sign in front of a bracket, the sign of each term inside the bracket changes when the bracket is expanded. For example, $-3(x - 2y) = -3x + 6y$.

5 For each identity, find the values of a, b and c.

a $5(2x - y) - 3(x + 2y) \equiv ax + by$

b $3(5x - y + 3z) - 3(3x - 2y + z) \equiv ax + by + cz$

c $4(3x - y) -(y - 2z) \equiv ax + by + cz$

d $2(3x + 2y - z) - 6(x - 2y + 3z) \equiv ax + by + cz$

6 Factorise fully.

a $3x^5y + 9x^4y^3$ b $12a^3b^2 + 16a^2b^3$ c $12x^6y^5z^4 + 18x^4y^5z^6 + 24x^5y^4z^6$

7 Expand each bracket and simplify each expression.

a $3a^2(a^4 + b^2) - 5b^2(a^2 - 1)$ b $x^4(2x - 5) + 2x^2(7 + 3x^2 - 2x^3)$

c $7s^3(2s - t) - st(3s + 5s^2)$ d $3x^2(x - 2x^2) - 2x(x^2 - 4x^3)$

Level 7

7c I can factorise an expression by taking out an algebraic common factor

7b I can simplify expressions involving brackets and negative signs in front of the brackets

7a I can simplify expressions involving brackets and negative signs in front of the brackets to find constants in an identity

7a I can factorise using the laws of indices

7a I can simplify expressions using laws of indices

Now try this!

A Play simplifications!

Work in a small group. In 3 minutes, each write as many different expressions as you can that can be simplified to $4x + 3y$. Pool and discuss the expressions made. How can you check if they are correct?

B Factorisation rotations

Play in a group of at least five, sitting in a circle.

- Each create an expression that involves a single bracket and indices. Write it at the top of a piece of paper. Pass the paper to the person sitting on your right.
- Now each multiply out the expression you have been given, writing the result underneath. Fold over the paper so the original expression can't be seen and pass it to the person sitting on your right. They now have to factorise it.
- Keep folding the paper to hide all but the last result, and passing it to the person on your right who expands or factorises the expression as appropriate. When the paper returns to the start, that person checks the results.

identity like terms

11.2 Working with double brackets

- Multiply out double brackets
- Use the difference of two squares to do mental calculations

What's the BIG idea?

Why learn this?

You can use algebra to work out how mental arithmetic 'tricks' work.

- Diagrams can help you to understand which terms to multiply in an expression with double brackets.

  For example, this is a diagram for the expression $(x + 3)(x + 4)$.

  Total area $= (x + 3)(x + 4)$
  $= x^2 + 4x + 3x + 12$
  $= x^2 + 7x + 12$

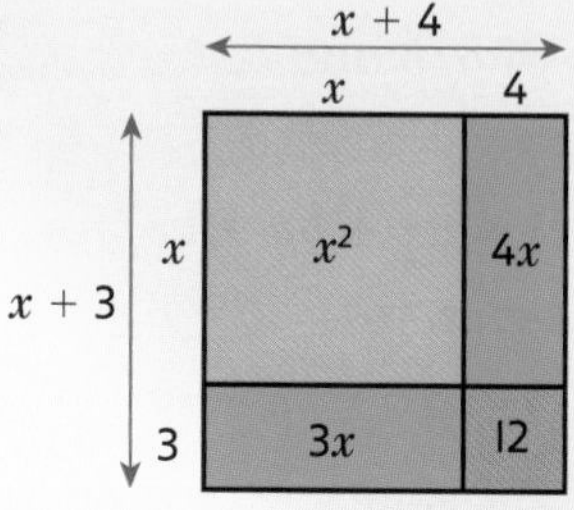

Tip

You can work out the total area from the four separate areas. Remember to simplify by collecting like terms.

  When you multiply expressions with negative signs, remember to look at the sign in front of each term. **Level 7 & Level 8**

- To **square** an expression, multiply it by itself.

  For example, $(x + 3)^2 = (x + 3)(x + 3)$
  $= x^2 + 3x + 3x + 9$
  $= x^2 + 6x + 9$

  $(2y - 3)^2 = (2y - 3)(2y - 3)$
  $= 4y^2 - 6y - 6y + 9$
  $= 4y^2 - 12y + 9$ **Level 8**

- The **identity** $(x + y)^2 \equiv (x + y)(x + y) \equiv x^2 + 2xy + y^2$ is called a perfect square. **Level 8**

- The identity $(x + y)(x - y) \equiv x^2 - y^2$ is called the difference of two squares. You can use it to simplify algebraic expressions and to help do mental arithmetic.

  For example, $67 \times 73 = (70 - 3)(70 + 3) = 70^2 - 3^2 = 4900 - 9 = 4891$

  Level 8

Practice, practice, practice!

1 **a** Write down the side lengths of the blue rectangle.

b Write down the area of the blue rectangle using brackets.

c Write down the area of the blue rectangle without using brackets.

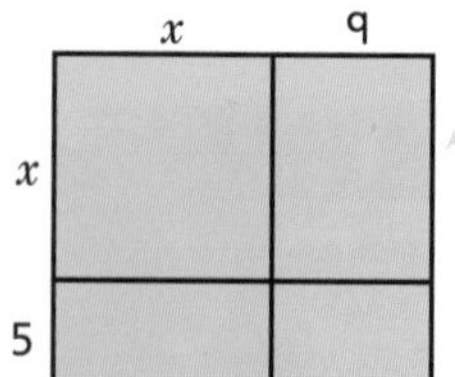

Level 7

7a I can link finding the area of a rectangle to expanding brackets

2 Multiply out the brackets for each expression and then simplify.

a $(x + 2)(x + 4)$ **b** $(x + 3)(x + 5)$

c $(x + 3)(x - 5)$ **d** $(x - 3)(x - 5)$

Level 8

8c I can multiply out brackets and collect like terms

identity

3 Multiply out the brackets for each expression and then simplify.

a $(x + 4)(x + 4)$ b $(2x + 15)(2x + 15)$ c $(x - 7)(x - 7)$
d $(3x - 7)(3x - 7)$ e $(x + y)(x + y)$ f $(x - z)^2$

4 Multiply out the brackets for each expression and then simplify.

a $(x + 8)(x - 8)$ b $(x + 11)(x - 11)$ c $(x + y)(x - y)$

5 a Multiply out $(10x + 5)(10x + 5)$.

b Write your result from part a in the form $ax(x + b) + c$, where a, b and c are positive whole numbers. State the values of a, b and c.

c Substitute $x = 4$ into the expressions in parts a and b.

d Work out 65^2 in your head.

6 a Multiply out $(a + b)(a - b)$.

b Use your result from part a to help you to work out $(a + b)(a - b)$, when $a = 25$ and $b = 2$.

c 43×37 can be written in the form $(a + b)(a - b)$.
Write down the values of a and b.

d Use your result from part a to help you to work out 43×37.

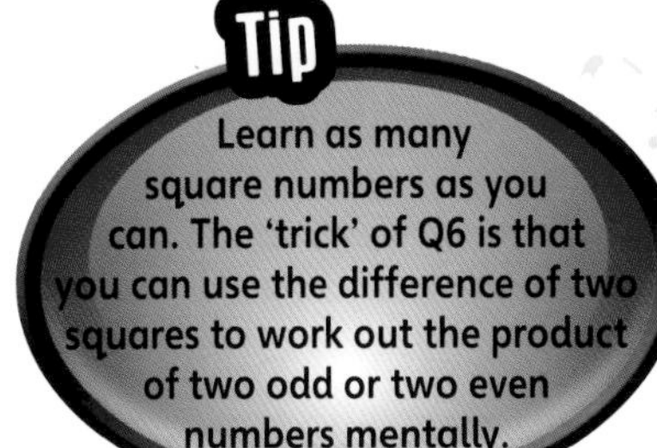

Tip
Learn as many square numbers as you can. The 'trick' of Q6 is that you can use the difference of two squares to work out the product of two odd or two even numbers mentally.

7 Use the techniques from Q5 and Q6 to work out each of these multiplications mentally or using rough jottings.

a 68×72 b 75^2 c 73×77
d 75×85 e 12.5^2 f 8.2×8.8

8 Multiply out the brackets in these expressions and simplify.

a $\dfrac{(x + y)^2 - (x - y)^2}{4}$ b $\dfrac{(s + t)^2 - (s - t)(s + t)}{2}$

Level 8

8b I can square a linear expression and collect like terms

8b I can multiply out brackets that are in the form $(a + b)(a - b)$

8a I can square a linear expression and collect like terms in a proof/problem-solving context

8a I can derive and use the difference of two squares

8a I can use identities to solve arithmetic problems mentally

8+ I can simplify complex algebraic expressions

Now try this!

A Rectangular areas

Copy the diagram.

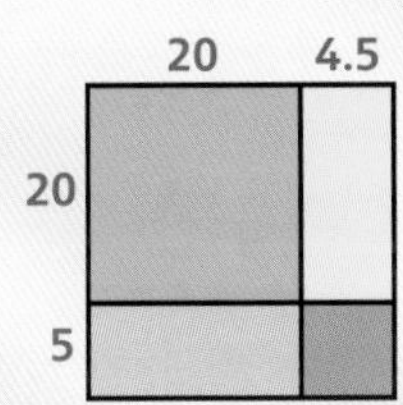

Work out the total area. Now change each number in red to another number, or a letter, and work out the area. Swap diagrams with a partner. Work out the area of each other's diagrams and check your results.

B Keep it in your head!

Work with a partner, each using a mini whiteboard. Agree on an amount of time to make up a multiplication which can be worked out mentally using the difference of two squares. Write the question on one side of your board and the answer on the back. Swap boards with your partner, question side up. Answer each other's questions and turn the boards over. Check your answer using a calculator. Score 1 point for the correct answer. If the 'author' of the question gave the incorrect answer, they lose 1 point.

square

11.3 Factorising quadratics

- Factorise a quadratic expression
- Use factorisation to simplify algebraic fractions

What's the BIG idea?

- **Factorising** is the reverse of **expanding**.
  For example, $(x + 1)(x + 2) = x^2 + 3x + 2$, so $x^2 + 3x + 2$ factorises to $(x + 1)(x + 2)$. **Level 7 & Level 8**
- Factorising and expanding change the form of an expression but not its meaning. For example, the area of this pattern can be expressed as

$$(x + 1)^2 = x^2 + 2x + 1$$

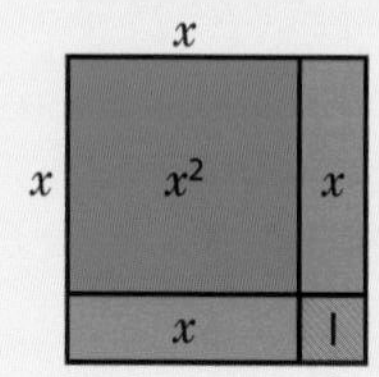

Level 7 & Level 8

- When you are factorising an expression:
  - Look for perfect squares, $(x + a)^2 = (x + a)(x + a) = x^2 + 2ax + a^2$
    For example, $x^2 + 12x + 36 = (x + 6)^2$
  - Look for the difference of two squares $x^2 - a^2 = (x + a)(x - a)$
    For example, $x^2 - 36 = (x + 6)(x - 6)$
  - Look for two numbers which add together to give the number in front of x and which multiply together to give the constant at the end. These numbers go in the brackets.
    First find **factor** pairs of the number at the end. Then choose the pair that add together to give the number in front of x.
    For example,
    $4 + 7$ 4×7
    $x^2 + 11x + 28 = (x + 4)(x + 7)$
    Expand to check: $(x + 4)(x + 7) = x^2 + 4x + 7x + 4 \times 7$
  - Watch out for negative signs.
    For example,
    $-4 + 9$ -4×9
    $x^2 + 5x - 36 = (x - 4)(x + 9)$
    Expand to check: $(x - 4)(x + 9) = x^2 + -4x + 9x + -4 \times 9$ **Level 8**

Why learn this?

You could describe the area of this pattern using a factorised or unfactorised expression. Being able to write expressions in more than one way helps you to work with them in the easiest way.

Watch out!

$x^2 - a^2$ factorises as $(x - a)(x + a)$, but $x^2 + a^2$ does *not* factorise.

Super fact!

St Stephen's chapel, in the Palace of Westminister, had the earliest known English example of a tiled floor. It was laid in 1237. The chapel was largely destroyed by fire in 1834.

Practice, practice, practice!

1 **a** Write the area of the rectangle without using brackets.
b Write down the length and width of the rectangle.
c Write the area of the rectangle using brackets.

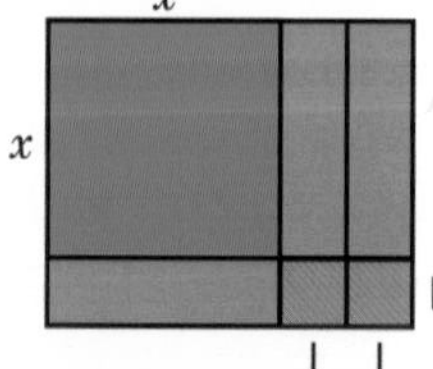

Level 7

7a I can use a diagram to help write down the unfactorised and factorised forms of a quadratic expression

expanding   factor

2 Copy and complete these equations.

a $x^2 + 6x = x(x + \square)$

b $x^2 + 6x + 5 = (x + 1)(x + \square)$

c $x^2 + 6x + 8 = (x + 2)(x + \square)$

d $x^2 + 6x + 9 = (x + \square)(x + \square)$

e $x^2 + 13x + 12 = (x + 1)(x + \square)$

f $x^2 + 8x + 12 = (x + \square)(x + \square)$

g $x^2 + 7x + 12 = (x + \square)(x + \square)$

h $x^2 + 9x + 14 = (x + \square)(x + \square)$

3 Factorise each of these expressions.

a $x^2 + 8x + 15$ b $x^2 + 8x + 16$ c $x^2 + 10x + 21$ d $x^2 + 10x + 25$

4 Each of these expressions can be written in the form $(x + a)^2$, where a is a whole number that is either positive or negative. Find a for each expression. Multiply out the brackets to check your answer.

a $x^2 + 10x + 25$ b $x^2 + 20x + 100$

c $x^2 - 8x + 16$ d $x^2 - 26x + 169$

Tip

To factorise a quadratic expression, check first whether it is the difference of two squares or a perfect square.

5 Each of these expressions can be written in the form $(x + a)(x - a)$, where a is a positive whole number. Find a for each expression.

a $x^2 - 9$ b $x^2 - 64$ c $x^2 - 144$ d $x^2 - 400$

6 Copy and complete these equations.

a $x^2 + 3x - 10 = (x + 5)(x - \square)$

b $x^2 - 3x - 10 = (x + \square)(x - \square)$

c $x^2 + x - 12 = (x + \square)(x - \square)$

d $x^2 - 10x + 16 = (x - \square)(x - \square)$

7 This fraction is made up of two different quadratic expressions.

$$\frac{x^2 + 3x}{x^2 + 5x + 6}$$

a Factorise the top and bottom of the fraction separately.

b Simplify by dividing the top and bottom by the common factor.

Level 8

8c I can begin to factorise a simple quadratic expression

8b I can begin to factorise a simple quadratic expression of the form $x^2 + bx + c$, where b and c are positive

8a I can factorise a perfect square

8a I can factorise the difference of two squares

8a I can factorise a quadratic expression of the form $x^2 + bx + c$, where b and c can be negative

8+ I can use factorisation to simplify fractions

Now try this!

A Template factorisation

Use templates (Resource sheet 11.3) to build a rectangle. Write down the side lengths of your rectangle. Write down the area of each piece and use this to help you find the total area. Swap with a partner. Build a rectangle from each other's pieces. Write down the side lengths.

B The memory factor!

Work in groups of six. You each need two blank cards. On one write an unfactorised expression and on the other write its factorised form. Mix up the cards and spread them out face down. Take turns to pick two cards. Show them. If they are factorised and unfactorised forms of the same expression keep them; otherwise return them. The winner is the player with the most cards when none are left on the table.

factorising   quadratic expression

11.4 Using and writing formulae

2.10

- Substitute integers into formulae and use the correct order of operations
- Write expressions and construct formulae
- Use a formula and find the value of a letter which is not the subject of a formula

Why learn this?

Formulae let us find out all sorts of information using methods that others have developed. For example, you can work out the capacity of a storage container.

What's the BIG idea?

→ To use a formula, **substitute** in the values of the letters you are given. Make sure you use the correct **order of operations**. Level 6 & Level 7

→ To write a formula, you need to **generalise**.

- Start by writing down the calculation for one or two examples using numbers. You don't need to work out the final answer.
- Work out which numbers always stay the same. Leave these numbers as they are. Replace each number which can change with a letter or expression.
- Don't use more letters than you need. So, if the length of a rectangle is double its width w, then the length is $2w$.

For example, this window is made up of a rectangle and a semicircle. The rectangle's height is two thirds its width.

To find its area …

first work with numbers.	Then work algebraically.
When width = 7, then height = $\frac{2}{3} \times 7$	Width = w, so height = $\frac{2}{3}w$
Area = $7 \times \frac{2}{3} \times 7 + \frac{1}{2} \times \pi \times \left(\frac{1}{2} \times 7\right)^2$	Area = $w \times \frac{2}{3} \times w + \frac{1}{2} \times \pi \times \left(\frac{1}{2} \times w\right)^2$
$= \frac{2}{3} \times 7^2 + \frac{1}{2}\pi \times \frac{1}{4} \times 7^2$	$= \frac{2}{3} \times w^2 + \pi \times \frac{1}{8} \times w^2$
$= 7^2\left(\frac{2}{3} + \frac{\pi}{8}\right)$	$= w^2\left(\frac{2}{3} + \frac{\pi}{8}\right)$

Level 7

→ If the unknown value is not the **subject** of the formula, substitute the known values as before, then solve the equation to find the unknown value. Level 7

Learn this

Brackets
Indices
Division and
Multiplication
Addition and
Subtraction

Practice, practice, practice!

Level 6

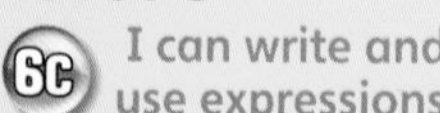

6C I can write and use expressions

1 A rectangle has a width of x cm. Its length is 3 cm more than its width.

a The width is 10 cm. Find the area.

b The width is 100 cm. Find the area.

c Write down an expression for the rectangle's length.

d Write down an expression for its area.

e Find the area when the length is 10 cm.

generalise · in terms of · order of operations

2 A mail-order company charges S for a pair of standard tights, L for a pair of luxury tights and P for postage and packing. Write a formula for the price C of x pairs of standard tights and y pairs of luxury tights including postage and packing.

3 A laundry charges £s for cleaning and ironing one shirt and £j for dry cleaning a jacket. Delivery is charged at £3 plus £1 per mile.
Write a formula for the cost C for laundering p shirts and q jackets, delivered to an address d miles from the laundry.

4 The volume V of a sphere is given by the formula $V = \frac{4}{3}\pi r^3$, where r is its radius.

a Find the volume when $r = 3$ cm.

b Find the volume when $r = 6$ cm.

Give your answers in terms of π.

Level 6

6a I can write formulae involving several different variables

6a I can substitute positive integers into algebraic formulae involving algebraic terms with small powers

Watch out!

Take care when you substitute a negative value into a formula. For example, when $x = 4$ and $y = -3$,
$4x + 5y = 4 \times 4 + 5 \times -3 = 16 - 15 = 1$.
Don't forget that you must multiply by the negative value.

5 A straight line has the equation $3x + 5y = 30$.

a Find x, when $y = -3$.

b Find y, when $x = 25$.

c Find out whether the point (18, −4) is on the line. Show your working.

6 The surface area S of a cylinder is given by the formula $S = 2\pi r(r + h)$, where r is its radius and h is its height.

a Work out the surface area when $r = 5$ cm and $h = 12$ cm.

b Work out the height if $S = 192$ cm^2 and $r = 3$ cm.

c Work out the radius to the nearest centimetre if the cylinder has a height of 7 cm and a surface area of 1000 cm^2.

7 The diagram shows a semicircle.

a Write a formula for its perimeter P in terms of its radius r.

b Write a formula for its area A in terms of its radius r.

c If the area is 50 cm^2, use both your formulae to find the perimeter.

8 The volume V of a cone is given by $V = \frac{1}{3}\pi r^2 h$, where r is the radius of the cone at the base and h is its height.

a Find the volume of a cone with $r = 5$ cm and $h = 18$ cm.

b Find the radius of a cone with a capacity of $\frac{1}{2}$ litre and a height of 25 cm.

Tip

1000 ml = 1 litre;
1 ml occupies 1 cm^3.

Level 7

7b I can substitute integers into a formula to form an equation to solve

7b I can write formulae and find any variable given values to substitute for the others by solving an equation

7a I can solve problems involving substituting integers into a formula to form an equation to solve

Now try this!

A VAT mark up

The standard rate of VAT is 17.5%. Write a formula to find the price of something including VAT given its price excluding VAT.

Tip

What do you multiply the original amount by?

B What's inside?

Write down a formula for the area of a triangle in the diagram. Give your own values to a and b. Work out the area of the middle square. Repeat the process in algebra, using the lengths a and b as on the diagram. What is the side length c of the middle square in terms of a and b?

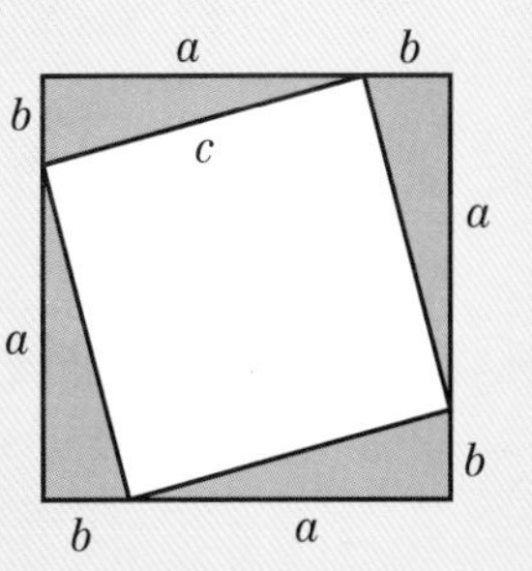

subject substitute

11.5 Working with formulae

➪ Change the subject of a formula

Why learn this?

A formula lets you work out one variable from others. If you want the formula written in terms of a different variable, you need to rearrange it. You could rearrange the formula that describes one dimension of a building to help work out another dimension.

What's the BIG idea?

→ If you apply the same operation to both sides of a formula, you get an **equivalent** formula.

For example,

	$3x + 6y = 21$		$3x + y = z$
Subtract $6y$ from both sides	$3x = 21 - 6y$	Subtract y from both sides	$3x = z - y$
Divide both sides by 3	$x = 7 - 2y$	Divide both sides by 3	$x = \frac{z - y}{3}$
Level 7		**Level 8**	

→ If the letter you want to make the **subject** of a formula appears in more than one **term**, you need all those terms to be on the same side. You then need to factorise.

For example, making x the subject of these formulae

	$4x - 2y = 12$		$ax + 3 = bx + c$
Add $2y$ to both sides	$4x = 12 + 2y$	Subtract bx from both sides	$ax - bx + 3 = c$
Divide both sides by 4	$x = 3 + \frac{y}{2}$	Subtract 3 from both sides	$ax - bx = c - 3$
		Factorise	$x(a - b) = c - 3$
		Divide both sides by $a - b$	$x = \frac{c - 3}{a - b}$
Level 7		**Level 8**	

→ More complex examples may involve fractions or taking the square root of both sides. **Level 7 & Level 8**

Practice, practice, practice!

Level 7

7b I can complete steps to rearrange a simple formula

1 **a** Solve the equation $3x + 4 = 25$. Write what you did to both sides to solve it.

b Rearrange this formula by copying and completing.

$3x + 4 = y$

$3x =$ ______

$x =$ ______

Super fact!

The Millennium Bridge in London was closed two days after it opened in 2000 because of excessive wobbling. Test results revealed a relationship between the movement of the bridge (V) and the force on the bridge by pedestrians walking across it (F). The formula $F = k \times V$ was derived where k is a constant.

2 **a** Solve the equation $5x + 7 = 2x + 22$.

b Rearrange this formula by copying and completing.

$6x + 7 = 4x + y$

$2x + 7 =$ ______

$2x =$ ______

$x =$ ______

7a I can rearrange a simple formula

3 Write each of these in the form $y = mx + c$.

a $2y = 4x + 6$ **b** $3y + 12x = 15$

c $3y - 12x + 15 = 0$ **d** $4 - 4x + 2y = 0$

e $3y - x = 5x + 18$ **f** $4y + 2x = 2y - 8$

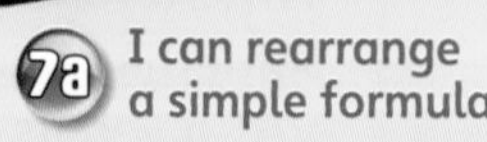

equivalent in terms of

4 Make r the subject of each formula.

a $A = \pi r^2$ b $V = \pi r^2 h$

c $V = 4\pi r^3$ d $V = \frac{4}{3}\pi r^3$

5 A sector of angle $\alpha°$ is shaded in this circle.

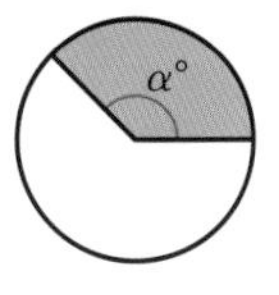

a What fraction of the circle is shaded?

b Write a formula for the shaded area A in terms of α and the radius r.

c Rewrite the formula to make r the subject.

Tip

Work out the area of a circle then multiply by the fraction that is shaded.

Level 7

7a I can change the subject of a formula involving the square or cube root

7a I can write formulae involving at least two steps

6 This formula has x on both sides. $ax + 4 = bx + c$

a Rearrange it by copying and completing the following. ______ $= c - 4$

b Factorise the left-hand side.

c Make x the subject of the formula.

7 This is the formula for the area of a trapezium.

$A = \frac{1}{2}(a + b)h$

a Make h the subject of the formula.

b Make a the subject of the formula.

8 This is the equation of a circle with its centre at (0, 0) and radius r.

$x^2 + y^2 = r^2$

a A circle has a radius of 5 units.
Find both possibilities for y when $x = -3$.

b Make x the subject of the formula.

9 Make x the subject of each formula.

a $3x + \frac{y}{4} = z$ b $3x + y = \frac{xz}{2}$

c $\frac{1}{x} = y + 4$ d $(x + a)^2 = \frac{b}{c}$

e $\frac{1}{x} + 3 = \frac{1}{y}$ f $\frac{2x^2}{3} + 4y = \pi^5$

Watch out!

If you take the square root of both sides, the square root sign goes over everything. Think carefully whether to put put ± with it or not; would a negative value make sense? (A radius, for instance, cannot be negative.)

Level 8

8c I can use factorisation to make a given letter the subject of a formula

8c I can change the subject of a formula which includes a fraction

8c I can change the subject of a more complex formula that involves the square root

8a I can change the subject of a complex formula that involves fractions

Now try this!

A Tall triangles

Sketch two right-angled triangles where the base is half of the height. Work out their areas. Write a formula for the area A of the triangle in terms of its base b. Rearrange your formula to make b the subject. Check by substitution that your formula works for the two triangles you sketched.

B Outside the inner circle

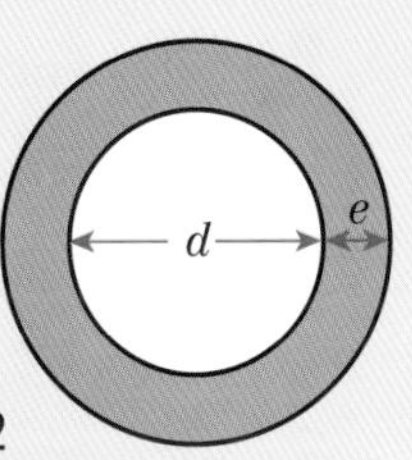

This ring is made of two circles which have the same centre. Write a formula for the ring's area A in terms of d and e. Write your formula in its simplest form. Check that your formula gives 45π, when $d = 12$ and $e = 3$. Rearrange your formula to make d the subject. Check it with the values you used before.

Tip

Begin by writing an expression for the radius of each circle.

subject term

11.6 Inequalities

- Understand inequality signs
- Identify and represent inequalities on a number line
- Solve inequalities
- Show inequalities graphically and link to practical problems

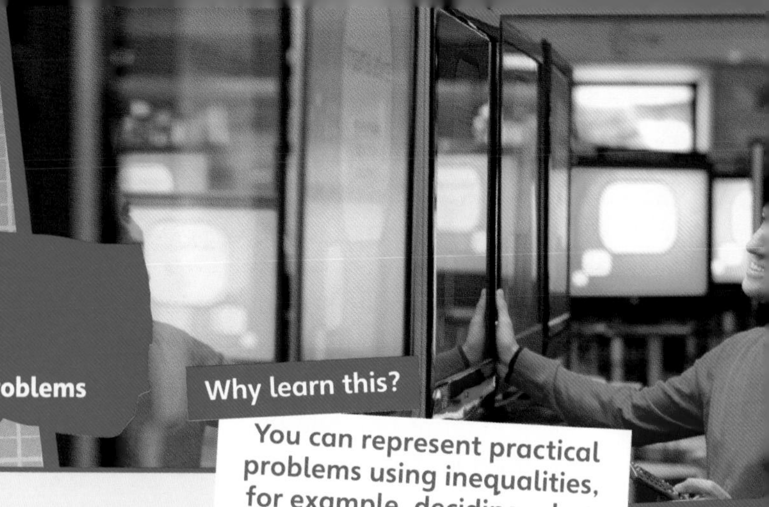

Why learn this?

You can represent practical problems using inequalities, for example, deciding what to sell to maximise profits given certain constraints.

What's the BIG idea?

→ You can show **inequalities** on a number line.
For example,

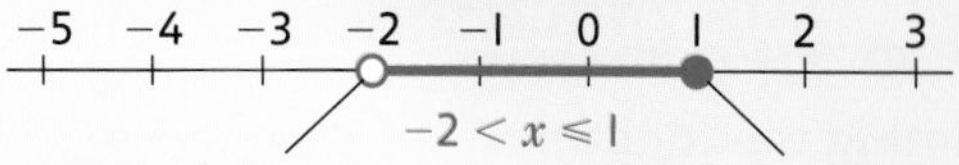

an open circle means the number at the circle is *not* included

a closed circle means the number *is* included **Level 7**

→ You can solve an inequality in one variable in almost the same way as solving an equation – by doing the same to both sides – but if you multiply or divide by a negative number, you must reverse the inequality sign. **Level 7 & Level 8**

→ You can show an inequality in two variables as a shaded region on a graph. On this graph of $y = 3$ and $y = 2x - 3$,

- the blue and green areas show $y \leqslant 3$
- the yellow and green areas show $y > 2x - 3$
- the green area shows $y \leqslant 3$ and $y > 2x - 3$
- the dotted red line indicates that the pairs on the line are not incuded. **Level 8**

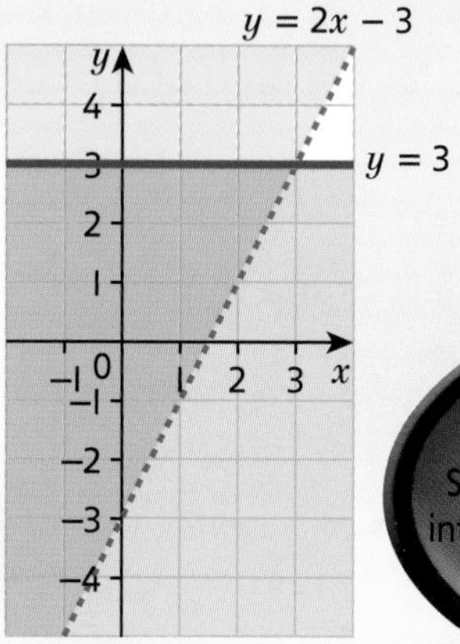

Learn this

$<$ less than
$\leqslant$ less than or equal to
$>$ greater than
$\geqslant$ greater than or equal to

Tip

To check which side of a line to shade on a graph, choose a point that is *not* on the line. Substitute the x and y values of this point into your inequality. If the inequality is true, shade the side of the line where the point lies, otherwise shade the other side.

Practice, practice, practice!

Level 7

7c I can identify and show inequalities on number lines

1 Copy this number line.

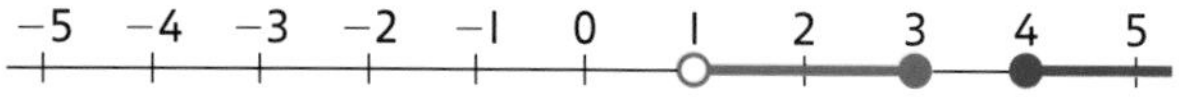

a Write down the inequality shown by the blue line.
b Write down the inequality shown by the red line.
c Draw a number line to show the inequality $x < -4$.
d Draw a number line to show the inequality $-3 \leqslant x < 0$.

Tip

An open circle means the number at the circle is not included.

7c I can solve inequalities and represent the solution on a number line

2 Solve each inequality and show your solution on a number line.

$2x + 1 > 9$ $2x > 8$, so $x > 4$

a $3x - 5 \leqslant 22$ **b** $8x - 1 < 3$ **c** $-6 < 2x < -4$
d $x^2 < 16$ **e** $x^2 \leqslant 36$ **f** $3 \leqslant \frac{1}{2}x < 5$

7b I can find integer solutions satisfying an inequality

3 Write down all the integer solutions to each of these inequalities.

a $-3 < x \leqslant 2$ **b** $-0.9 \leqslant x < 1$ **c** $-7.5 < x \leqslant -6$
d $-6 \leqslant 2x < 3$ **e** $3x + 2 < 11$ and $2x - 1 > 1$

7a I can represent an inequality on a graph

4 Draw the line $y = 2x + 1$. Shade the region $y \geqslant 2x + 1$.

greater than   greater than or equal to   inequality

5 a Write down the equation of each coloured line.

b Write down three inequalities which are satisfied in the shaded region.

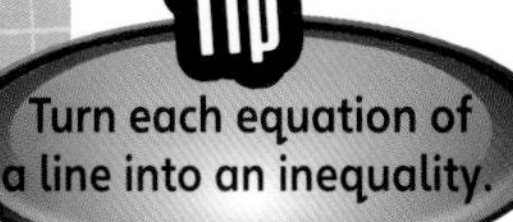

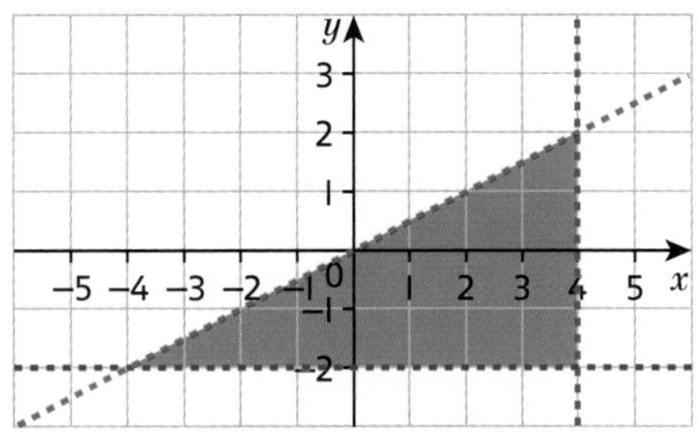

6 a Write an inequality to express the distance, x, from the centre of the circles to any point in the purple ring.

b Represent your solution on a number line.

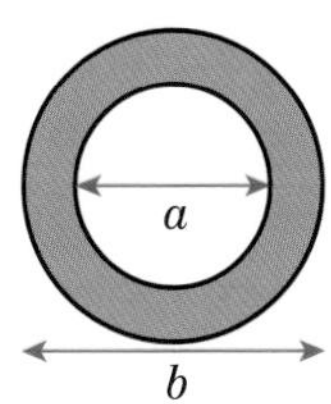

7 Show on a graph the region enclosed by the inequalities.

a $x + y > 4$, $y < x$ b $x + y \leqslant 7$, $y \geqslant 2$ c $y \leqslant 2x$, $3x + 2y < 6$

8 a Sketch $y = x^2$ and $y = x$ on the same graph.

b Shade the area of your graph which satisfies $y \geqslant x^2$ and $y \leqslant x$.

c Show the values of x within the shaded region on a number line and express this as an inequality.

9 a Write down the equation of each coloured line.

b Write down three inequalities to describe the shaded region.

c Write down a statement for each of your inequalities to express what is true in terms of the number of boys and the number of girls that can be invited to the party.

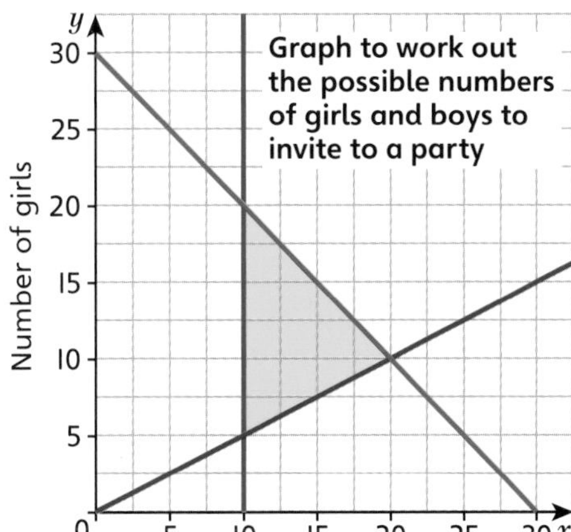

Level 8

8c I can identify a set of inequalities which define a shaded region on a graph

8c I can use inequalities to express practical relationships and represent the solution on a number line

8c I can identify the region on a graph which satisfies inequalities in two variables

8b I can solve more complex inequalities in two variables by using linear and quadratic graphs

8a I can link the solution to practical problems to the graphical representation of inequalities in two variables

Now try this!

A Where does it lie?

Use a mini whiteboard. Show an inequality on a number line. On the reverse, write the inequality. Swap boards with a partner and challenge them to work out what's on the other side of the board. Check and discuss.

B Refreshment planner for your party

Plan a party with a partner. Agree on the maximum number of bottles of lemonade and cola to buy. Plot on a graph (e.g. $x + y = 20$, where x is the number of bottles of lemonade, y the number of bottles of cola, and the maximum number of bottles is 20). Agree on two more conditions. Plot each as a straight line, shading the appropriate side of the line to represent the inequality. For example, if you decide the number of bottles of cola must be less than twice the number of bottles of lemonade, plot $y = 2x$ and shade below it. Use the graph to find different solutions to the problem.
Check each solution against your inequalities.

less than less than or equal to

maths!

Algebra from the edge

Mathematics can help us describe and understand physical situations at the edges of human experience.

Black holes

In his general theory of relativity, Albert Einstein laid out a set of equations that explained gravity as a consequence of curved spacetime. In 1916, the physicist and astronomer Karl Schwarzschild found a solution to these equations for a spherical body – but the mathematics went wild when the radius of the sphere had a particular value.

Physicists have since realised that if a sphere is compressed so that its radius is smaller than this value, known as the Schwarzschild radius, it becomes a black hole. The surface at the Schwarzschild radius is called the 'event horizon' of the black hole – the point beyond which nothing, not even light, can escape from the black hole's gravitational pull.

The Schwarzschild radius of a body with a mass of m kg is given by the formula

$$r = \frac{2Gm}{c^2}$$

where c, the speed of light, is 3×10^8 m/s and G is a constant equal to 6.67×10^{-11} Nm^2/kg^2.

- The mass of the Sun is approximately 1.99×10^{30} kg.
  What radius would the Sun have to collapse down to, in order to produce a black hole?
  Give your answer to the nearest km.
- The mass of the Earth is approximately 5.97×10^{24} kg.
  What radius would the Earth have to collapse down to, in order to produce a black hole?
  Choose an appropriate unit for your answer.
- Rearrange the formula for the Schwarzschild radius to make
  a m
  b c
  the subject of the formula.

Time dilation

One of the surprising results of the general theory of relativity is that time itself seems to '**stretch**' for observers travelling at incredibly high speeds.

Imagine two events happening t seconds apart at the same place on Earth. A spaceship travelling at v m/s will see the events happening T seconds apart, where

$$T = \frac{t}{\sqrt{1 - \frac{v^2}{c^2}}}$$

and c, the speed of light, is 3×10^8 m/s.

- An observer on a spaceship travelling at $0.5c$ sees two events that take place 10 seconds apart on Earth. How far apart in time do they appear to take place as seen by this observer?
- An observer on a spaceship travelling at $0.9c$ sees two events that take place 10 seconds apart on Earth. How far apart in time do they appear to take place as seen by this observer?
- Rearrange the formula to make
  a t
  b v
  the subject.
- Use your rearranged formula to find out at what speed a spaceship would need to travel so that events would appear to take twice as long as they would for a stationary observer.

Accessorised with algebra

Shapes are a good way of modelling algebra. You can write a formula for an area in terms of lengths, or write a formula for a length in terms of areas and other lengths. Seeing the relationships helps you to understand how to rearrange formulae.

1 The earrings below are made of a central square surrounded by four identical rectangles to make a larger square. The edges of the rectangles are finished with silver.

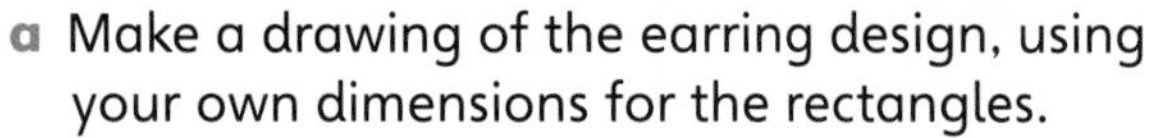

a Make a drawing of the earring design, using your own dimensions for the rectangles.

b Write down the side length and the area of the outer square and the inner square. **Level 6**

2 The rectangles in the earring design have width p and length q. Write a formula for the visible length of edging, E, needed. **Level 6**

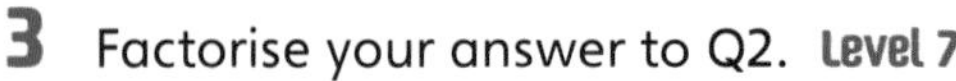

3 Factorise your answer to Q2. **Level 7**

4 Rearrange your answer to Q2 to make p the subject. **Level 7**

5 Find p when $E = 10$ cm and $q = 1$ cm. Sketch a diagram and check that your answer works. **Level 7**

6 A jewellery maker has 40 cm of silver edging material. She uses the material to produce a *pair* of the earrings. q is 2.2 cm. What is the maximum possible value of p? **Level 7**

7 Write a formula for the area of the front face of one earring in terms of p and q. **Level 7**

8 Write a formula for the area of the inner square of one earring in terms of p and q. **Level 7**

9 A different pair of earrings is based on the same design. The front of one earring has an area of $p^2 + 8p + 16$.

a Factorise this expression.

b Use your answer to find the dimensions of one rectangle in terms of p.

c The width is p. Could the area be $p^2 - 8p + 16$? Give a reason for your answer. **Level 8**

The BIG ideas

- → To use a formula, **substitute** in the values of the letters you are given. Make sure you use the correct **order of operations**. Level 6 & Level 7
- → To write a formula, you need to **generalise**. Level 7
- → If the unknown value is not the **subject** of the formula, substitute the known values as before, then solve the equation to find the unknown value. Level 7
- → If you apply the same operation to both sides of a formula, you get an **equivalent** formula. Level 7 & Level 8
- → When you are factorising an expression of the form $x^2 + bx + c$:
  - Look for perfect squares, $(x + a)^2 = (x + a)(x + a) = x^2 + 2ax + a^2$
    For example, $x^2 + 12x + 36 = (x + 6)^2$
  - Look at the difference of two squares, $x^2 - a^2 = (x + a)(x - a)$
    For example, $x^2 - 36 = (x + 6)(x - 6)$
  - Look for two numbers which add together to give the number in front of x and which multiply together to give the constant at the end. These numbers go in the brackets.

    $3 + 4 = 7$ $\quad$ $3 \times 4 = 12$

    For example, $x^2 + 7x + 12 = (x + 3)(x + 4)$ Level 8

Find your level

Level 6

Q1 Look at this equation.

$2a + 10 = 4a + b$

a If $a = 10$, what is the value of b?

b If $a = -10$, what is the value of b?

Q2 A rectangle has an area of $18x^2$ units and a perimeter of $18x$ units.
What are the dimensions of this rectangle?

Level 7

Q3 Look at these expressions.

$3a - 2 \qquad 5a + 4$

What value of a makes these two expressions equal in value?

Q4 Rearrange these equations.

a $x + 2 = y \qquad x =$ ________

b $3a = b \qquad a =$ ________

c $3(2 + s) = r \qquad s =$ ________

Q5 Julie uses thin wire to make a design based on two squares, one inside the other. The shape has rotational symmetry of order 4.

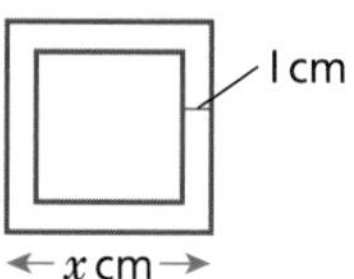

a Write a formula for the length, L cm, of wire required.

b Rearrange your formula to make x the subject.

Level 8

Q6 A square has an area of $x^2 + 10x + 25$.
Find the side length of the square.

Q7 Solve these inequalities.

a $6 - 5x \geqslant 2 - 3x$

b $4(x - 3) \geqslant 3 - x$

Q8 Difference of two squares:
$(x + a)(x - a) = x^2 - a^2$

a Write 43×37 in the form $(x + a)(x - a)$.

b Work out 43×37 using the difference of two squares.

12 No problem?

This unit is about the techniques that you can use to solve problems.

James is a race engineer for a Formula 1 team. Before going to a circuit, he decides on the set-up of the car for the race. For example, should the car have big wings (for slow corners) or small wings (for fast corners)? Other things that can be changed include the type of tyres, tyre pressures, fuel levels and gear ratios.

Before each race, the team carries out simulations in a factory to test the choices they have made. They also use mathematical models to calculate what speeds they will be able to achieve at various corners.

During the race, James is in constant contact with the driver. He needs to be able to make split-second decisions that may affect the outcome of a race. Fractions of a second are vital – the difference between first and tenth places may be as little as 0.6 seconds.

Activities

A If a corner is very tight (small radius) then the car can only go round it slowly. Match the four speeds below with the labelled sections of the track.

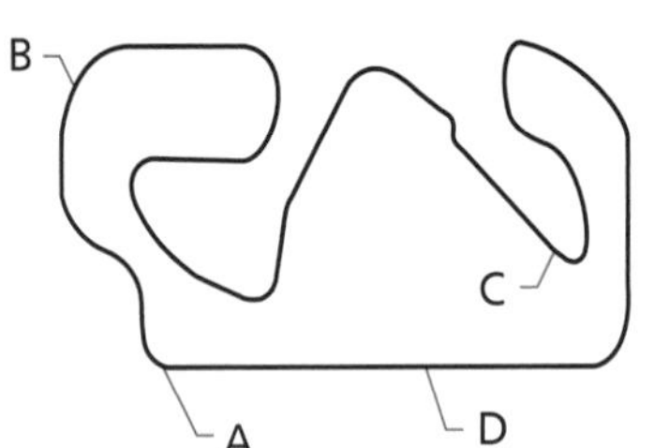

234 km/h 80 km/h 340 km/h 153 km/h

B For a racing car to go as fast as possible, it must be as light as possible.

- 1 kg of extra fuel slows a car down by 1.9 seconds over a race. A car needs 195 kg of fuel to complete the race. Suppose it takes 200 kg of fuel instead of 195 kg. How much time will it lose?
- Fuel consumption is about 54 kg per 100 km. Estimate the distance of the race.
- The race is 65 laps. How much fuel is used per lap?
- In a race, the engine's computer requests 85 333 pulses of fuel per lap. How much fuel is delivered in each pulse?

Did you know?
Every Formula 1 car on the grid is capable of going from 0 to 100 mph and back to 0 in less than 5 seconds.

Before you start this unit...

1 Draw a tree diagram to show all the different meals possible with this two-course menu.

page 170

Starters	Soup Salad
Main courses	Chicken pie Risotto Meatballs

2 Write these numbers using standard form.

page 130

a 7 654 000 **b** 0.203 458
c 507 000 **d** 1.5934

3 Solve these equations.

page 24

a $4x - 1 = 9$
b $2x + 2 = x + 7$

4 **a** Increase £200 by 25%.

page 46

b Decrease £1050 by 10%.
c Find $\frac{1}{5}$ of £470.

12.1 Data problem solving
12.2 Number problem solving
12.3 Algebra problem solving
12.4 Geometrical reasoning: lines, angles and shapes
12.5 Percentage and proportion problems
12.6 Functions and graphs
maths! Unsolved problems
Unit plenary: Cars, cars, cars!

Did you know?
Cornering speeds are so high that Formula 1 drivers have strength training routines just for their neck muscles.

12.1 Data problem solving

➩ **Understand and use the data handling cycle to solve problems**

Why learn this?

Statistical investigations can reduce waiting times.

What's the BIG idea?

→ You can use the **data** handling cycle to help with statistical problems.
- Specify and **plan**: state the aims of your investigation. This may include using a **hypothesis**.
  - What are you going to try to find out?
  - What contexts will you **investigate** within?
  - What types of data will you need?
  - How will you collect the data?
- Collect data that is relevant to your investigation.
  - How will you avoid **bias**?
- Process your data: do calculations and draw appropriate graphs and charts.
  - Do your calculations and graphs relate to the aims of your investigation?
- **Interpret** and discuss.
  - What do your graphs and calculations tell you?
  - Do they confirm your hypothesis?
- **Evaluate** and extend: after evaluating your results, you may extend the aims of your investigation. This would take you back to the beginning of the cycle.

Level 6, Level 7 & Level 8

Practice, practice, practice!

Level 6

1 Plan an investigation, for example, the differences between newspapers.

Newspapers investigation
- *What are the differences that exist between newspapers? Do these differences relate to the audience they target?*
- *Why do people choose to read a certain newspaper?*
- *Do people tend to read the same newspaper every day?*
- *Which newspapers are the most popular?*

Clearly state the aims of your investigation.

6b I can suggest a problem to explore using statistical methods

2 **a** State what data you will need to collect to test the hypothesis of the investigation you planned in Q1.
b How does the data you have chosen to collect fit the aims of your investigation?
c How much data will you need to make your conclusions valid?
d How will you record the outcomes of your data collection?
e What degree of accuracy will you need for your investigation?

6a I can plan how to collect data, determine an appropriate sample size and degree of accuracy needed

Tip
Appropriate graphs might be bar charts and pie charts. Appropriate calculations might include finding the mean, median and mode.

3 **a** Collect the data that you need for your investigation.
b Use the data to draw appropriate graphs and carry out appropriate calculations.

6a I can draw appropriate graphs and carry out appropriate calculations

bias data evaluate evidence hypothesis

4 Here are a doctor's consultation times for a week.

Consultation time (min)	0	1	2	3	4	5	6	7	8	9	10
Number of patients	0	10	15	19	15	13	10	6	4	3	2

Consultation time (min)	11	12	13	14	15	16	17	18	19	20
Number of patients	1	0	0	0	0	0	1	0	1	0

a **i** Draw a graph to show this data. **ii** Describe your graph

b Suggest a suitable appointment system for the surgery to adopt. Think about the best time for patients and doctors.

5 Morse code represents letters of the alphabet with patterns of up to four dots and dashes.

Letter	A	B	C	D	E	F	G	H	I	J	K	L	M
Code	• –	– • • •	– • – •	– • •	•	• • – •	– – •	• • • •	• •	• – – –	– • –	• – • •	– –

Letter	N	O	P	Q	R	S	T	U	V	W	X	Y	Z
Code	– •	– – –	• – – •	– – • –	• – •	• • •	–	• • –	• • • –	• – –	– • • –	– • – –	– – • •

a **i** Draw a tree diagram to find all the possible patterns with up to four dots and dashes.

ii How many possible patterns are there with up to four dots and dashes?

b **i** Dots take one unit of time to transmit and dashes take three. Do you think Morse code is an efficient code to use in terms of time?

ii Which letters' codes might you change?

Tip

Start with two branches, dot and dash. Stop when you have extended a further three branches along each route.

Did you know?

Codes can be cracked by going through books, newspapers and magazines to find how often letters appear, and then comparing the results with how often a symbol appears in a piece of code. This method is called 'frequency analysis'.

Level 7

7a I can use statistics to analyse a situation or problem

6 **a** Look at the graphs and calculations you produced in Q3. What do they tell you in relation to the aims of your investigation?

b Do you think your conclusions are reliable? Why?

c How could you make your findings more reliable?

Level 8

8c I can review my findings and discuss how reliable my evidence is

Now try this!

A Reaction times

- Work with a partner. Hold a ruler vertically from 30 cm mark. Your partner holds their hand ready by the 0 cm mark and tries to catch the ruler when you drop it. Record how far the ruler falls. Swap roles and repeat.
- Include more pupils in your investigation.
- What if someone is distracted or guesses when the ruler will be dropped? How can you avoid bias or the effect of outliers (extreme results, for example, very small or very large)?

B Capture and recapture

'Capture and recapture' is a method used to find out how big a population is. You can use it to find out how many counters there are in a bag. Begin by 'capturing' a number of counters from the bag and marking them in some way. Put them back in the bag to 'mix' with the rest of the population and then take out another sample. Use proportion and the sizes of your samples to work out the size of the whole population. Check your answer by counting all the counters in the bag.

Learn this

In statistics, the word 'population', means the set from which any statistical analysis is taken.

interpret investigate justify plan sample size

12.2 Number problem solving

- Solve complex problems involving number
- Break down substantial tasks to make them simpler
- Justify solutions to problems

Why learn this?

By being able to solve problems, we can work out how much bigger the Sun is than all the planets combined.

What's the BIG idea?

→ When solving problems, follow the same procedure each time.
- Read the question.
  - What do you need to find out?
  - Highlight the most important information.
- Which maths is relevant to the problem?
- Break down the problem into a series of steps if the task is complex.
- Carry out calculations and/or draw graphs and diagrams.
- Give your results in an appropriate form and to an appropriate degree of accuracy.
- Check your results – does your solution seem reasonable?

Level 6, Level 7 & Level 8

→ Very large and very small numbers can be written in **standard form**. For example 9.109×10^{-31} kg is read as 'nine point one zero nine times ten to the power of minus thirty-one'. Level 8

→ When a number has been rounded, there are many values it could originally have been. The **lower bound** is a limit on the lowest possible value of the original number. The **upper bound** is a limit on the highest possible value of the original number. Level 8

Super fact!

The Sun's radius is 696 000 million km, 109 times bigger than the Earth's radius.

Practice, practice, practice!

Level 6

6c I can conjecture and generalise in simple cases

6b I can identify exceptional cases or counter examples

6a I can use appropriate mathematical procedures

1 A three-digit number is multiplied by a two-digit number. What are the minimum and the maximum numbers of digits that the answer could have? Give full reasons for your answer.

2 'Squaring numbers always makes them bigger.' Is this statement true? Justify your answer.

Tip

Use a variety of numbers to test the statement.

3 Sarah likes cheesecake but to make it last she decides to limit the amount of cheesecake she eats each day. On the first day she eats $\frac{1}{10}$ of a whole cheesecake, on the second day $\frac{1}{9}$ of what is left, and so on.

a Sarah ate 150 g of cheesecake on the first day. What is the weight of the whole cheesecake?

b How many days will the cheesecake last?

counter example exceptional case generalise justify

4 A space shuttle is connected by a 6 m length of cable to a repair device.
An astronaut can shorten the connecting cable at a rate of 1.3 cm per hour.
Repairs to the shuttle can be made when the connecting cable is 1.48 m long.
After how many days can the repairs be done?

5 In the old monetary system, Great Britain used pounds and shillings and pence.
There were 20 shillings in £1.
One gallon of petrol cost 7 shillings in 1970 and it now costs £3.80. How much has the price of a gallon of petrol risen by? Give your answer as a percentage.

Level 7

7b I can break down substantial tasks to make them simpler

6 The table shows the mass of the major planets in the Solar System.
The mass of the Sun is 1.989×10^{30} kg.

a Which mass is not in standard form? Convert this mass to standard form.

b About how many times bigger than the total mass of the planets is the Sun? Show your working.

Planet	Mass (kg)
Mercury	3.3×10^{23}
Venus	4.87×10^{24}
Earth	5.98×10^{24}
Mars	6.5×10^{23}
Jupiter	1.9×10^{27}
Saturn	5.7×10^{26}
Uranus	8.7×10^{25}
Neptune	100×10^{24}

Level 8

8c I can justify solutions to problems set in an unfamiliar context

7 Ramesh is a central-heating installer. For a house on a new estate, he needs 22 m, to the nearest half metre, of copper piping. Ramesh buys 250 m of copper piping, to the nearest 10 m.

a Ramesh needs to install central heating in 11 houses. Will he have enough copper piping? Use calculations to justify your answer.

b What is the greatest length of piping that Ramesh could have left?

8a I can generate fuller solutions using reasoned argument

Learn this

If 5 m is accurate to the nearest half metre, then the lower bound is 4.75 m and the upper bound is 5.25 m.

Now try this!

A Standard form snap

Work in a group of four.
Cut out 40 playing cards and put them in 10 groups of four. On one card in each group write a number. On the others write it in standard form and two other variations of it. For example

4900	4.9×10^3	0.49×10^4	70^2

When you have filled in all the cards, shuffle, deal and play snap – a match occurs when two cards show equivalent numbers.

B Number conundrum

Work with a partner. Use standard form, square and cube numbers, roots and other calculations to write a list of numbers from smallest to largest. Think of a mathematical word of the same length as your list and assign a letter to each number. Mix your numbers and challenge your partner to put them in order, to work out what the word is. For example

U	B	E	C
$(0.1)^2$	$2.56 \div 8$	$0.9 - 3 \div 10$	1×10^{-3}

becomes

C	U	B	E
1×10^{-3}	$(0.1)^2$	$2.56 \div 8$	$0.9 - 3 \div 10$

lower bound   standard (index) form   upper bound

12.3 Algebra problem solving

- Represent problems in algebraic form
- Solve complex problems and evaluate solutions
- Justify generalisations, arguments and solutions

Why learn this?

Simultaneous equations can be used to work out how far a car has travelled at different speeds.

What's the BIG idea?

→ **Equations** are true for particular values.
For example, $3x + 7 = 13$ is only true for $x = 2$. Level 6

→ You can construct an equation from the facts that you are given. Some facts are given **explicitly**, others **implicitly**. Level 6 & Level 7

→ You can use the problem-solving process (see Lesson 12.2) for algebraic problems. Level 6, Level 7 & Level 8

→ A pair of **simultaneous equations** consists of two equations each with two unknowns usually, x and y. The solution is the pair of values that satisfies both equations. Level 8

→ To solve simultaneous equations
- If necessary, multiply one or both equations so that they have either the same number (ignoring the sign in front) of x or the same number of y.
- Add or subtract the equations to eliminate one of the variables, then solve the resulting equation to find the other variable.
- Use **substitution** to find the remaining variable. Level 8

Practice, practice, practice!

Level 6

6b I can represent mathematical problems using symbols

6a I can represent more demanding mathematical problems using symbols

Tip Start by calling the first integer x.

1 **a** The sum of two consecutive integers is 91.
Use an algebraic method to find the value of these integers.

b The sum of two consecutive even numbers is 102.
Find these numbers using an algebraic method.

2 Each face of a cube has an area of $(9x + 1)$ cm². The surface area of the cube is 249 cm².

a Write an equation connecting the area of each face and the surface area of the cube.

b Solve the equation to find x and so find the area of one face of the cube.

Level 7

7c I can solve substantial problems by breaking them down into smaller tasks

3 A cube has faces with an area of $16y^2$ cm². Four such cubes are used to make a cuboid as shown.

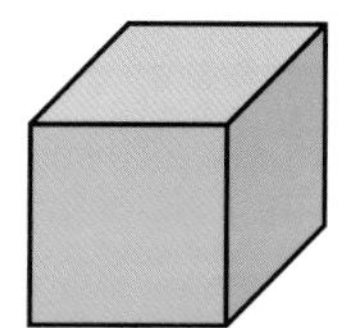

a Write an expression for the surface area of the cuboid.

b Write an expression for the side length of each cube.

c Write an expression for the volume of the cuboid.

d The actual surface area of the cuboid is 1024 cm². Write an equation using the expression you formed in part **a** and solve it to find y.

e Recombine the cubes to make a different cuboid and write down expressions for their volume and surface area. What do you notice?

algebraic expression counter example equation exceptional case

4 This rectangle has an area of 200 cm². Use an algebraic method to find the height and width of the rectangle.

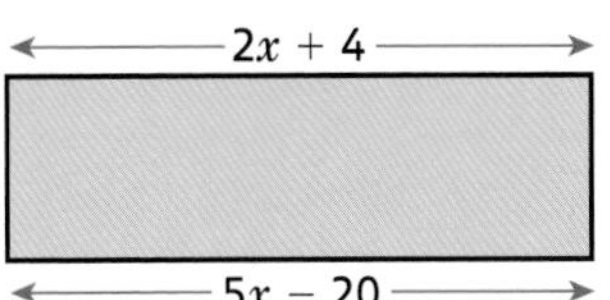

Level 7

7c I can solve substantial problems by breaking them down into smaller tasks

5 For each mathematical statement decide whether it is always true, sometimes true or never true. For each statement that is sometimes true, state under what circumstances it is true.

a $x^y \div x^z = x^{y-z}$
b $x^2 > 4$
c $(a + b)(a + c) = a^2 + ab + ac + bc$
d $x^2 = 5x$
e $x^4 = (x^2)^2$
f $x^{-3} = (-x)^3$

Level 8

8c I can identify exceptional cases or counter-examples and explain why

6 Michael and Lyla are downloading music.
Michael buys two albums and four tracks; he spends £11.
Lyla buys four albums and nine tracks; she spends £22.50.

a Form a pair of simultaneous equations and solve them to find the price of one album and the price of one track.

b Lyla likes a lot of tracks from one album. How many tracks must she like to make buying the whole album cheaper than buying separate tracks?

8b I can generate fuller solutions to increasingly demanding problems

7 A 2 by 2 box is drawn on a 10 by 10 numbered grid.
The numbers in the opposite corners are multiplied and then the difference is found.
For the box shown the calculation is $(2 \times 11) - (1 \times 12) = 10$.

1	2
11	12
21	22

a Investigate other 2 by 2 boxes on the same size grid.

b Use algebra to represent the problem and prove that the result will always be 10.

c How can the problem be extended? Investigate with a range of grid sizes and introduce a new variable where necessary.

8b I can justify generalisations, arguments or solutions and investigate whether particular cases can be generalised further

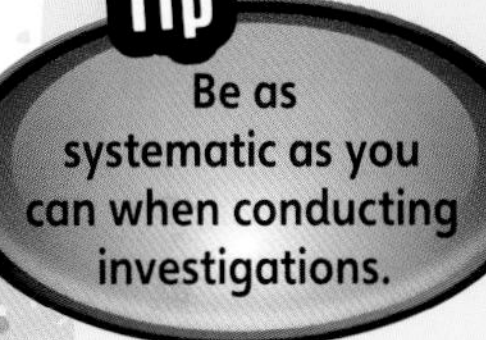

Tip
Be as systematic as you can when conducting investigations.

Now try this!

A Mobile phone packages

Investigate the range of mobile phone packages available. Model prices using formulae. Use your formulae to justify why your family and friends should use certain packages.

B Investigating L shapes

Draw some L shapes on a 10 by 10 numbered grid. Add up the numbers in each L.

Use algebra to model this situation. Start with *L* as the value of the top square in each L shape.

Now use algebra to model the situation for any grid size.

1	2	3	4	5
11	12	13	14	15
21	22	23	24	25
31	32	33	34	35

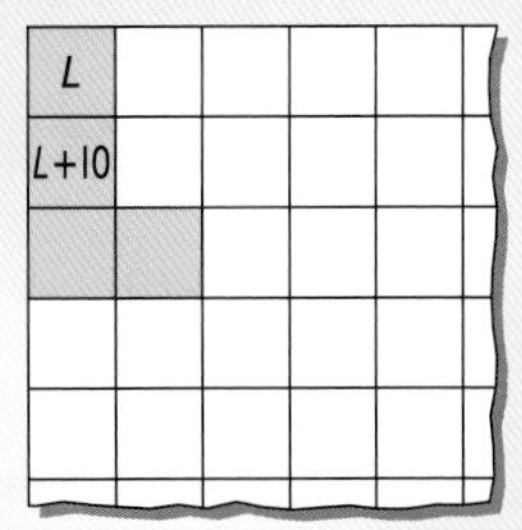

explicitly implicitly justify proof simultaneous equations substitution

12.4 Geometrical reasoning: lines, angles and shapes

- Solve problems using properties of angles and parallel and intersecting lines
- Solve problems using properties of triangles and other polygons
- Use diagrams and calculations to justify arguments and solutions

Why learn this?

Geometrical reasoning is essential for architecture – without it, the spectacular bridges and buildings in the world would not exist.

What's the BIG idea?

- Always begin with the problem-solving process (see Lesson 12.2). Level 6, Level 7 & Level 8
- **Corresponding angles** are equal. **Alternate angles** are equal. Level 6
- **Pythagoras' theorem** is $h^2 = a^2 + b^2$, where h is the length of the **hypotenuse** and a and b are the lengths of the shorter sides in a right-angled triangle. Level 7
- A **surd** is a square root that cannot be simplified further, for example $\sqrt{3}$. Level 8
- In any right-angled triangle,

$$\sin\theta = \frac{\text{opp}}{\text{hyp}} \qquad \cos\theta = \frac{\text{adj}}{\text{hyp}} \qquad \tan\theta = \frac{\text{opp}}{\text{adj}}$$

You can rearrange these formulae to find the lengths of sides in right-angled triangles when you know some sides and angles. Level 8

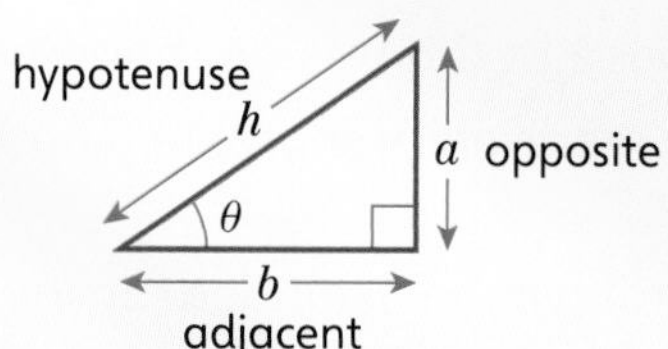

Did you know?

The corners of the Bermuda triangle – an area where many planes and boats have mysteriously disappeared – are accepted to be in Florida, Puerto Rico and Bermuda.

Practice, practice, practice!

Level 6

6b I can solve problems using angle properties of parallel and intersecting lines and isosceles triangles

6a I can solve problems using properties of angles, parallel and intersecting lines and triangles and polygons

1 Find the size of angle x.
State the angle properties you are using at every stage of your working.

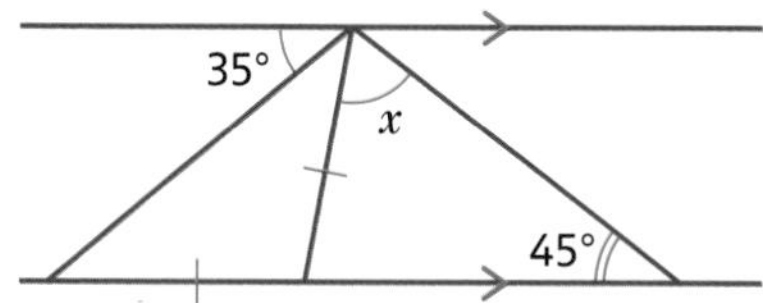

2 Use alternate angles to show that the opposite angles in a parallelogram are equal.

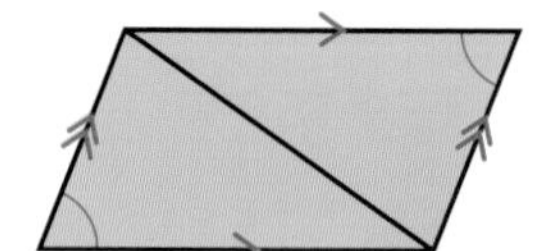

Level 7

7b I can solve substantial problems by breaking them down into smaller tasks

3 The diagram shows the plan of a theatre. The semicircle represents the stage and the square section holds the seating. Find the area of the theatre, given that the seating section has an area of 110.25 m^2.

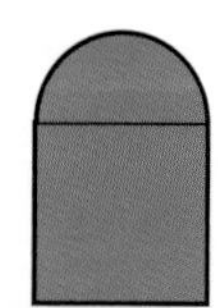

alternate angles bisect corresponding angles deduce hypotenuse

4 A 32-inch widescreen TV has a width of 27.89 inches and a diagonal of 32 inches.
Work out its viewable area in square centimetres.

Tip
1 in = 2.54 cm

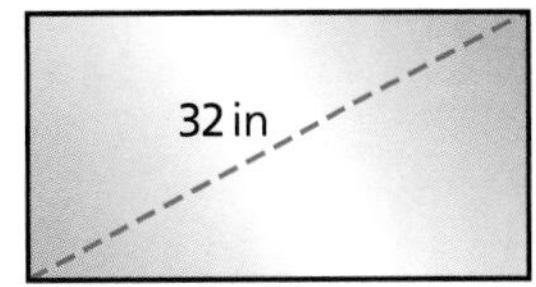

Level 7

7a I can use connections with related contexts to improve the analysis of a problem

5 This rhombus has a longer diagonal of 60 cm and an angle of 60° as marked.
Use this information to find

a the length of the diagonal connecting the other two corners

b the side length of the rhombus.

Justify each of your answers.

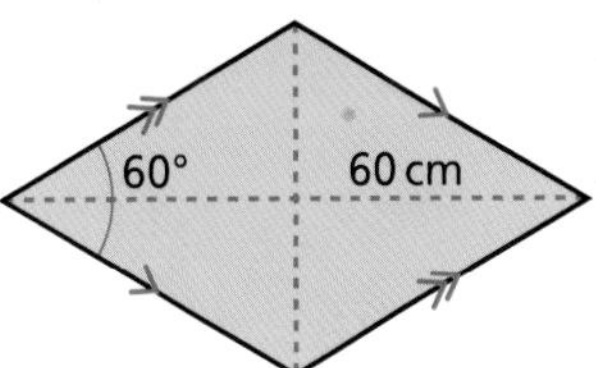

Learn this
The diagonals of a rhombus are perpendicular and bisect each other.

Level 8

8b I can generate fuller solutions by presenting a concise and reasoned argument

6 The diagram shows an equilateral triangle with sides of length $4x$ units.

a Show that the length of AB is $2\sqrt{3}x$ units.

b Give the area of the triangle in terms of x.

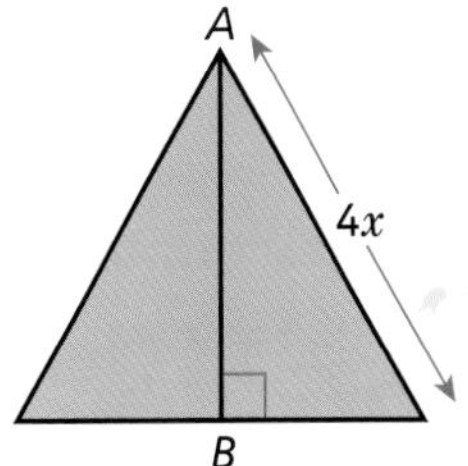

8a I can present a concise and reasoned argument using surds

c This regular hexagon has sides of length 8 cm.
Use your result from part **b** to calculate the area of the hexagon.

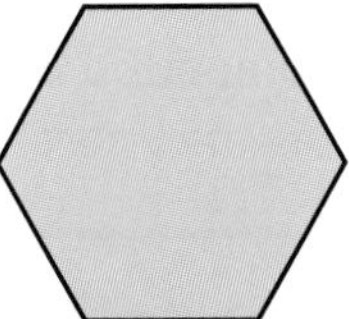

Now try this!

A Shape modelling

Make as many different shapes as you can from six multilink cubes.
Find the model with the largest … the smallest surface area.
Experiment with different numbers of cubes.

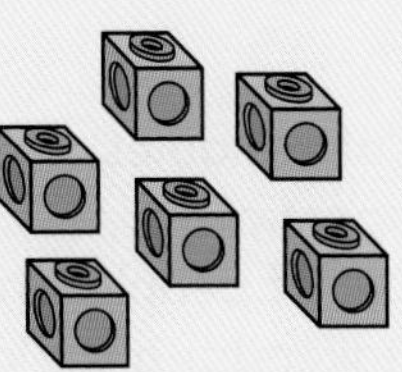

B Pythagorean triples

Pythagorean triples are sets of three whole numbers that will form a right-angled triangle, for example, 3, 4 and 5.
Pythagorean triples can be found by taking k as any number and substituting into the formulae $a = 2k + 1$, $b = 2k^2 + 2k$ and $c = b + 1$, where a, b and c are the side lengths of the triangle.
Find three or four right-angled triangles using the formulae. Represent the information on a poster entitled 'Pythagorean triples'.

justify parallel perpendicular Pythagoras' theorem surd

12.5 Percentage and proportion problems

- Identify, describe and recognise when values are in direct proportion
- Solve problems involving percentage change
- Use graphs and equations to solve problems involving direct proportion
- Work with repeated proportional change

Why learn this?

You need to be able to solve problems involving percentages or ratio to get value for money from bank accounts, mortgages and in the shops.

What's the BIG idea?

- Two **variables** are in **direct proportion** when they increase or decrease in the same **ratio**. Level 7
- Two variables in direct proportion will always produce a graph where all the points can be joined to form a straight line that goes through the origin. Level 7
- The symbol for 'is **proportional to**' is $\propto$.
  For example, if y is directly proportional to x, then you write $y \propto x$. Level 7
- If two sets of numbers are in direct **proportion**, you can use a constant k to set up a formula.
  For example, if a is proportional to b, you can say that $a = kb$. If you know a pair of values for a and b, you can substitute them into the formula to find the value of k. Level 7
- You can make solving problems involving percentage change easier by changing the percentage into a decimal multiplier.
  For example, the decimal multiplier for an increase of 25% is (100 + 25)% ÷ 100% = 1.25. Level 7
- **Compound interest** is worked out on the principal (original amount) plus any interest already earned. You can calculate compound interest by finding the multiplier and raising it to a power equal to the number of times the interest is added. Level 8

Practice, practice, practice!

Level 7

7c I can make convincing arguments to justify solutions

7b I can identify examples of direct proportion and recognise when two values are in direct proportion

1 Which offer is the better value? Give a full and clear answer.

Special offer!
50 blank CDs
£21
+ VAT (17.5%)

SALE! $\frac{1}{3}$ OFF!
Packs of ten CDs
Normally £7.50
Delivery £1 per pack

2 These pairs of variables are sometimes in proportion, always in proportion or never in proportion.
- the number of people who live in a country and its land area
- the time taken for a car journey and the distance travelled
- the temperature of a saucepan of soup and how long it has been cooking
- the number of hours a salesperson works and their earnings
- the weight of a pile of money and its value

Sort the pairs of variables into the table. For each pair you class as 'sometimes in proportion', state when they would be proportional.

Always	Never	Sometimes

compound interest direct proportion justify logic

3 Choose one of the 'sometimes in proportion' examples from the table in **Q2** and sketch a graph for it.
Use an arrow to indicate where you think the variables are in proportion.

4 A bottle of diet cola shows that the energy content of 100 m*l* is 0.4 calories.
The amount of energy e is directly proportional to the volume in m*l* of cola v.

a Find a formula connecting the energy content e to the amount of cola v.

b How much energy would you get from 230 m*l* of cola?

c How much cola would you need to drink to get 4 calories of energy?

d Use a graph to represent the information.
How can you tell that the values are in direct proportion?

e Use your graph to find out how much energy you would get from a 550 m*l* bottle of cola.

f What are the advantages and disadvantages of using a graph compared with the algebraic method used in parts **a** to **c**?

Level 7

7b I can recognise when two values are in direct proportion by using a graph

7b I can use graphical and algebraic methods to solve problems involving direct proportion

5

Savings account 1	*Savings account 2*	*Savings account 3*
Receive £100 bonus when you invest £2000 or more for a year **5.1% interest**	5.1% on the first £3000 then 6.4% on any further investment	£50 set-up fee Invest a maximum of £3600 in a year **6.4% interest**

a Which type of bank account would be suitable for each of these people.
Justify your answers with appropriate calculations and explanations.

i Dinesh has £3500 to invest; these are his only savings so he may need to withdraw them at short notice. He wants to invest for about two years.

ii Marian has £5000 to invest over a long term; she is unlikely to need to withdraw these savings without good notice.

b Review your recommendations and use this to help you make general recommendations for different types of investor.
Give a full mathematical justification for your answer.

Level 8

8a I can explore the effects of varying values and make convincing arguments to justify generalisations

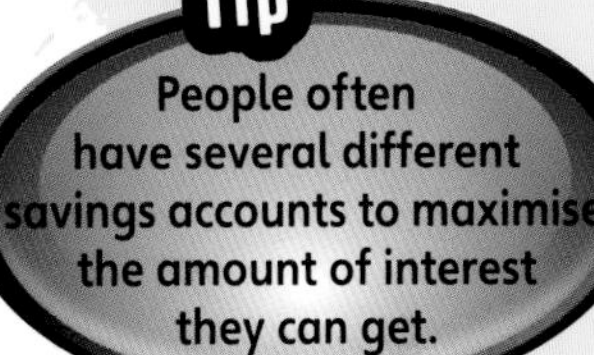

Tip
People often have several different savings accounts to maximise the amount of interest they can get.

Now try this!

A Photographs

You can enlarge or stretch a photograph on a computer. Measure the original photo and divide the height by the width. Do the same for the two images. What is the ratio for each image? Which image has been enlarged and which has been stretched?

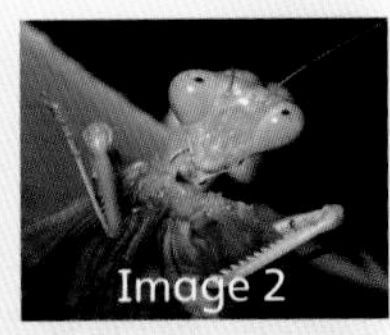

Practise making similar enlarged and stretched images of your own pictures.

B The golden face

The golden ratio is about 1.618 : 1 and is said to contain great beauty.
Look in magazines for images of different faces.
Measure these lengths and then work out the ratios.

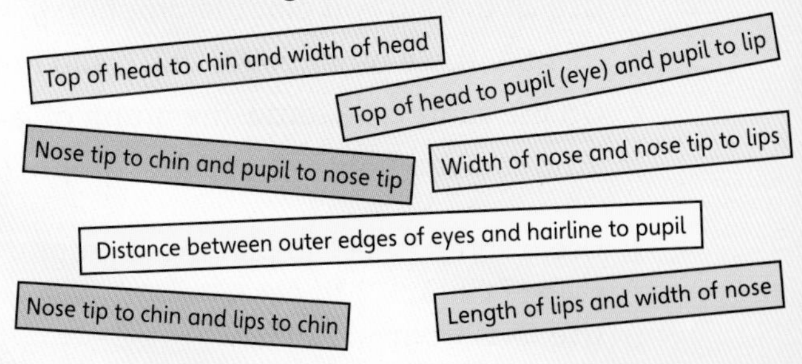

Top of head to chin and width of head
Top of head to pupil (eye) and pupil to lip
Nose tip to chin and pupil to nose tip
Width of nose and nose tip to lips
Distance between outer edges of eyes and hairline to pupil
Nose tip to chin and lips to chin
Length of lips and width of nose

Are any of your values close to the golden ratio?

proportion   proportional to (∝)   ratio   variable

12.6 Functions and graphs

- Generate points and plot the graphs of linear functions on paper and using ICT
- Construct models of real-life situations by drawing graphs and constructing algebraic equations
- Solve complex problems using graphs

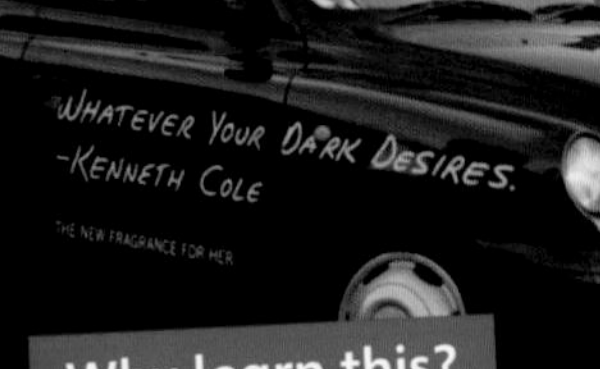

Why learn this?

You can use graphs and algebra to work out which taxi firm to use.

What's the BIG idea?

- You can generate coordinate pairs for a **linear function** by substituting numbers into its equation. **Level 6**
- A linear function has the form $y = mx + c$. Its graph has **gradient** m and it **intercepts** the y-axis at c. **Level 7**
- A linear function can be written **explicitly** in terms of x.
  For example, $y = 3x + 2$.
  Or y can be given **implicitly** in terms of x.
  For example, $y - 3x - 2 = 0$. **Level 7**
- You can use **spreadsheets** to draw graphs.
  - If necessary, rearrange your equation so that it begins '$y =$'.
  - Name two columns x and y.
  - Put some numbers in the 'x' column and write the formula in the 'y' column.
  - Select the 'graphs and charts' option and choose an XY type graph.

  Level 6, Level 7 & Level 8

	A	B
1	x	y
2	1	=A2*2+3
3	=A2+1	=A3*2+3
4	=A3+1	=A4*2+3
5	=A4+1	=A5*2+3
6	=A5+1	=A6*2+3

A	B
x	y
1	5
2	7
3	9
4	11
5	13

Watch out!

When writing formulae in a spreadsheet you must use cell references (e.g. A1) and start the formula with an '=' sign.

Practice, practice, practice!

1 **a** Copy and complete the table.

x	−3	0		2	3
$y = 2x - 2$			0	2	

b Write the values in the table as coordinate pairs.

2 **a** Use Q1 to help you plot the graph of $y = 2x - 2$.

b Identify three other points you could have used to plot the same graph.

c On the same axes, draw the graph of $y = 3x - 1$.

d What do you notice about your graphs?

3 **a** Use a spreadsheet to plot these graphs.

i $y = 7x - 3$ **ii** $y = 4 - 3x$ **iii** $y = 10 - 2x$ **iv** $y = \frac{1}{2}x - 1$

Use axes from −10 to +10.

b What are the advantages of using a spreadsheet to draw graphs?

Level 6

6c I can generate coordinate pairs for simple linear functions

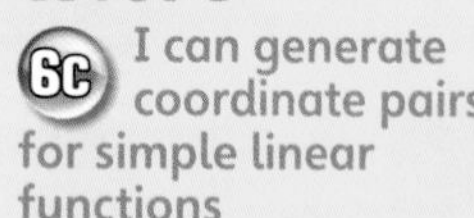

6b I can plot the graphs of simple linear functions

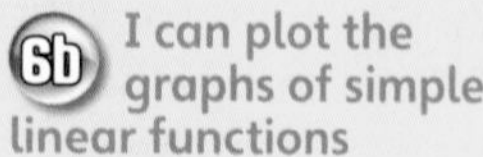

6a I can plot the graphs of linear functions using a spreadsheet

explicitly gradient implicitly intercept

4 a Use the graph to complete the table.

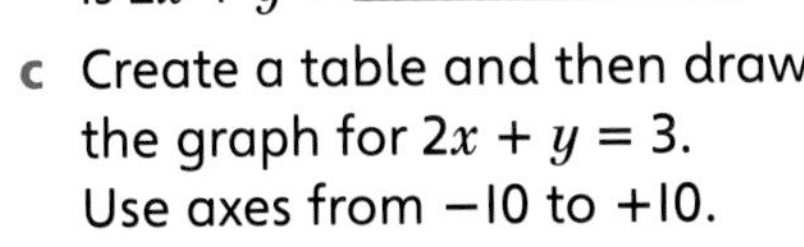

x	2	1	0	−1	−2
y	1	3			
$2x$	4				

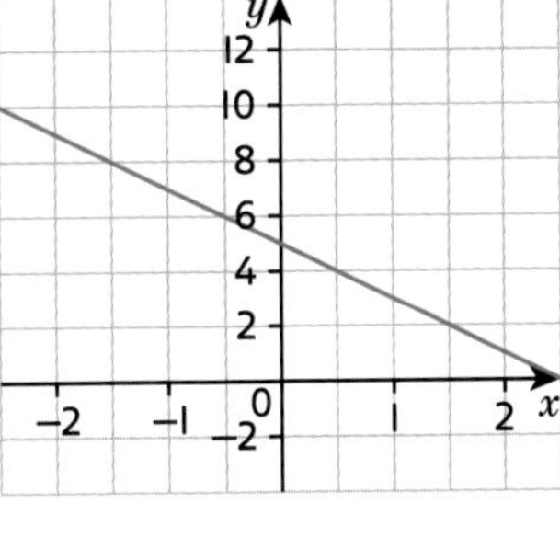

b Copy and complete:
The equation of the line is $2x + y =$ ______________.

c Create a table and then draw the graph for $2x + y = 3$.
Use axes from −10 to +10.

d Rearrange $2x + y = 3$ so that it begins '$y =$'.

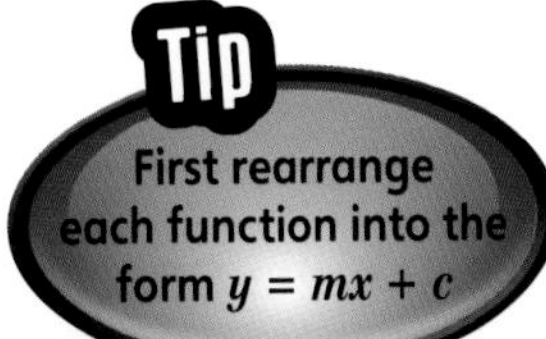

Tip
First rearrange each function into the form $y = mx + c$

5 Use a spreadsheet to plot these graphs.

a $2x + 3y = 6$ b $x - 2y = 2$

Level 7

7b I know that equations of lines can be rearranged

7a I can use a spreadsheet to plot the graph of a linear function given in the form $ax + by = c$

6 Rachel needs a taxi for a distance of 8 km. She needs to decide which taxi firm to use.

Crazy cabs	***Terrific Taxis***
£2.20 fixed charge **+ £1 per km**	No fixed charge! **£3.30 per mile**

a Represent the information graphically by working out each firm's price for distances between 1 km and 10 km. Use your graph to make a recommendation for Rachel's journey.

b Represent the information algebraically by writing a pair of simultaneous equations and solving them.

c When should customers use Crazy Cabs and when should they use Terrific Taxis?

d What are the advantages of using a graphical method?

e What are the advantages of using an algebraic method?

Tip
Convert all values to the same unit before you start.
1 mile = 1.6 km

7 The net of an open box is cut out of a square with side length 18 cm. A smaller square with side length x is cut from each corner.

a Find the capacity of the box when $x = 2$ cm.

b Use a spreadsheet to find the length of x, to the nearest millimetre, for the box to have a capacity of 350 cm^3.

c Use the graph sketching function on the spreadsheet program to find when the capacity of the box will be at its maximum.
Comment on the accuracy of your answer.

Level 8

8b I can construct models of real-life situations by drawing graphs and constructing algebraic equations

8a I can use a spreadsheet to solve increasingly demanding problems

Now try this!

A Many rectangles

Work with a partner.
Use a set of axes from −10 to +10. Take turns drawing vertical lines, $x = a$, and horizontal lines, $y = b$. The aim of the game is to make as many rectangles as possible when drawing your line. You score 1 point for each rectangle you make. The game ends when each player has taken five turns.

B Draw that shape!

Use a graph sketching package or a graphical calculator to draw shapes using linear equations.
Try drawing a rectangle, a square and a parallelogram.
Can you draw a rhombus and a kite?

linear function simultaneous equations spreadsheet

maths!

Unsolved problems

Goldbach conjecture

UNSOLVED

Every even integer greater than 2 can be written as the sum of two primes.

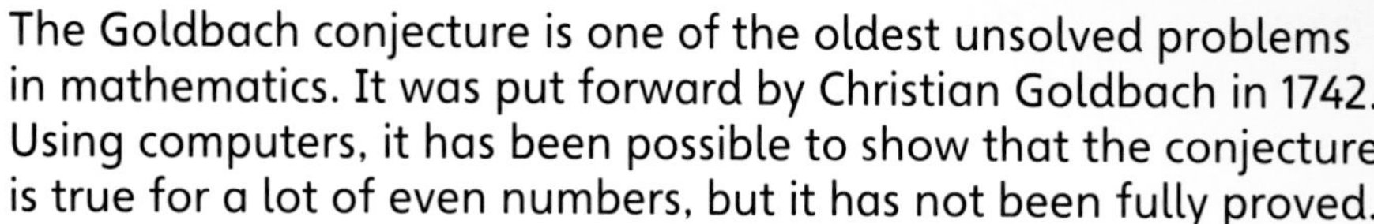

The Goldbach conjecture is one of the oldest unsolved problems in mathematics. It was put forward by Christian Goldbach in 1742. Using computers, it has been possible to show that the conjecture is true for a lot of even numbers, but it has not been fully proved.

$$4 = 2 + 2 \qquad 6 = 3 + 3 \qquad 8 = 3 + 5$$

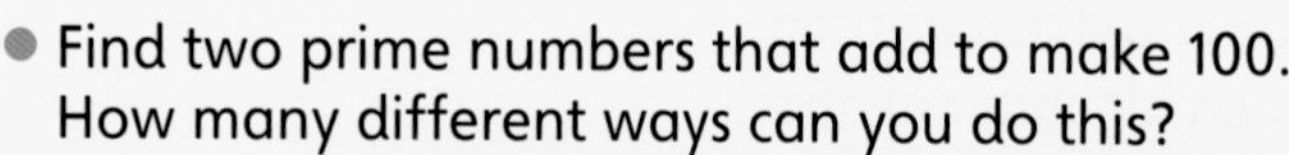

- Find two prime numbers that add to make 100.
  How many different ways can you do this?
- What's the biggest even number you can show to be equal to the sum of two primes?
  Make sure you know that the numbers really are primes!

$170 = 67 + 103$ $\qquad$ $280 = 131 + 149$

$438 = 197 + 241$

Four colour problem

Any map split into regions can be coloured so that no adjacent regions are the same colour using four colours or fewer.

UNSOLVED

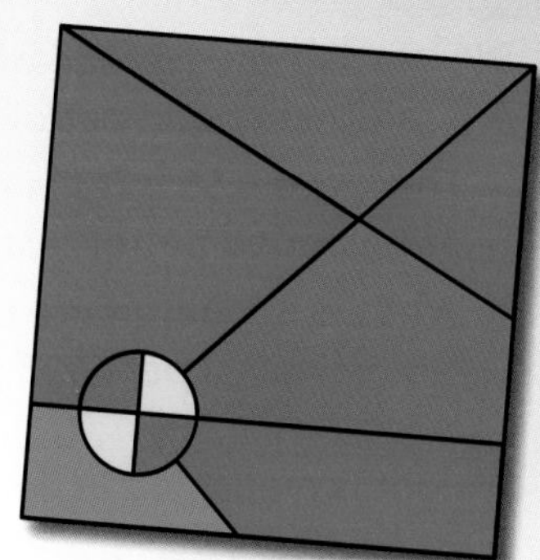

The four colour problem was first put forward in 1853 by Francis Guthrie, a botanist, who had been trying to colour a county map of England.

Many people have attempted to prove this problem and it was the first theorem to be proved with the help of computers, in 1977. However, many mathematicians do not accept this as a proof because it would be impossible for a human to prove 'by hand'.

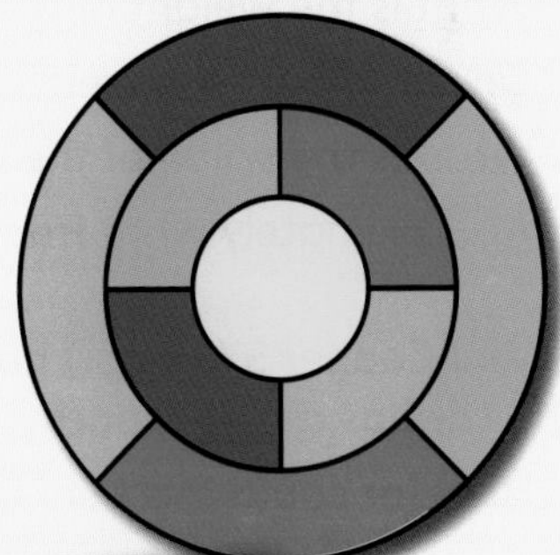

Over the years there have been many false disproofs or counter-examples. Regions have been drawn where it is claimed that more than four colours are needed to colour them.

- Show that this is a false disproof and that it can be coloured with only four colours.

UNSOLVED

Millennium Prize Problems

In 2000 the Clay Mathematics Institute announced seven problems called the Millennium Prize Problems.

For each problem they are offering \$1 000 000 for a correct solution. By 2008 only one of these problems (known as the Poincaré conjecture) had been solved, so there is still \$6 000 000 up for grabs!

Cars, cars, cars!

Numerous technical regulations are involved in Formula 1 racing, for example the minimum weight and maximum overall bodywork height. Car designers must use maths to develop a car that meets all of these regulations. But maths is also used when buying and running cars on a day-to-day basis.

1 Diesel costs an average of £4.49 a gallon.
A litre of unleaded petrol costs an average of 87.2p.

- **a** There are 4.55 litres in a gallon. How much does a gallon of unleaded petrol cost?
- **b** What is the difference between the cost of diesel and the cost of petrol *per litre*? **Level 6**

2 Weight is measured in newtons and is calculated using this formula.

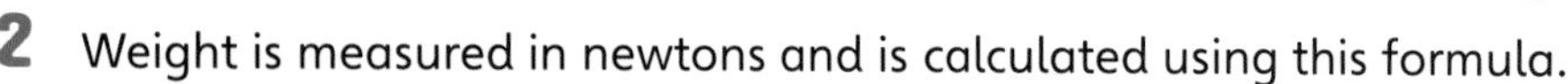

weight = mass × gravity

- **a** An F1 car (with driver) has a mass of 605 kg.
  What is the weight of the car (and driver)? (Use gravity = 9.81 m/s^2)
- **b** On the Moon, the weight of the same car (and driver) would be 989.18 newtons.
  What is the value of the Moon's gravity?
- **c** Approximately what fraction of the Earth's gravity is the Moon's gravity? **Level 7**

3 A Mini Cooper S (3 door) has a fuel consumption of 46.5 miles per gallon (mpg).
The car has a fuel capacity of 50 litres.

- **a** How far could this Mini travel on one tank of petrol?
- **b** The average yearly mileage of a driver is 12 000 miles. Unleaded petrol costs 87.2p per litre.
  What is the fuel cost of the Mini per year? **Level 7**

4 During the 2008 Monaco Grand Prix, the fastest practice lap was recorded by Lewis Hamilton at 1 minute 15.140 seconds.
The circuit length at Monaco is 3.340 km and the race distance is 260.520 km.

- **a** What was Lewis's average speed?
- **b** Based on his average speed, how long would it take him to complete the full race? **Level 8**

5 A Lexus IS 200 SE has a basic price of £18 395.

- **a** The list price of the car is the basic price plus VAT.
  Given that VAT is 15%, work out the list price.
- **b** At the end of the first year, the Lexus depreciates (decreases in value) by 22%.
  What is the value of the car after 1 year?
- **c** What fraction of the original list price is the car worth after 1 year?
- **d** Another Lexus model depreciates by 27 per cent in the first year, and 9 per cent in the second year.
  Is it correct to say that the total depreciation after 2 years is 36%?
  Give full reasons for your answer. **Level 8**

The BIG ideas

- → If the unknown value in a **formula** is not the **subject**, substitute the known values, then solve the equation to find the unknown value. **Level 7**
- → You can make solving problems involving **percentage** change easier by changing the percentage into a decimal multiplier.
  For example, the decimal multiplier for an increase of 25% is (100 + 25)% ÷ 100% = 1.25. **Level 7**
- → Compound measures combine measures of two different types. For example, speed measures distance and the time taken to travel it.
  $\text{Speed} = \frac{\text{distance}}{\text{time}}$ **Level 7**
- → $\text{Average speed} = \frac{\text{total distance travelled}}{\text{total time taken}}$ **Level 8**
- → When solving problems, follow the same procedure each time:
  - Read the question – what do you need to find out?
  - Which maths is relevant to the problem?
  - Break down the problem into a series of steps if the task is complex.
  - Carry out calculations and/or draw graphs and diagrams.
  - Give your results in an appropriate form and to an appropriate degree of accuracy.
  - Check your results – does your solution seem reasonable? **Level 6, Level 7 & Level 8**

Find your level

Level 6

Q1 Paul is 14 years old. His sister is exactly 6 years younger. This year, the ratio of Paul's age to his sister's age is 14 : 8. This can be simplified to 7 : 4

a When Paul is **21**, what will be the ratio of Paul's age to his sister's age? Write the ratio as simply as possible.

b When his sister is **36**, what will be the ratio of Paul's age to his sister's age? Write the ratio as simply as possible.

Q2 a

> When you multiply a number by 3, the answer is always greater than 3.

Give an example to show that this is *not* correct.

b

> When you subtract a number from 3, the answer is always less than 3.

Give an example to show that this is *not* correct.

Level 7

Q3 In 2007 a company made a profit of £1.36 million. In 2008 the company's profit was £1.47 million.
What was the company's percentage increase in profit?

Q4 This rectangle has an area of 330 cm^2. Find the height and width of the rectangle.

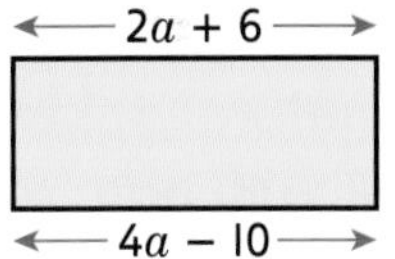

Level 8

Q5 38 755 people attended an Aston Villa home game. This was 8% fewer than at the previous home game.
How many attended the previous home game?

Q6 a Show that in this right-angled triangle $a = \frac{b}{\sqrt{2}}$

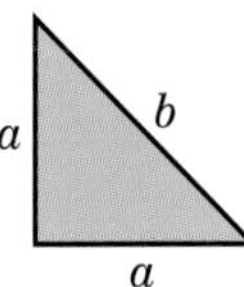

b What is the area of the triangle in terms of b?

c Calculate the area of this regular octagon.

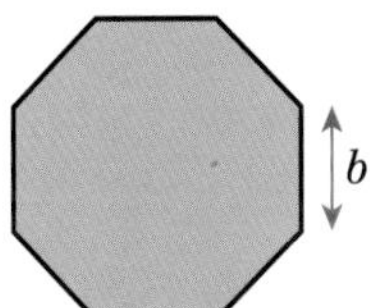

Hint: Start by drawing a square around the outside of the octagon.

13 Data day statistics

This unit is about dealing with data, including looking for correlation – how sets of data values are related.

Amaia Ruiz de Alegria is a research student in the Department of Geography at Plymouth University. She collects data on how sand and shingle are moved around beaches in storms, and on wave heights during those storms. By comparing the two sets of data, she can see if there is a relationship between the height of the waves and the movement of the sand. Amaia uses a GPS system to measure the heights of the ground from the top to the bottom of the beach.

'A few years ago the road that ran along the top of the beach was washed away. In the future we expect sea levels to rise, so these kinds of problems will only get worse,' says Amaia. 'I am using the data to find out about the natural adjustment the beach makes after storms. This means we can work with nature to engineer better coastal protection for the future.'

Activities

A Look at the page from Amaia's field notebook below.

Height of waves (m)	Height of top of beach (m)	Height of bottom of beach (m)
0.25 (calm)	6.0	0.3
2.00 (stormy)	5.9	0.61
2.5 (bad storm)	5.3	0.8
0.75 (moderate)	5.9	0.5
0.2 (calm)	6.1	0.25

- What happens at the top and bottom of the beach when the waves are high?
- What may happen to the beach when waves are high?

B This graph shows global temperatures and carbon dioxide levels over the years.

- What was the global average temperature in 1900, 1950 and 2000?
- What was the carbon dioxide concentration in 1900, 1950 and 2000?
- What has happened to global temperatures and carbon dioxide concentrations over time?

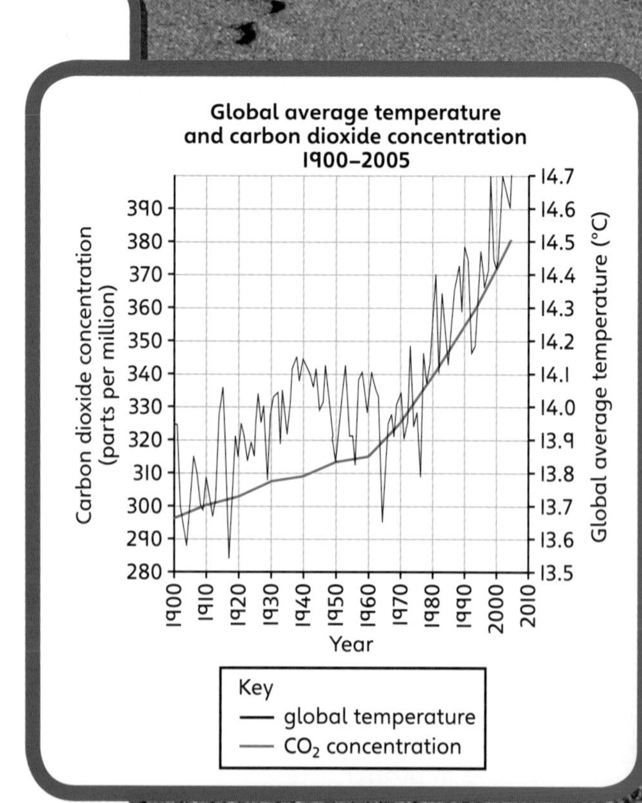

Before you start this unit...

1 Aisha is doing a survey to find out about what people eat. What is wrong with this question?
'Do you eat a lot of vegetables?'

page 84

2 Here are the times that Bill has to wait for a bus at his bus stop.
12, 15, 10, 12, 35, 14, 12, 13, 11, 10
Work out the mean, median, mode and range.

page 88

3 This table shows the number of people in Year 9 classes. Find the mean class size.

Class size	Frequency
27	1
28	3
29	3
30	1

page 88

4 Here are the eye colours of 60 pupils: 10 blue, 25 brown, 10 green, 15 hazel. Draw a pie chart for this data.

page 90

13.1 Questionnaires and samples
13.2 Averages from grouped data
13.3 Classroom challenge
13.4 Scatter graphs and correlation
13.5 Misleading graphs and charts
13.6 Comparing distributions
Unit plenary: There's a storm coming

World's Greatest Maths!

Plus digital resources

13.1 Questionnaires and samples

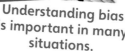

- ➪ Recognise sources of secondary data
- ➪ Recognise sources of bias
- ➪ Identify ways to reduce bias in questionnaires and samples

What's the BIG idea?

Why learn this?

Understanding bias is important in many situations.

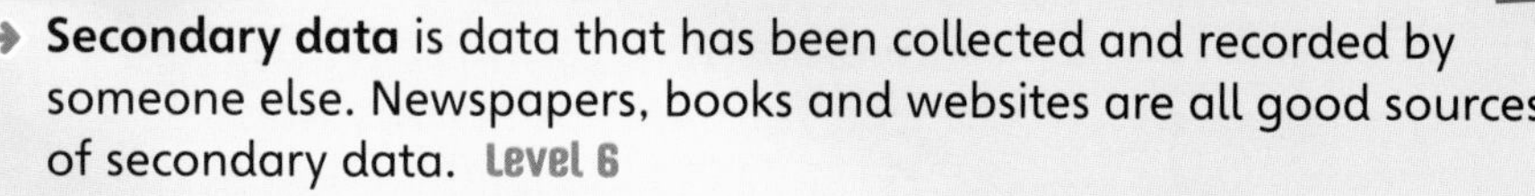

- → **Secondary data** is data that has been collected and recorded by someone else. Newspapers, books and websites are all good sources of secondary data. **Level 6**

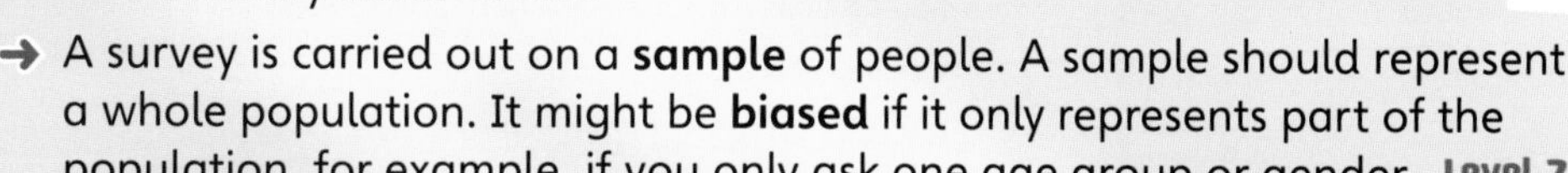

- → A survey is carried out on a **sample** of people. A sample should represent a whole population. It might be **biased** if it only represents part of the population, for example, if you only ask one age group or gender. **Level 7**
- → The questions in a **questionnaire** should
  - be easy to understand
  - be easy to answer with options to choose from, particularly for numerical answers, for example use boxes 0–5 ☐ 6–10 ☐ 11–15 ☐ etc.
  - not lead to bias. **Level 7**
- → Badly written questions can lead to bias. For example, questions that start 'Do you agree ...' might push the respondent to say 'Yes' **Level 7**

Practice, practice, practice!

Level 6

6b I can interpret data in a table from a secondary source

1 Look at Table A on Resource sheet 13.1a and answer these questions.

- **a** What percentage of households owned a home computer in 1985?
- **b** What percentage of households owned a home computer in 2000?
- **c** Which appliance showed the greatest change in percentage ownership between 1978 and 2000?
- **d** What was the mostly widely owned appliance in 2000?
- **e** When was satellite, cable or digital TV first introduced in the UK?
- **f** Describe what happened to television ownership between 1981 and 2000.

2 Look at Table B on Resource sheet 13.1a and answer these questions.

- **a** How much did an average family spend per week on clothing and footwear in 1982?
- **b** What did an average family spend per week on fuel and power in 1998–99?
- **c** Which item has increased the most in spending between 1982 and 2001?
- **d** What items have decreased in spending between 1982 and 2001?
- **e** What percentage of the total average family spending was on food and non-alcoholic drinks in
  i 1982 **ii** 2001?
- **f** What percentage of the total average family spending was on leisure goods and services in **i** 1982 **ii** 2001?
- **g** Write a short paragraph explaining the differences in average family spending between 1982 and 2001.

bias questionnaire

3 You are investigating which of two schools has, on average, the better sprinters. You are planning to pick a sample of pupils from each school and to record their times in the 100 m sprint.
What factors should you consider in picking your sample and recording the data?
Here are some suggestions. Pick the ones you would use and explain why.

age | on the athletic team or not | hair colour | weight | gender | direction of track | location

4 You are conducting a survey about people's diets.
Where in the town centre should you position yourself in order to carry out your survey?
Explain your reasons for choosing or not choosing each one of these locations.

a outside the butcher's shop
b in the library
c by the sports centre
d outside the fish and chip shop

5 Describe what is wrong with each of these questions.

a How much do you spend on clothes each week?
b It's important to look good. Do you buy new clothes each month?
c Frivolous fashions are inconsequential to the gross spending per household. Do you agee or disagree?
d How much do you spend on average on a skirt?

6 Rewrite each of the questions in Q5 so they are suitable for a questionnaire.

Level 7

7c I can identify bias in a sample

Super fact!
In 1992, a badly chosen sample caused the wrong result to be predicted in the general election. Samples must accurately represent the population!

7b I can identify bias in a questionnaire

7a I can plan to minimise bias in a questionnaire

7 In a random sample, every item in the population has an equal chance of being selected. Write whether each sample is random. If not, explain why.

a Take the first 20 names in the year list to choose a sample of pupils in your year.
b Randomly select from the inhabitants of Glasgow, to obtain a sample of the population of the UK.
c Consider all the cars sold last year by five randomly selected showrooms to obtain a sample of cars on the UK roads.
d Obtain random number between 1 and 10, say 6. Select the sixth name on the judo club members list and every tenth name after that, to select a sample of members.

Level 8

8c I can identify a random sample

Now try this!

A What's wrong?

Look at the healthy-eating questionnaire (Resource sheet 13.1b). In pairs, discuss what is wrong with the questions, then write a better set of questions.

B Deliberate mistakes!

Write a badly-worded questionnaire that aims to find out how often people take part in a sporting activity. Challenge a partner to improve your questionnaire.

sample secondary data

13.2 Averages from grouped data

Estimate the mean and median for large sets of grouped data

What's the BIG idea?

→ To estimate the mean of grouped data:

- calculate the mid-point of each interval $\left(\text{mid-point} = \frac{\text{start point} + \text{end point}}{2}\right)$
- multiply each mid-point by the **frequency** for that interval

Estimated mean = $\frac{\text{total of all the values of (mid-point} \times \text{frequency)}}{\text{total frequency}}$ **Level 7**

→ To estimate the median of grouped data:

- calculate the position of the median value $\left(\text{the } \frac{n+1}{2}\text{th data value}\right)$
- find the interval that contains the median value
- calculate how far along that interval the median value lies
- use this information to estimate the median. **Level 8**

→ You can also make estimates of the mean and the median from a **frequency polygon** in the same way. **Level 8**

Why learn this?

Grouping data and then calculating an estimate of the mean or median saves valuable time when there is a lot of data.

Practice, practice, practice!

Level 7

7a I can calculate an estimate of the mean of a set of grouped data

1 The table shows the ages of the owners of a new model of mobile phone. Calculate an estimate of the mean age of an owner.

Age	Frequency	Mid-point of interval	Frequency × mid-point
$10 \leq \text{age} < 20$	13	15	195
$20 \leq \text{age} < 30$	4	25	100
$30 \leq \text{age} < 40$	2	35	70
$40 \leq \text{age} < 50$	1	45	45
Totals	**20**		**410**

Watch out! Remember to divide by the total number of data values and not the number of groups.

Level 8

8c I can estimate the median of a set of grouped data from a table

2 These are the ages of the users of a new MP3 player.

13 16 31 36 27 14 16
30 18 26 27 16 43 17
33 31 34 28 24

a Copy and complete this grouped frequency table.

Age	Frequency
$10 \leq \text{age} < 20$	
$20 \leq \text{age} < 30$	
$30 \leq \text{age} < 40$	
$40 \leq \text{age} < 50$	

b In which group will the median value be?

c Estimate the median age of the MP3 users.

estimated mean estimated median

3 This table shows the number of minutes (to the nearest minute) that a sample of Year 11 pupils spent on phone calls in one week.

a Estimate the median length of time spent on phone calls.

b Estimate the mean length of time spent.

Time spent on the phone in a week	Frequency
$0 \leqslant$ minutes < 10	28
$10 \leqslant$ minutes < 20	25
$20 \leqslant$ minutes < 30	16
$30 \leqslant$ minutes < 40	9
$40 \leqslant$ minutes < 50	5
$50 \leqslant$ minutes < 60	0
$60 \leqslant$ minutes < 70	2
$70 \leqslant$ minutes < 80	0

Level 8

8c I can estimate the mean and the median of a set of grouped data

4 This frequency polygon shows the ages of people who attend a fitness club.

Use the frequency polygon to calculate an estimate of the mean age of the members of the club.

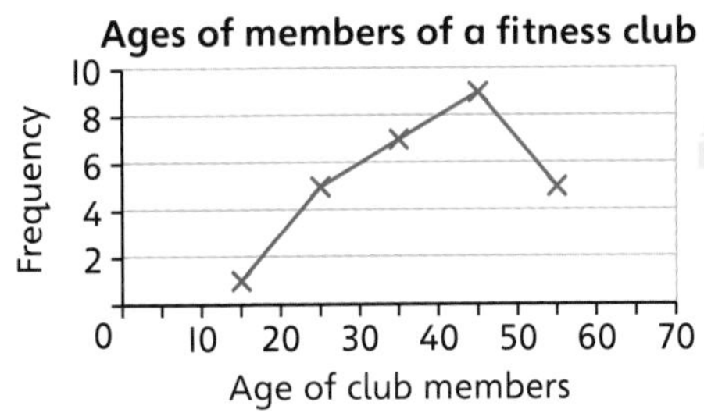

8c I can estimate the mean from a frequency polygon

5 These two frequency polygons show the lengths of time members spent in the gym per visit during the week and at weekends.

a Estimate the mean time per visit during the week and at weekends.

b Compare the two means.

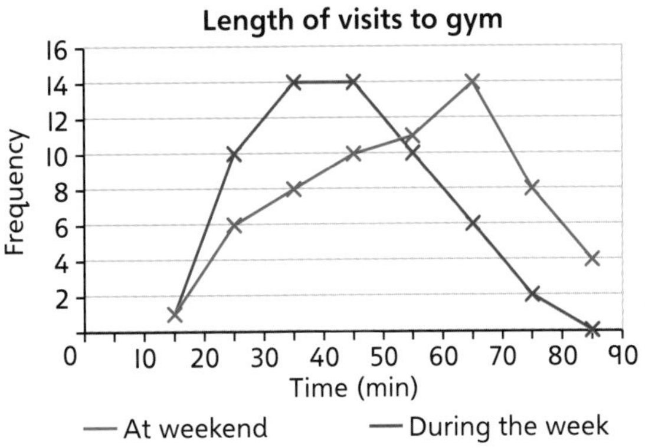

6 Use the frequency polygon in Q4 to estimate of the median age of the members of the fitness club.

8a I can estimate the median from a frequency polygon

7 **a** Use the frequency polygons in Q5 to estimate the median time per visit during the week and at weekends.

b Compare the median times. When do people spend more time in the gym?

Now try this!

A It's a goal!

Look up the results of 20 premiership football matches and 20 lower league matches. Group the first goal times in a pair of frequency tables and then find an estimate of the mean first goal time in each league.
Write a short report to compare the results.

B Homework

Find out how long everyone in your class spent doing homework last night.
Design a pair of grouped frequency tables to collect the results.
Using estimated means, compare the boys' homework time with the girls' homework time.
Write a report to compare the results.

frequency frequency polygon

World's Greatest Maths!

13.3 Classroom Challenge

Stopwatch Stoppers

In Stopwatch Stoppers, contestants have to stop a timer as close to 1 second as possible.

This contestant was 8 hundredths of a second too quick!

Result t (hundredths of a second)	Frequency	Cumulative frequency
$0 \leq t \leq 5$	2	2
$5 < t \leq 10$	7	
$10 < t \leq 15$	17	
$15 < t \leq 20$		34
$20 < t \leq 25$		38
$25 < t \leq 30$		40

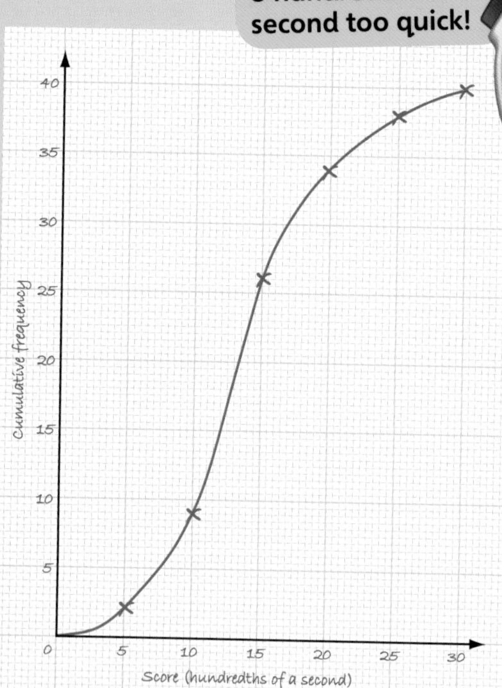

The results for 40 pupils were recorded in a cumulative frequency table and diagram.

1 Copy and complete the cumulative frequency table.

You can use a copy of Resource sheet 13.3 to help you.

2 Use the cumulative frequency diagram to estimate the number of pupils who were closer than 8 hundredths of a second.

3 Estimate the median of this data.

4 Estimate the upper and lower quartiles, and find the interquartile range.

Practice makes Perfect

After practising, the same class played Stopwatch Stoppers again.
Their results are shown in this tally chart.

Result, t (hundredths of a second)	Tally
$0 \leq t \leq 5$	𝍸 I
$5 < t \leq 10$	𝍸 𝍸 𝍸
$10 < t \leq 15$	𝍸 𝍸 III
$15 < t \leq 20$	𝍸
$20 < t \leq 25$	I
$25 < t \leq 30$	

1 Draw a cumulative frequency table to show the data.

2 Use your table to draw a cumulative frequency diagram.

3 Write down estimates for the median, upper and lower quartiles and interquartile range of this data.

4 Compare these results with the class's scores before practising.
What conclusions can you draw from your statistics?

You can draw your cumulative frequency diagram on a copy of the earlier one to compare the distributions.

Cliff Hanger

The aim of this game is to shove a 2p piece from one side of the table to the other. How close can you get to the edge of the table? Can you score a 'cliff hanger'?

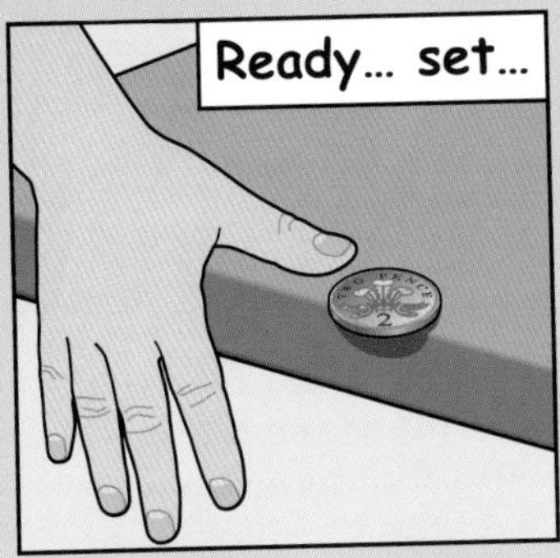

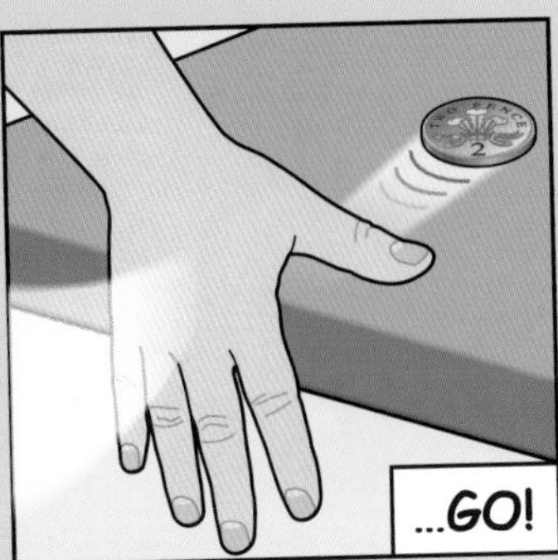

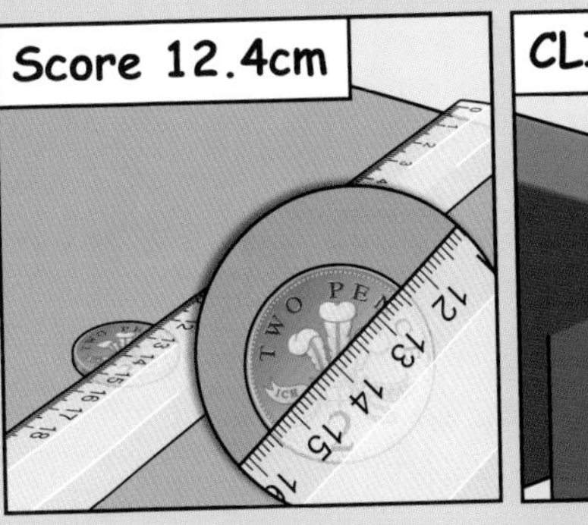

- Start with a 2p piece half over the edge of a table.
- Shove the 2p piece with the palm of your hand.
- Measure the distance from the coin to the other edge of the table (to the nearest millimetre). This is your score.

Have a few practice 'shoves' before you start your experiment. Once you have started, make sure you record every shove.

1 Play Cliff Hanger in pairs. Measure each other's distances and record them in separate lists. Also keep count of how many times your coin falls off the table. Continue until you have ten distance scores each.

2 Write your distance scores in a table like this one.

Distance, d cm	Tally	Frequency
$0 \leqslant d \leqslant 5.0$		
$5.0 < d \leqslant 10.0$		
$10.0 < d \leqslant 15.0$		

3 Collect together the scores of all the members of the class.
Keep the boys' and girls' results separate.

4 Draw cumulative frequency diagrams of the boys' and girls' scores on the same axes.

5 Calculate the median, upper and lower quartiles and interquartile range for the boys and the girls.

6 Compare the scores of the boys and the girls.
Who performed better?
Use your results to justify your conclusions.

You could also calculate the average number of times the coin fell off the table before ten scores were recorded.

Reaction Times

- Use a reaction time program from the internet to record reaction times to the nearest millisecond.
- Repeat your experiment 20 times.
- Design a data collection sheet for this experiment, and choose appropriate class intervals for the data.
- Compare the reaction times of girls and boys.
- Investigate whether you have a better reaction time using your right or your left hand.

Nearest the Mark

- Mark a starting line on your desk using masking tape.
- Stick another small piece of tape 1 m away from the starting line, and draw a cross on it.
- Place a coin behind the starting line and flick it towards the cross. How close can you get?
- Design a table to record data from this experiment.
- Compare the results of girls and boys.
- Investigate whether you can achieve better results if you use your right or your left hand.

13.4 Scatter graphs and correlation

- Construct and interpret scatter graphs
- Identify key features presented in the data
- Identify correlation in a scatter graph

Why learn this?

A road engineer might use a scatter graph to see if there is a correlation between a road surface's time to failure and how many vehicles drive on it.

What's the BIG idea?

→ A **scatter graph** shows whether there is a relationship or **correlation** between two sets of data.

Miles travelled / Age of car

Positive correlation: as one value increases, so does the other.

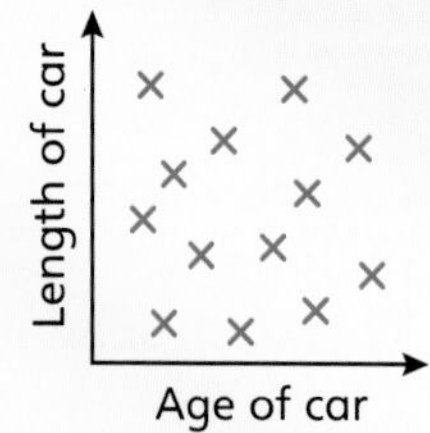

No correlation: the values are independent of each other.

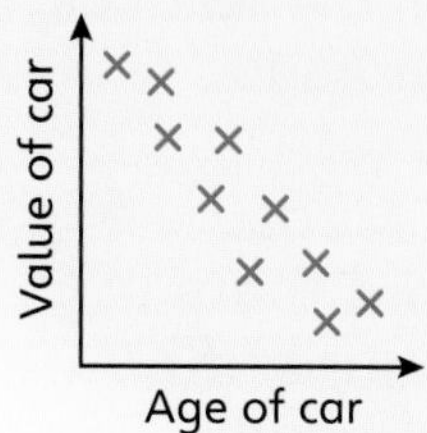

Negative correlation: as one value increases, the other decreases. **Level 6**

→ If there is correlation, you can draw a **line of best fit** to represent the relationship between the two variables. There should be the same number of crosses on each side of the line. **Level 7**

→ You can use the line of best fit to predict unknown values. **Level 7**

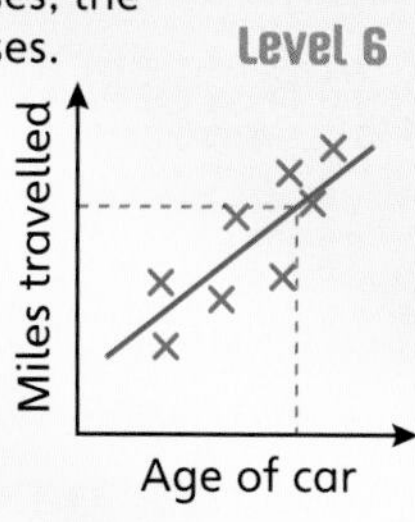

Practice, practice, practice!

1 Which type of correlation does each of these newspaper headlines describe? Explain your answers.

- **a** Watching too much TV lowers your IQ.
- **b** Taller people are more intelligent than shorter people.
- **c** Taking more exercise each day increases your metabolic rate – which means you burn up the calories in the food you eat more quickly.

2 Les owns a mobile café.
He recorded sales of different items and average daily temperature.
Describe the relationship shown in each of his scatter graphs.

a

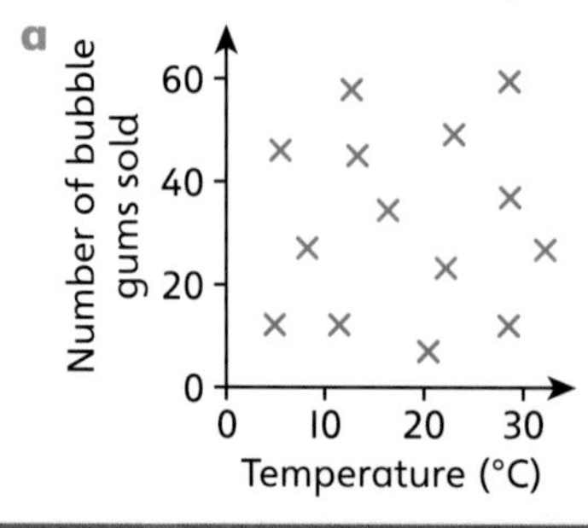

b

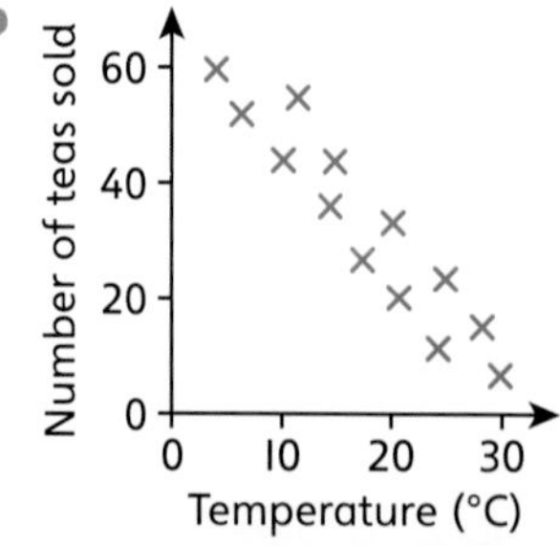

c

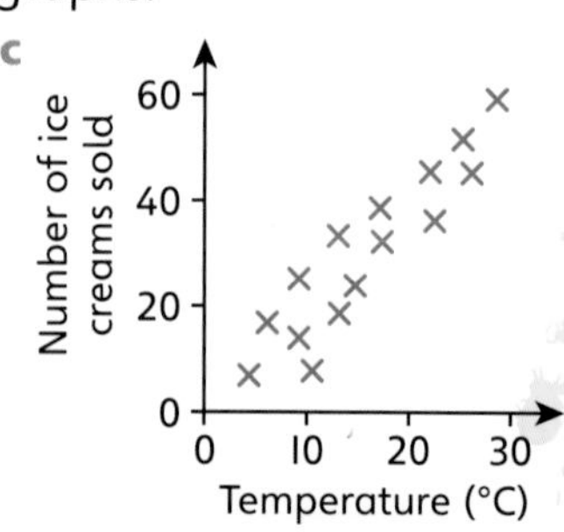

Level 6

6c I can use correlation to describe relations between sets of data

6b I can use correlation to describe relations between sets of data in simple cases

correlation **line of best fit**

3 The table shows the times that 15 runners took to complete two events in a school sports day.

Runner	A	B	C	D	E	F	G	H	I	J	K	L	M	N	O
100 m time (s)	14.9	17.6	18.1	15.3	16.4	16.0	17.2	16.9	20.3	15.8	15.7	17.0	18.2	19.0	14.6
400 m time (s)	54.3	67.1	82.0	65.3	59.5	53.0	71.2	75.0	73.0	59.5	64.2	81.5	76.0	95.5	70.0

a Draw a pair of axes with the vertical axis from 40 to 100 and labelled '400 m time (s)', and the horizontal axis from 10 to 22 and labelled '100 m time (s)'.

b Plot the data from the table to give a scatter graph.

c Describe the connection between the 100 m times and the 400 m times.

d Use a transparent ruler to draw a line of best fit on the scatter graph.

e Two pupils each ran in only one event. Use your line of best fit to estimate their probable times in the other event.

Runner	100 m time (s)	400 m time (s)
P	16.1	
Q		54.0

4 The scatter graph shows the gestation period (length of pregnancy) and the lifespan of different animals.

a Describe the correlation between the gestation period and the lifespan of an animal.

b Place a transparent ruler over the scatter graph to find the position of a line of best fit.

c Hyenas have a gestation period of 110 days. Use your 'line of best fit' to estimate the lifespan of a hyena.

d Identify a data value that does not follow the trend.

e Humans have a gestation period of 265 days. Use your 'line of best fit' to estimate the life span of a human. Is your estimate sensible? Explain your answer.

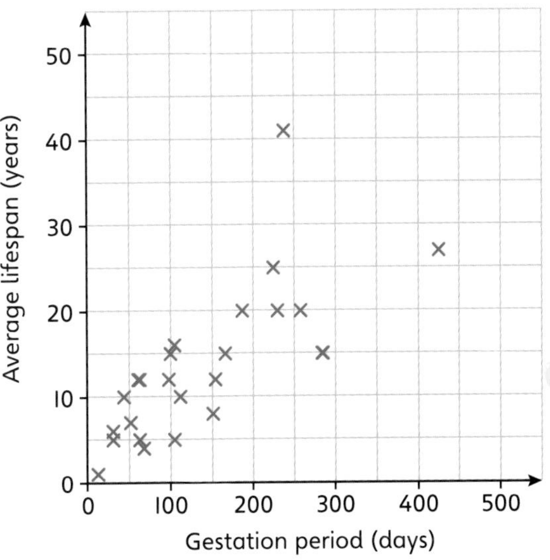

Level 7

7b I can draw a line of best fit by eye and use it to estimate the missing data

Now try this!

A Car performance

Use websites or a car magazine to find the engine sizes of 20 cars and the times the cars take to accelerate from 0 mph to 60 mph. Plot the data on a scatter graph. Describe the correlation.

B Correlation check

Repeat activity A, but choose your own data to investigate. Before you plot your scatter graph, predict what type of correlation you expect. Was your prediction correct?

scatter graph

13.5 Misleading graphs and charts

- Identify misleading graphs and statistics
- Interpret graphs and charts
- Draw conclusions from graphs and charts using statistical measures

Why learn this?

It is important not to be misled by advertisements. Do eight out of every ten people recommend the skin care product or just eight out of the ten people who were asked?

What's the BIG idea?

→ Graphs and charts can give a misleading impression if
- the scales on the **axes** are missing, or uneven, or do not start at zero
- dicrete data points are joined
- a small **sample** was used. **Level 7**

→ Graphs might not give enough information to answer all questions about the data. **Level 8**

Practice, practice, practice!

Level 6

6b I can identify which graphs are the most useful in the context of the problem

1 The table shows data on the prices of second-hand bikes in Mike's cycle shop. Which type of graph or chart would make it easiest for Mike to compare the numbers of bikes sold in each price range?

Bar chart | Pie chart | Line graph

Price of bike (£)	Frequency
0–99.99	12
100–199.99	15
200–299.99	9
300–399.99	7
400+	2

Level 7

7b I can identify misleading graphs and give reasons why they are misleading

2 This graph shows information about raw material costs.

a The cost of raw materials in 1985 was four times what it was in 1965.

Explain why the graph does not show this.

b Redraw the graph so that it is less misleading.

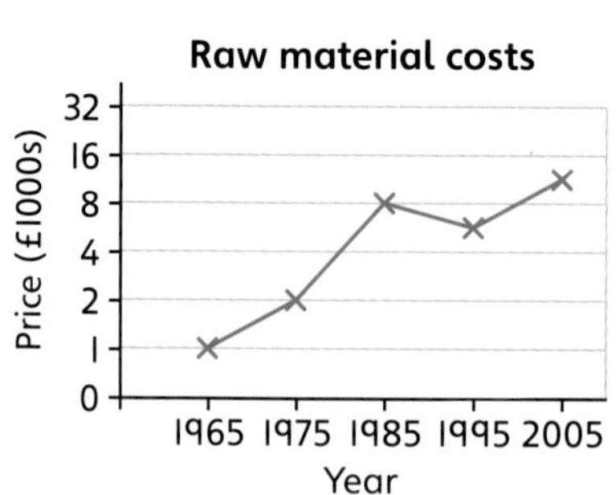

3 What is wrong with this pictogram?
How could you improve it?

Number of pets in Year 8

fish

cats

dogs

Key

= 10 fish

= 10 cats

= 10 dogs

axes proportion

4 In the 1960s and 1970s, the prize money for women's and men's singles was roughly equal at the Wimbledon tennis tournament.

a Look at the bar chart.
Do you agree or disagree with the statement?

b Look at this table, which shows the same information for the 1960s and 1970s.
Do you agree or disagree with the statement now?

Year	1968	1969	1970	1971	1972	1973
Men's singles (£)	2000	3000	3000	3750	5000	5000
Women's singles (£)	750	1500	1500	1800	3000	3000

Year	1974	1975	1976	1977	1978	1979
Men's singles (£)	10 000	10 000	12 500	15 000	19 000	20 000
Women's singles (£)	7000	7000	10 000	13 500	17 100	18 000

c Explain why the bar chart is misleading.

d In 1998, the prize money was £391 500 for women and £435 500 for men.
- **i** Sketch how the graph would look if you added this information.
- **ii** Would the graph be easy to read accurately? Explain your answer.

e Tristan says, 'Plotting the women's prize money as a percentage of the men's would give a clearer picture'. Plot the graph to see if you agree.

Level 7

7a I can interpret graphs, diagrams and tables and draw inferences to support/cast doubt on conjectures

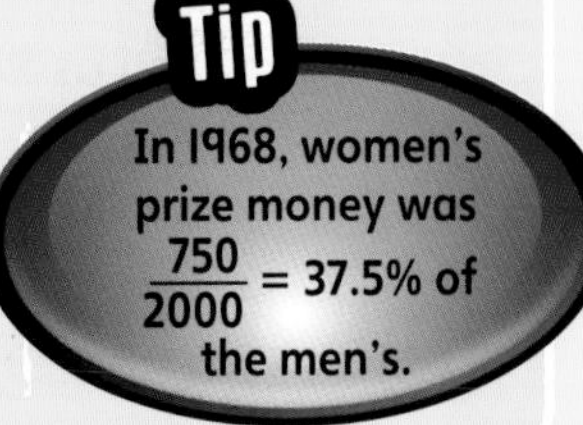

Tip
In 1968, women's prize money was $\frac{750}{2000}$ = 37.5% of the men's.

5 Dan tests his buzzer game on his class to see how long they take to complete it. His results are shown in this bar chart.
Use the bar chart to say whether each statement is true or false, or whether there is not enough information to tell. Explain your answers.

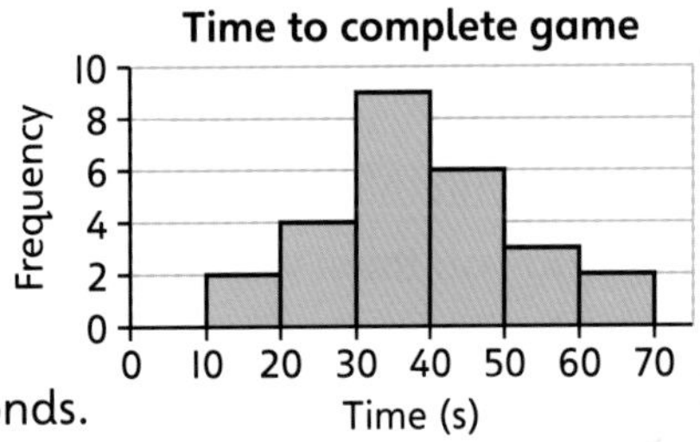

Time to complete game (Frequency against Time (s): 0–10: 0, 10–20: 2, 20–30: 4, 30–40: 9, 40–50: 6, 50–60: 3, 60–70: 2)

a The median time was between 30 and 40 seconds.

b No-one took less than 10 seconds.

c No-one took more than 65 seconds.

d The boys did better than the girls.

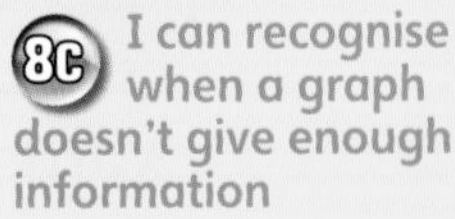

Now try this!

A Is it clear?

Look in magazines or newspapers and find five examples of graphs. Write a sentence on each graph, explaining whether it is clear or not, and why.

B Rising prices!

Look on websites for data about the rising prices of tickets to see a football match or a music concert, or make up some data of your own.
Draw two graphs to illustrate the data, one misleading and one not. Who might use each graph to support their case – the owner of the club or music company, or the editor of the fan magazine?

13.6 Comparing distributions

➪ **Compare two distributions using appropriate statistics and the shape of the distributions**

What's the BIG idea?

→ You can use the range and averages to compare sets of data and draw conclusions. **Level 6**

→ You can use two frequency polygons drawn on the same pair of axes to compare **distributions**. **Level 7**

Why learn this?

To answer questions in business you need to use graphs and averages to compare data and write a report.

Practice, practice, practice!

Level 6

6b I can compare two sets of data using the range and an average

1 The Smith family are deciding whether to buy a holiday apartment in Madrid or in Las Palmas. Look at the mean temperatures and ranges in this table.

	Mean temperature (°C)	Range (°C)
Madrid	13.9	19.7
Las Palmas	20.2	6.3

a Copy and complete these sentences.

i __________ is warmer on the whole because it has a higher __________.

ii __________ has a greater variety of temperatures because the __________ is bigger.

b Which resort do you think would be hotter in summer?

c Which resort do you think would be colder in winter?

d Which city do you think the Smith family should choose? Explain your choice

Watch out!

Just one really large value or really small value can have a big effect on the mean.

2 In Newtown there are two travel agencies, Sunnytours and Easibreak. 14 people work in each. Here are the statistics about wages in the two companies.

	Mean (£)	Median (£)	Range (£)
Sunnytours	243	252.20	185
Easibreak	252.21	220	500

Which company would you choose if you were trying to earn the most? Explain your reasons.

distribution

3 This frequency polygon shows the percentage scores in a science exam for three classes.

Look at the shape of the distributions. Write three sentences to compare the scores of the boys and the girls.

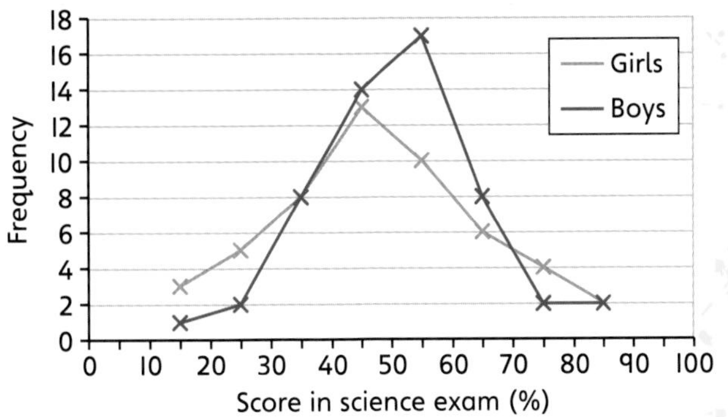

4 The sales team at Hillside Travel earn commission by selling holidays. This means that the more they sell, the more they earn. These graphs show the commission earned in March and in April.

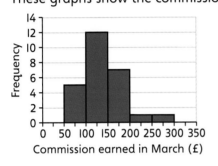

Frequency

0 50 100 150 200 250 300 350

Commission earned in March (£)

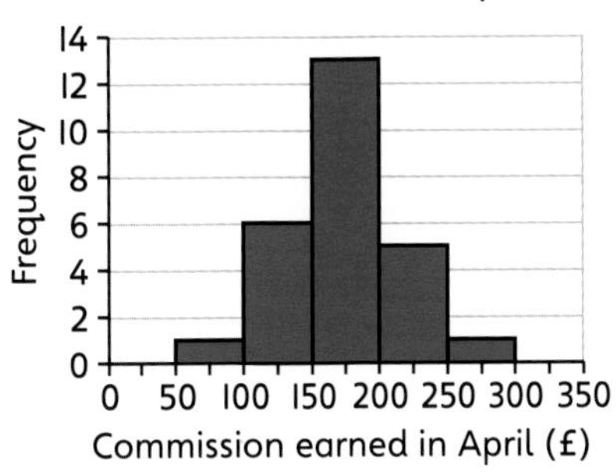

Commission earned by sales team

Commission earned in April (£)

Commission earned in March (£)

Write a report about the commission earned in each month.

> **Tip**
> Include average commission, commission earned by individuals, and the connection between commission earned in March and in April.

5 The graphs on Resource sheet 13.6 show information about how much fruit and vegetables people eat, and how many crisps and sweets they eat. Write a report describing what they tell you.

Level 7

7c I can compare two distributions using the shape of the distributions

7a I can interpret graphs and charts, draw conclusions by using a variety of statistical measures, and write a statistical report

Now try this!

A Climates

Choose two places in the world that you would like to visit and find weather graphs or charts for them on websites. Write a report to explain what the graphs tell you about the weather in each place.

B Your choice

What two related sets of data might be shown in these two grouped bar charts?

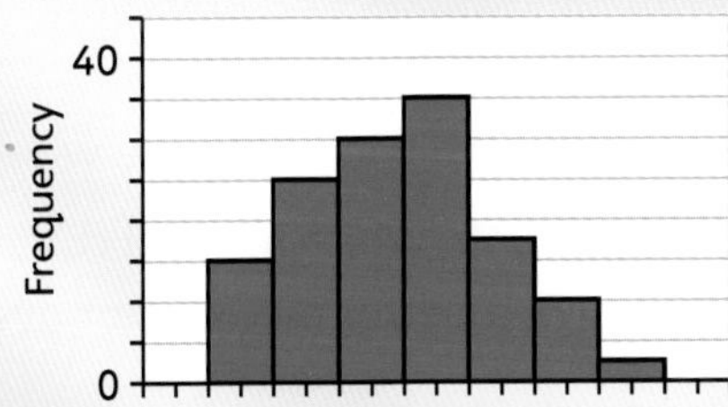

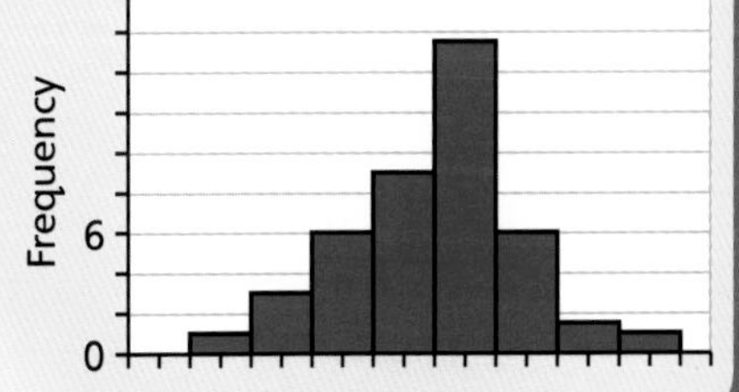

There's a storm coming

Hurricane Katrina hit New Orleans, Louisiana, in August 2005. Over 1800 people were killed in the hurricane itself and in the subsequent floods, making it one of the five deadliest hurricanes in the history of the US.
The table below shows all the hurricanes in the Atlantic in 2005.

Name	Wind speed (knots)	Wind pressure (millibars)	Category
Hurricane Cindy	65	992	1
Hurricane Dennis	130	930	4
Hurricane Emily	135	929	4
Hurricane Irene	85	975	2
Hurricane Katrina	150	902	5
Hurricane Maria	100	960	3
Hurricane Nate	80	979	1
Hurricane Ophelia	80	976	1
Hurricane Philippe	70	985	1
Hurricane Rita	150	897	5
Hurricane Stan	70	979	1
Hurricane Vince	65	987	1
Hurricane Wilma	150	882	5
Hurricane Beta	100	960	3
Hurricane Epsilon	75	979	1

1 What type of data is given in the table, primary or secondary? **Level 6**

2 What percentage of the hurricanes were category 3? **Level 6**

3 The category type is determined by the wind speed.
Using the information in the table, estimate the intervals for each category type, 1 to 5. **Level 6**

4 **a** Calculate the mean wind speed.

b Put the wind speed data into a frequency table.
Use the intervals $60 \leqslant w < 80$, $80 \leqslant w < 100$, and so on.
Calculate an estimate for the mean wind speed.

c Compare and discuss your answers to parts **a** and **b**.

d Repeat part **b** using the intervals $50 \leqslant w < 100$, $100 \leqslant w < 150$, and so on.

e Compare and discuss your answers to parts **d** and **b**. **Level 7**

5 Plot the wind and pressure data to produce a scatter graph.
Use wind speed on the x-axis and pressure on the y-axis.

a Is there a correlation between wind speed and pressure? Discuss.

b Draw a line of best fit on the scatter graph.

c Use your line of best fit to estimate the wind pressure when the speed is 110 knots. **Level 7**

6 **a** Look at your frequency table from Q4b. Which interval contains the median wind speed?

b Does the mean or the median give a better measure of average for this data?
Explain your answer. **Level 8**

The BIG ideas

- Secondary data is data that has been collected and recorded by someone else. Level 6
- A **scatter graph** shows whether there is a relationship or **correlation** between two sets of data.
  Positive correlation: as one value increases, so does the other.
  Negative correlation: as one value increases, the other decreases.
  No correlation: the values are independent of each other. Level 6
- To estimate the mean of grouped data:
  - calculate the mid-point of each interval $\left(\text{mid-point} = \frac{\text{start point + end point}}{2}\right)$
  - multiply each mid-point by the **frequency** for that interval
  - estimated mean $= \frac{\text{total of all the values of (mid-point} \times \text{frequency)}}{\text{total frequency}}$ Level 7
- To estimate the median of grouped data:
  - calculate the position of the $\left(\frac{n+1}{2}\right)$th data value (the median)
  - find the interval that contains the median value
  - calculate how far along the interval the median value lies
  - use this information to estimate the median. Level 8

Find your level

Level 6

Q1 Tom wants to investigate whether more people prefer practical or theory subjects. Here is his planning sheet.

> I am going to ask 30 people in my year group.
> I will ask each one: 'Do you prefer practical or theory subjects?', then I will write down their answers.

Write two different ways Tom could improve his survey.

Level 7

Q2 The scatter graph shows the average body length and average fin length of fish in an aquarium.

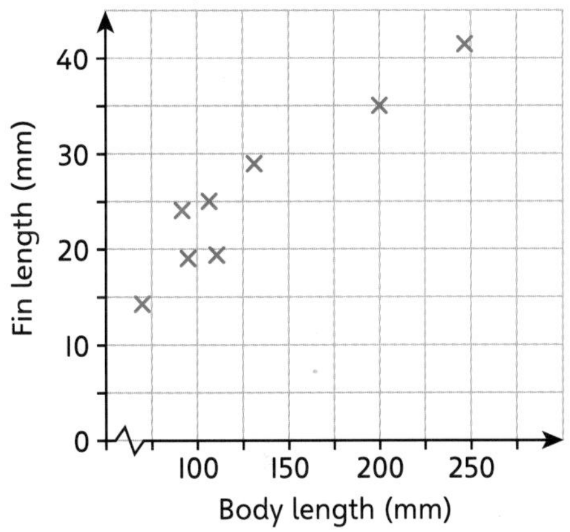

a What does the scatter graph tell you about the type of correlation between the body length and fin length?

b If body length increased by 50 mm, by approximately how many millimetres would you expect fin length to increase?

c A fish from another aquarium has a body length of 186 mm and fin length of 15 mm.
Is this fish likely to be the same type as those in the first aquarium? Explain your answer.

Level 8

Q3 The table shows how much time 50 students spent exercising yesterday.

Time spent exercising (minutes)	Frequency
$0 \leqslant \text{time} < 30$	7
$30 \leqslant \text{time} < 60$	14
$60 \leqslant \text{time} < 90$	19
$90 \leqslant \text{time} < 120$	10

a Show that an estimate of the mean time spent exercising is 64.2 minutes.

b Which interval contains the median?

14 Trig or treat?

This unit is about solving shape problems involving angles and lengths.

Matt is in the fourth year of his apprenticeship as a carpenter and joiner. He works for Lloyd Woodworking, who specialise in making windows, doors, staircases and roof timbers.

'When I first started this job, I didn't realise how much maths I would use every day,' says Matt. 'Obviously I need to measure and calculate accurately to make sure the windows and doors end up the right size. I also have to use Pythagoras' theorem to calculate the measurements for staircases. The vertical height between the ground floor and the first floor is called the "rise". The floor space available is called the "go". The staircase forms the hypotenuse of a right-angled triangle.

'Last week we needed to work out the angles for a triangular roof truss. We knew the vertical height of the triangle and the distance across its base. We split the triangle into two identical right-angled triangles and then used tan to find the base angles.'

Activities

A Here is a sketch showing the side elevation of a staircase.

Use Pythagoras' theorem to work out the length of wood needed to make the side of the staircase (the hypotenuse).

The staircase will have 14 steps like this:

Work out the rise and go for each step.

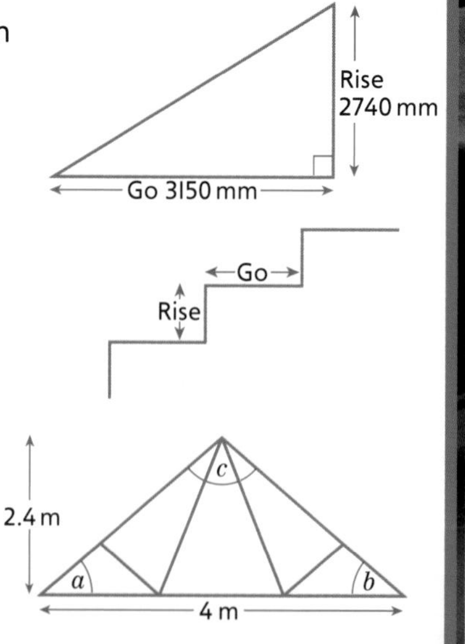

B Here is a diagram of triangular roof truss.

Split the triangle into two right-angled triangles. Use tan to work out the base angles (a and b).

Now calculate the angle at the top (c).

Did you know?
Building regulations state that for steps in a staircase the minimum go is 220 mm and the maximum rise is 220 mm.

Before you start this unit...

1 Draw the plan, front and side elevation of this 3-D shape.

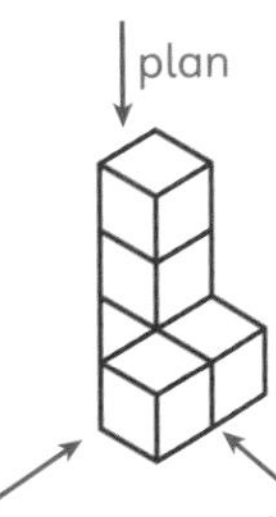

2 A cuboid has length 5 cm, width 2 cm and height 3 cm.

a Find the volume.
b Find the surface area.
c Sketch a possible net.

3 Find the size of angle x.

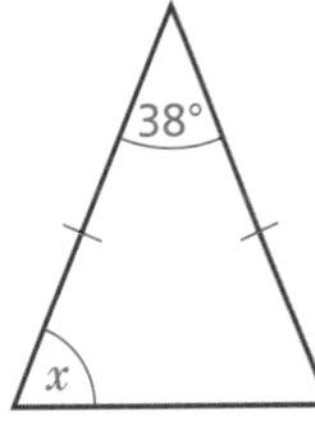

14.1 Solving geometrical problems
14.2 More geometrical problems
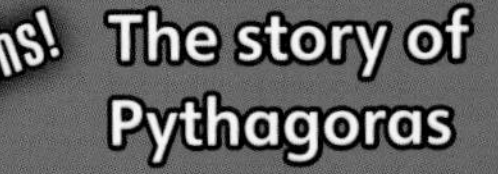

The story of Pythagoras
14.3 3-D shapes
14.4 Prisms and cylinders
14.5 Using trigonometry 2
14.6 Solving problems in trigonometry
Unit plenary: Accessibility problems

14.1 Solving geometrical problems

- Solve problems using properties of triangles and quadrilaterals
- Solve problems using Pythagoras' theorem

What's the BIG idea?

- You can use your knowledge of the properties of **triangles** and **quadrilaterals** to solve problems. Level 6
- **Pythagoras' theorem** is $h^2 = a^2 + b^2$, where h is the length of the hypotenuse and a and b are the lengths of the shorter sides. Level 7

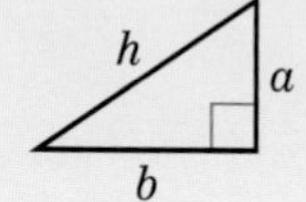

- You can use Pythagoras' theorem to solve problems. Level 7

Why learn this?

To make a perfect football a manufacturer needs to know the properties of shapes and angles to make sure the sections fit together correctly.

Practice, practice, practice!

Level 6

6c I can find the missing angle in a triangle given the other two angles

1 Find the size of the unmarked angles in these triangles.

a 82°, 23°

b

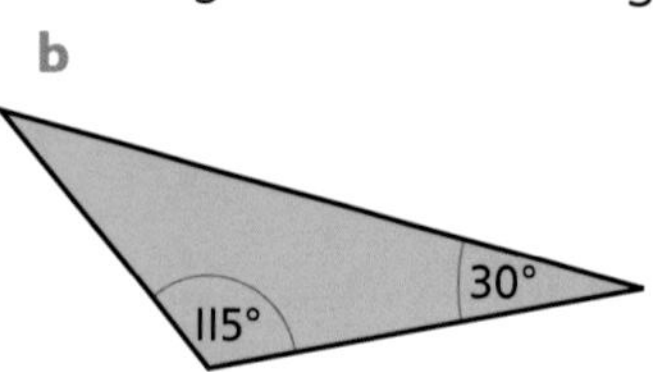

c

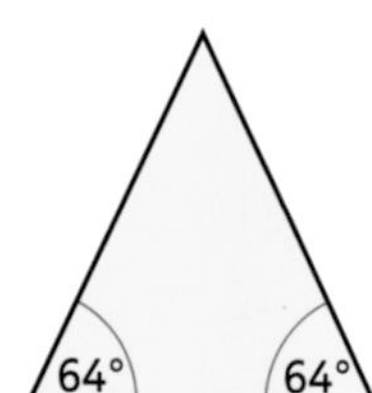

6b I can solve problems, using what I know about the properties of triangles

2 These five identical triangles fit round a point.
Work out the sizes of angles a, b and c.

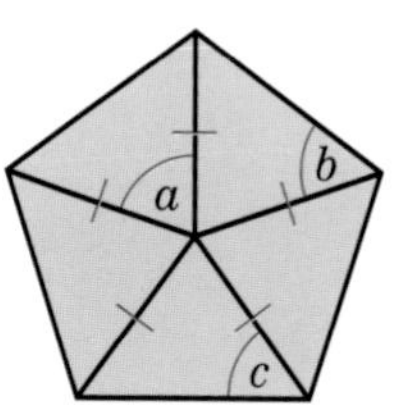

3 Which of these isosceles triangles can fit around a point, as in Q2?
If they do fit, how many of each triangle are used?

a

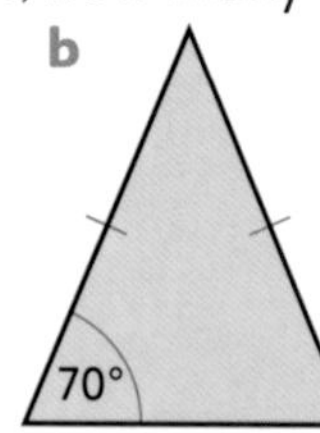

b 70°

c

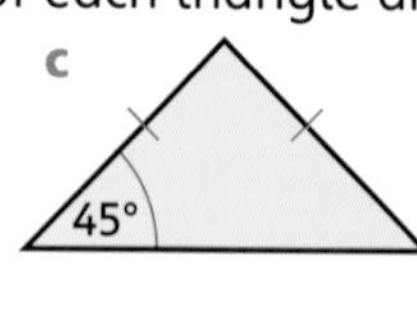

d

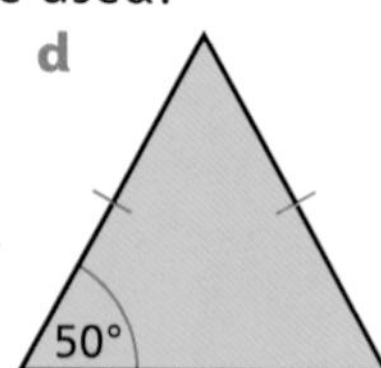

e

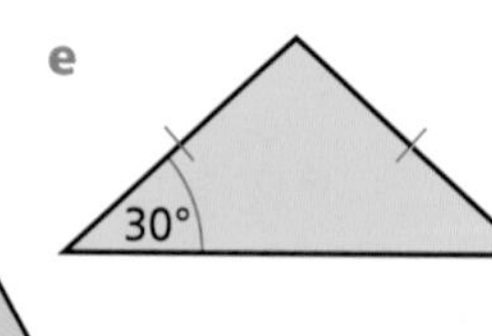

6a I can find and explain solutions to problems involving quadrilaterals

4 Copies of this isosceles trapezium can be fitted around a point in different ways.
Draw a sketch of each one you find.
Mark angles to show how the shapes fit.

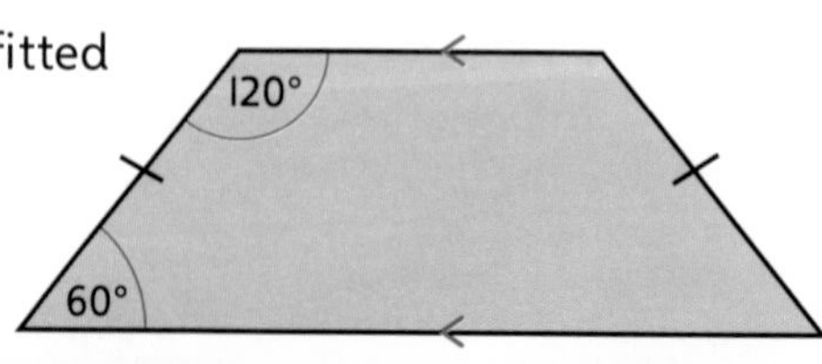

hypotenuse isosceles Pythagoras' theorem

5 Which of these quadrilaterals can fit around a point?
If they do fit, how many will fit? In each case, explain your reasoning.

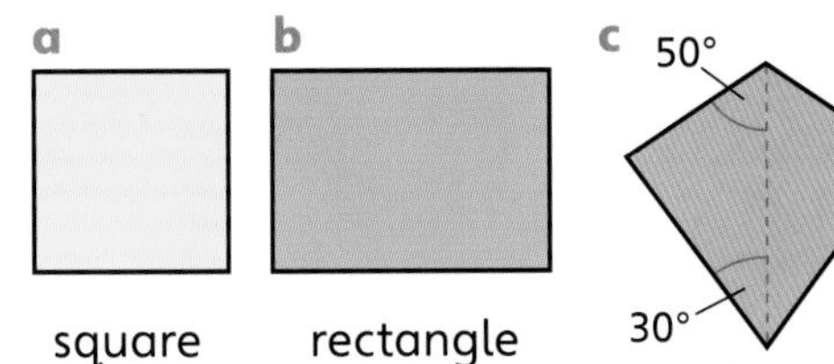

c kite (50°, 30°)
d parallelogram (150°)
e parallelogram (45°)

Level 6

6a I can find and explain solutions to problems involving quadrilaterals

6 Use Pythagoras' theorem to calculate the unmarked lengths.
Give your answers to one decimal place.

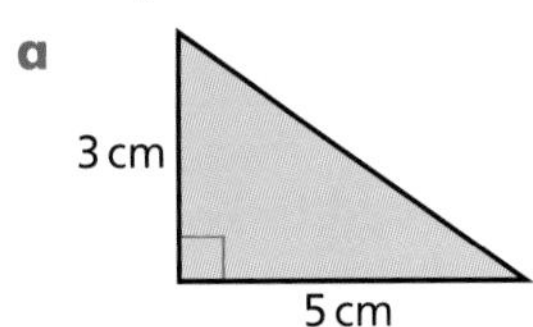

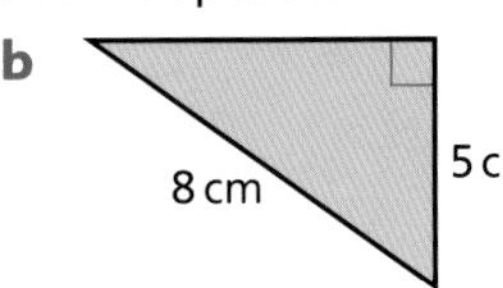

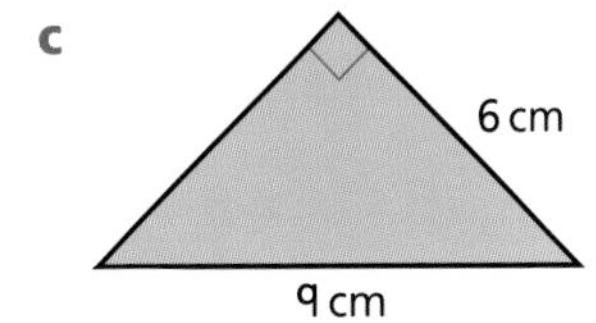

Level 7

7b I can use Pythagoras' theorem to find missing lengths

7 Alex is making a corner cupboard.
The door forms a triangle with the walls.
He has a 75 cm piece of wood.
Will it be long enough for the door?
Explain your reasoning.

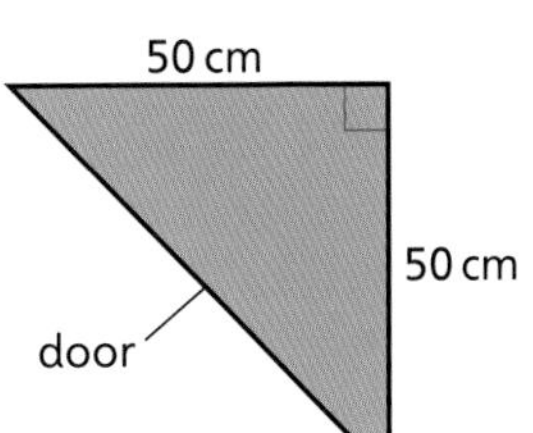

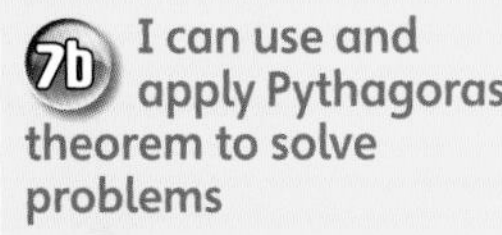

7b I can use and apply Pythagoras' theorem to solve problems

8 Alex wants to check whether the corner of a room is a right angle.
He measures lengths along two walls and the diagonal, as shown.
Is the corner a right angle?
Explain your reasoning.

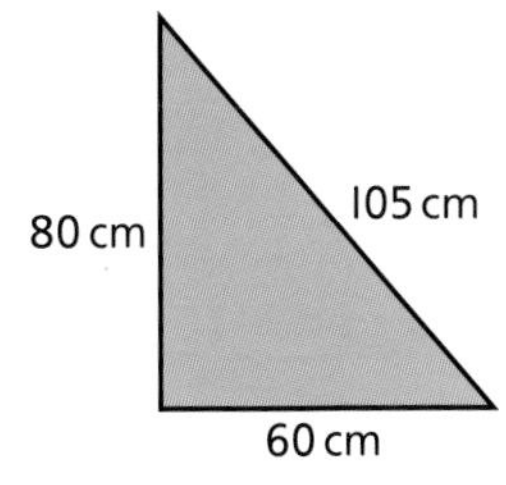

9 Work out the length of the long side of this trapezium.

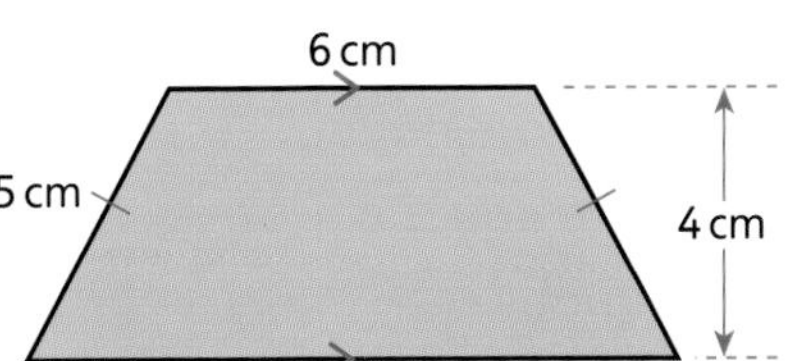

Now try this!

A Make an Escher tessellation

- Start with a square and cut out a shape from one side, then add it to the opposite side. Show how the new shape tessellates.
- Experiment by cutting from the top or bottom of the original square as well. Try to make a tessellation of creatures.

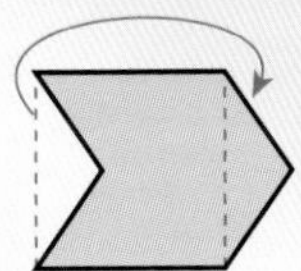

B Semi-regular tessellations

A semi-regular tessellation is formed by two or more different regular polygons. The arrangement of polygons at every vertex is the same.

- Find two ways of tessellating a regular triangle and hexagon.
- Which other semi-regular tessellations can you find?

quadrilateral   tessellate

14.2 More geometrical problems

- Solve problems using properties of angles, parallel and intersecting lines, and of triangles and other polygons
- Justify reasons and explain reasoning with diagrams and text

Why learn this?

Many patterns on fabrics, wallpaper and carpets are geometric, with repeating patterns. To design these patterns you need to know the geometrical properties of the lines and shapes.

What's the BIG idea?

- **Corresponding angles** are equal. Level 6
- **Alternate angles** are equal. Level 6

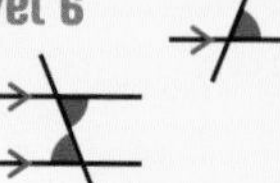

- You can use the angle rules to solve problems. Level 6 & Level 7
- To solve a geometrical problem with reasons or justification, you need to present **evidence** and a reason for each step. Level 7

Practice, practice, practice!

Level 6

6c I can understand the angle rules

1 For each of these angle rules, make up a question to test another pupil.

Angles on a line add up to 180°.

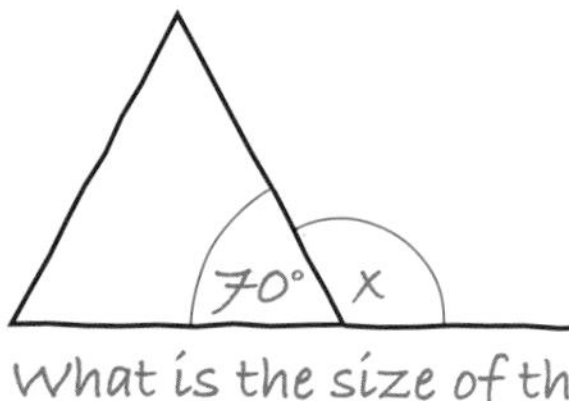

a Vertically opposite angles are equal.
b Corresponding angles are equal.
c Alternate angles are equal.
d An angle rule of your choice.

6c I can solve simple problems using alternate and corresponding angles and the properties of quadrilaterals

2 Find the size of each lettered angle. Give reasons for your answers.

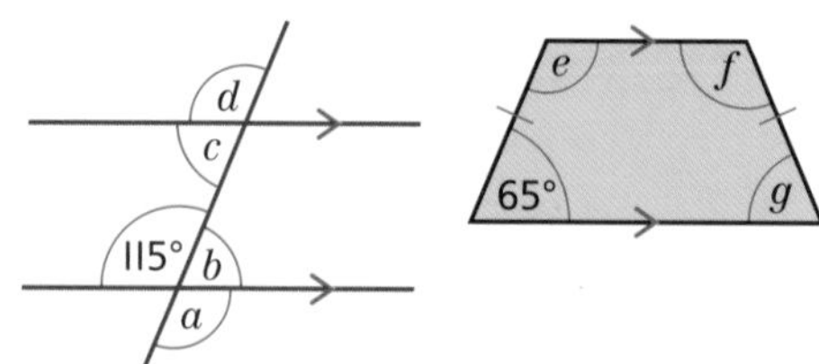

6b I can use reasoning to solve geometrical problems

3 Work out the sizes of the angles marked p and q. Give reasons for your answers.

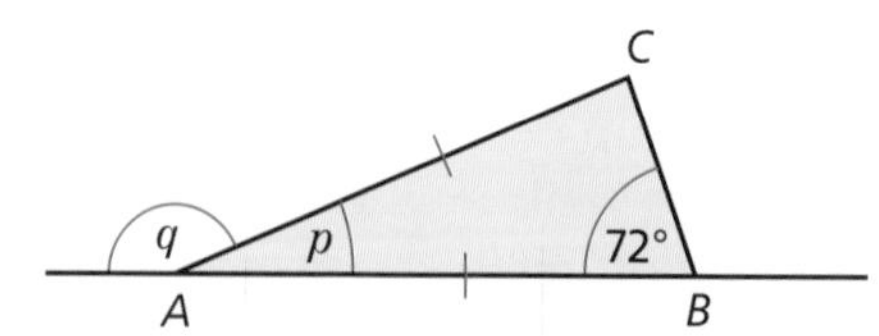

alternate angles corresponding angles

Level 7

7c I can solve more complex geometric problems

4 This diagram shows two squares.
Find the sizes of the lettered angles.
Explain how you found your answer.

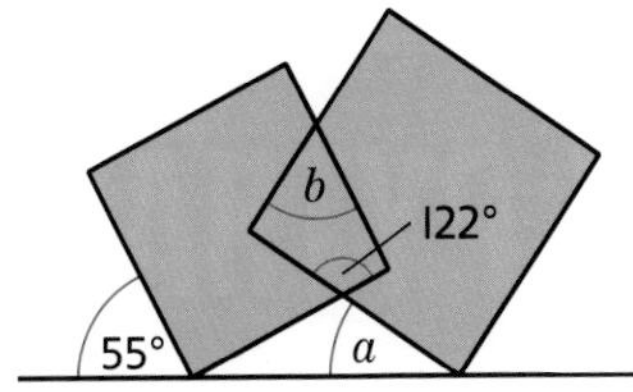

5 Find the values of c and d.
Justify your answer.

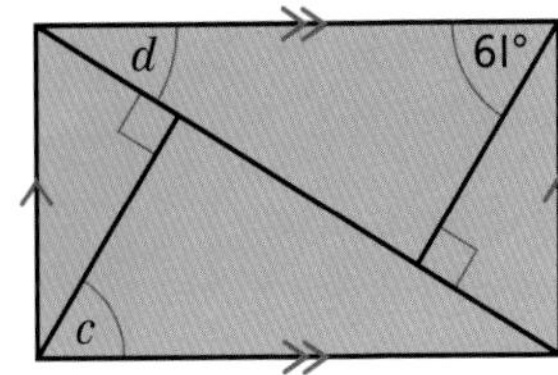

Super fact!
The Alhambra palace in Spain is famous for its geometric patterns.

6 This five-pointed star is made from a regular pentagon and five congruent triangles.

Find the values of e and f, giving reasons.

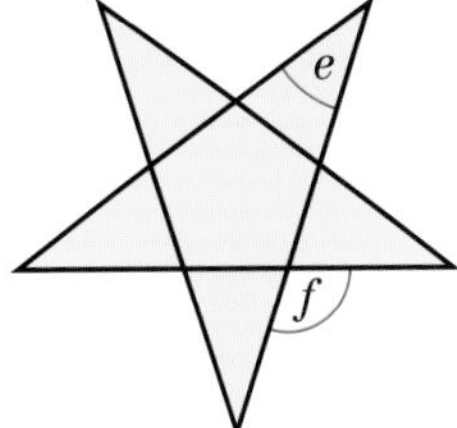

Tip
A regular pentagon has all sides and all angles equal.

7b I can justify inferences and explain reasoning using diagrams and text

7 Regular pentagons are put next to each other as shown in the diagram.

a Find the sizes of the angles marked a and b, explaining your reasons.

b How many more pentagons are needed to make a complete ring? Explain your reasoning.

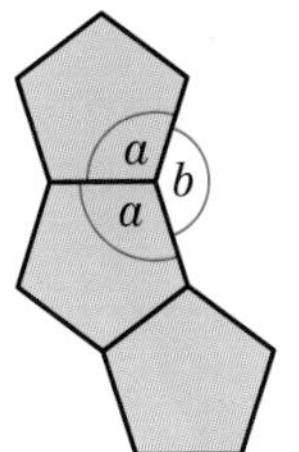

8 How many regular hexagons are needed to form a ring like the one in Q7?

9 This ring is made from squares and equilateral triangles.

a Find the sizes of angles around the point *A*.

b Sketch a ring made from squares and octagons. Label the angles at one point.

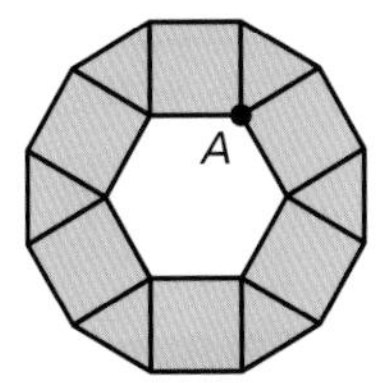

Now try this!

A Rotating pentagon

1 Draw a regular pentagon.
2 Rotate it about one vertex as shown.
3 Draw around the outside of the shape that is formed by the two pentagons.
4 What shape has been formed?
5 Repeat steps 2 to 4 until you have rotated the pentagon back to its starting point.
6 Find the sizes of the angles that form the final shape.

B Pentagon ring

You made a ring from pentagons in Q7.
Here is a different pentagon ring.

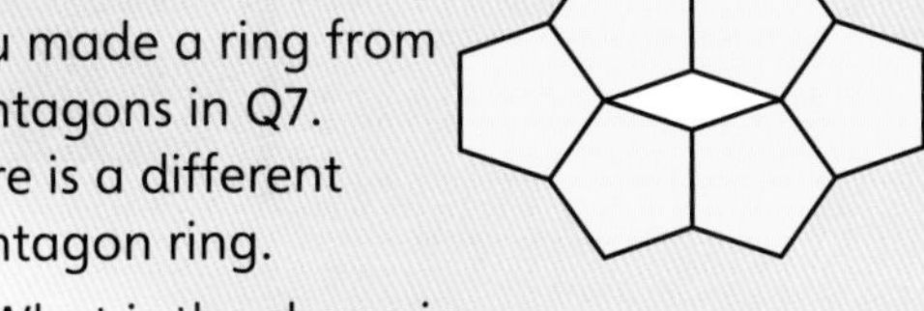

a What is the shape in the middle? Calculate the angles to justify your answers.

b Investigate the different ways to make a pentagon ring – how many pentagons are needed, and what is the shape in the middle?

evidence

maths! presents

the story of PYTHAGORAS

Prove it!

The diagram below is the basis of just one of the many ways in which you can prove Pythagoras' theorem.

Pythagoras was born around about 580 BC on the Greek island of Samos.

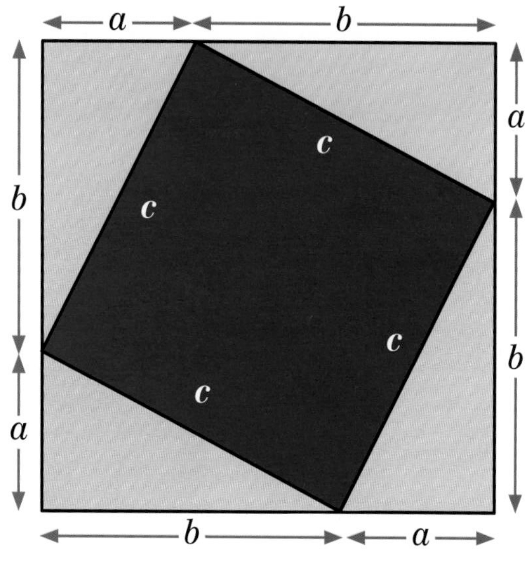

Rearrange the statements A to G into the correct order to form a proof.

A The area of the small square is c^2 and the area of each triangle is $\frac{1}{2}ab$.

B Therefore $a^2 + 2ab + b^2 = c^2 + 2ab$

C Divide a square as shown into four congruent right-angled triangles and a smaller square.

D Taking $2ab$ away from both sides gives $a^2 + b^2 = c^2$.

E The area of the big square is $(a + b)^2$.

F The total area of the small shapes is $c^2 + 2ab$.

G $(a + b)^2 = a^2 + 2ab + b^2$

Pythagoras led a group of men and women who studied under him and followed his teaching. They were known as Pythagoreans. Pythagoras' inclusion of women on equal terms with men was very unusual in Ancient Greece. Within the group there was an inner circle of students called the Mathematikoi.

Say not little in many words, but much in few.

Pythagoreans were not allowed to own any personal possessions and they followed a vegetarian diet.

He studied in Italy and in Egypt, where he might have come across the idea for his famous theorem.

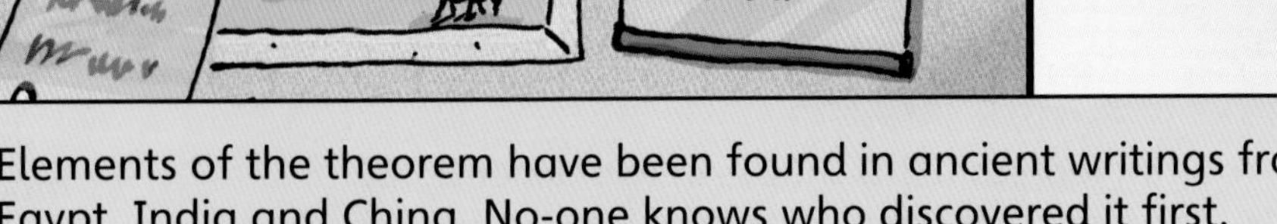

Elements of the theorem have been found in ancient writings from Egypt, India and China. No-one knows who discovered it first.

Beyond squares

Does Pythagoras' theorem work with other shapes?

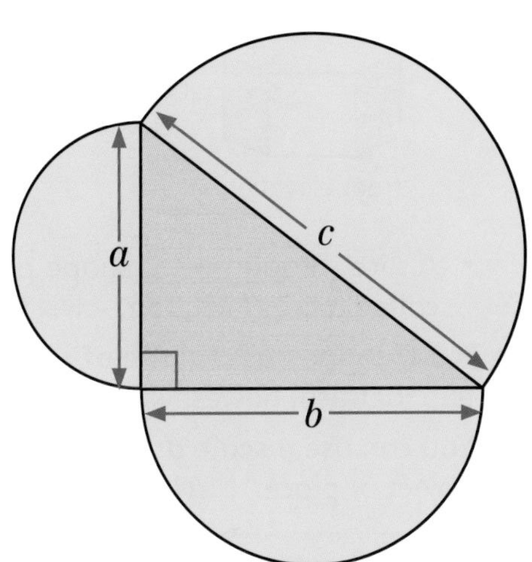

- Calculate the areas of the three semicircles when $a = 3$ cm, $b = 4$ cm and $c = 5$ cm.
- Is the area of the largest semicircle equal to the sum of the areas of the two smaller semicircles?
- Write down an algebraic expression for the area of each of the three semicircles.
- Use your expressions to prove that the area of the largest semicircle is always equal to the sum of the areas of the two smaller semicircles.

Pythagoras thought that everything should be neatly explained by whole numbers. One story says that when his student, Hippasus, proved that $\sqrt{2}$ could not be written as a whole-number fraction, Pythagoras had him thrown off a ship.

Pythagoras also studied the connection between mathematics and music. He died around about 500 BC.

14.3 3-D shapes

- Use plans and elevations
- Visualise and use a wide range of 2-D representations of 3-D objects
- Analyse 3-D shapes through informal 2-D representations, cross-sections, plans and elevations

Why learn this?

3-D shapes can be seen in boxes, buildings, designs and computer models. The architect of this building needed to know the properties of the shapes when designing the building.

What's the BIG idea?

→ You can represent 3-D shapes in two dimensions.
- The **front elevation** is the view looking from the front.
- The **side elevation** is the view looking from the side.
- The **plan view** is the view looking down from above.

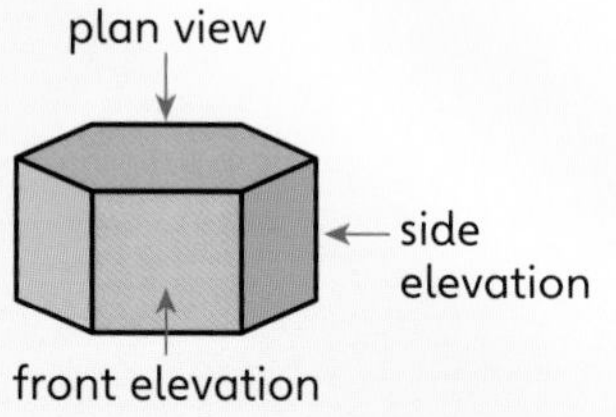

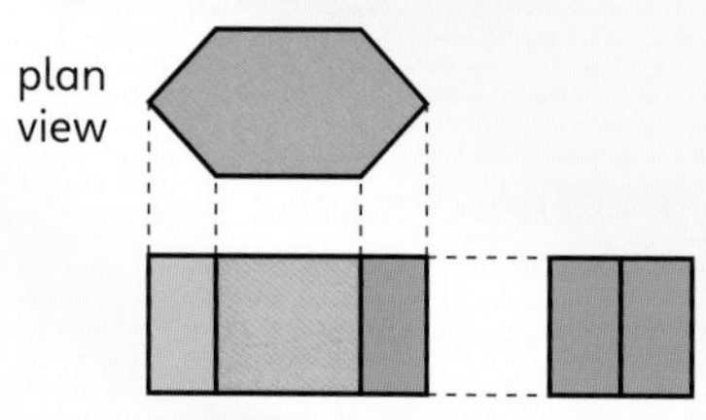

- A slice through a 3-D shape parallel to the base is called a **cross-section**. Level 6

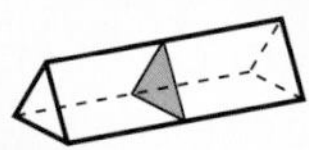

→ A 3-D shape has a **plane of symmetry** if the plane divides the shape into two identical halves, where one is the mirror image of the other. Level 7

→ You can use a **scale drawing** to work out the real-life dimensions of an object or place. Level 7

Super fact!

There are five regular 3-D solids. They are called the Platonic solids. The Ancient Greeks called them the atoms of the Universe.

Practice, practice, practice!

Level 6

6c I can visualise and use a variety of 2-D representations of 3-D objects

1 Sketch the front and side elevations and plan views of these shapes.

a

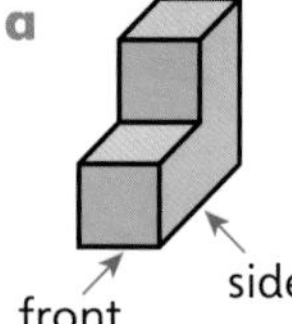

b

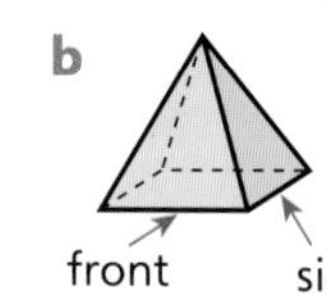

c

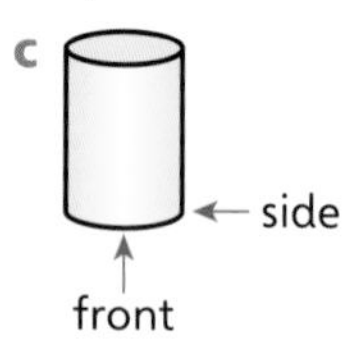

d

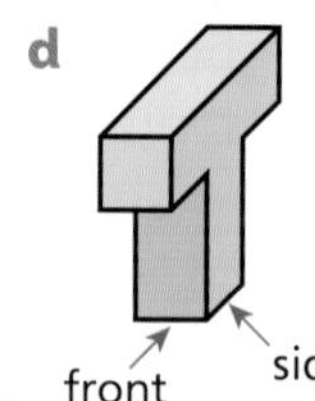

2 Sketch three solids which have this front elevation.

3 Sketch three solids which have this plan view.

6b I can analyse 3-D shapes by looking at 2-D representations

4 These are the shadows of shapes.
Describe and sketch the solid that makes each shadow.

A

B

C

Tip

There may be more than one solid that can make each shadow.

cross-section front elevation plan view

5 This is a plan of Willow Farm. Draw an accurate scale drawing using a scale of 1:20 000.

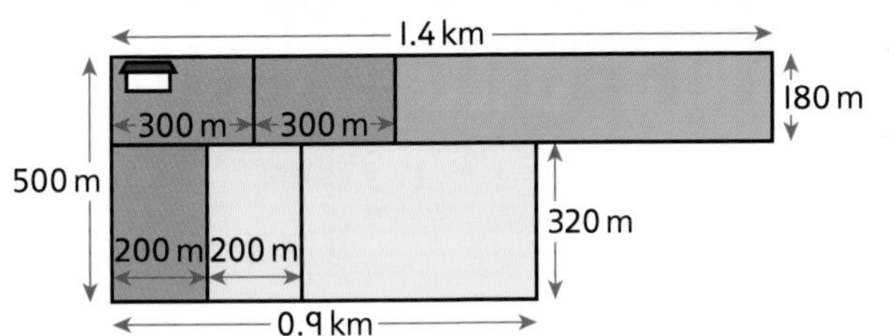

Level 6

6b I can draw diagrams to scale

6 For each set of views, draw the net and sketch the 3-D solid.

a

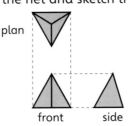

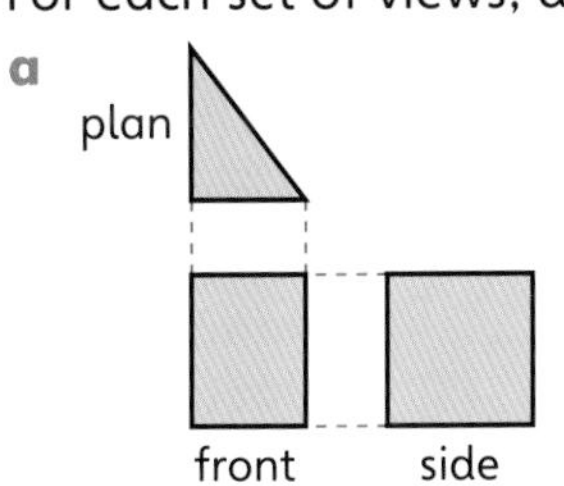

b

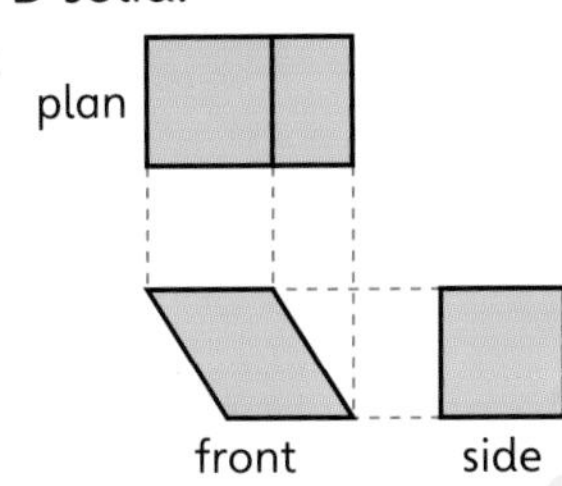

c

6a I can analyse 3-D shapes by looking at cross-sections, plans and elevations

7 Draw the cross-sections of these 3-D shapes.

a

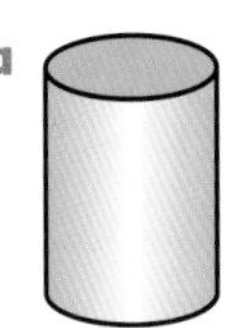

b

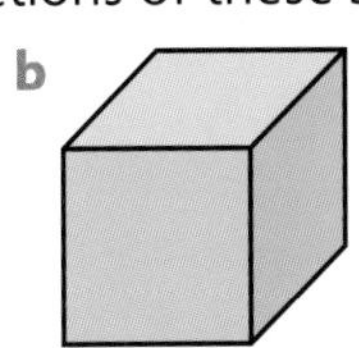

c

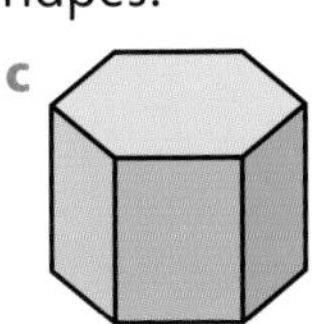

d

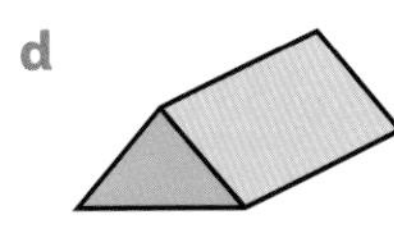

8 How many planes of symmetry does each of these solids have?

a

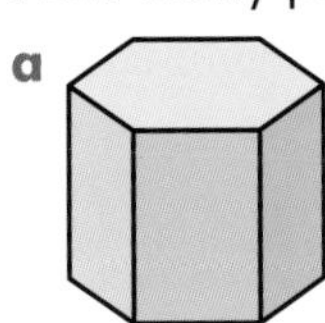

b

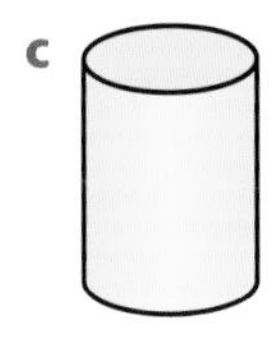

c

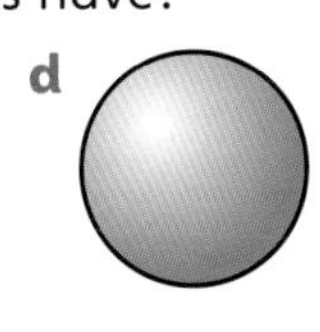

d

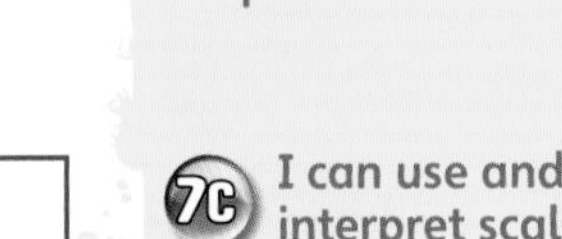

Level 7

7c I can identify planes of symmetry in 3-D shapes

9 Sam has made this scale drawing of a floor of his house.

a By measuring the scale mark, work out what length on the diagram represents 1 m in real life.

b What are the actual length and width of the living room?

c Sam wants to mark the position of the sofa in the living room. The sofa is 1.75 m long and 1 m wide. How big should it be on the scale drawing?

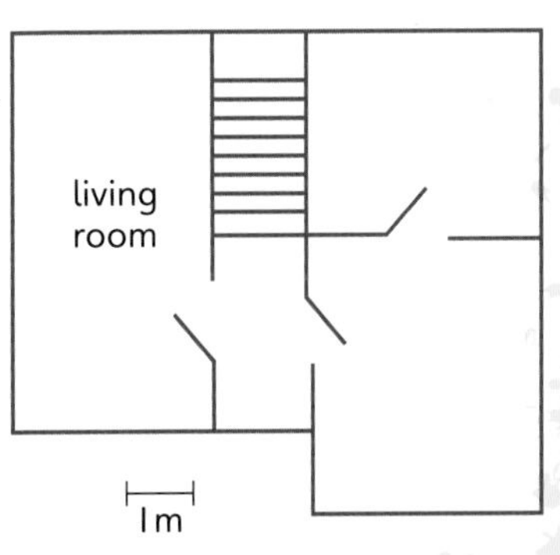

7c I can use and interpret scale drawings

Now try this!

A Shape sorting

The diagram show the top of a toy which has holes in it for a child to 'post' bricks into. Sketch the shapes of several different bricks that would fit through each hole.

B Creating 3-D shapes by reflection

Look at this blue 3-D shape. Imagine it reflected in two planes of symmetry at right angles. Sketch the new shape. How many planes of symmetry does it have? Draw your own 3-D shape and the new shape created when it is reflected in two planes of symmetry.

plane of symmetry scale drawing side elevation

14.4 Prisms and cylinders

- Calculate the volumes and surface areas of prisms
- Calculate the volumes and surface areas of cylinders
- Calculate the lengths and areas of prisms, given the volumes

Why learn this?

Many food items are sold in cylindrical tins. The manufacturers need to work out the area of metal required to make the tins.

What's the BIG idea?

→ To calculate the **volume** of a **prism**, find the area of its **cross-section** and then multiply by the length.

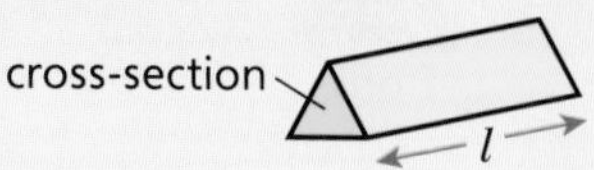

V = area of cross-section × length **Level 7**

→ To calculate the **surface area** of a prism, work out the area of each face and add them together. **Level 7**

→ The formula for the volume of a **cylinder** = $\pi r^2 h$. **Level 8**

→ The formula for the surface area of a cylinder = $2\pi rh + 2\pi r^2$. **Level 8**

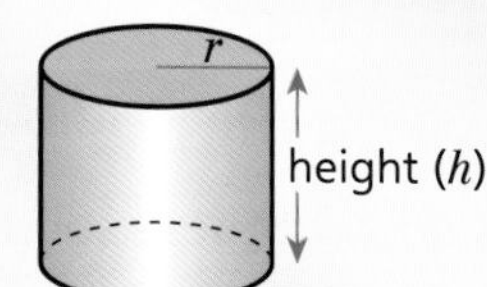

Tip

When working out the surface area of a solid, sketch a net to make sure that you have included all of the faces.

Practice, practice, practice!

Where necessary, use the π key on your calculator.

Level 7

7c I can calculate the volumes of prisms

1 Find the volume of each of these prisms. Give your answers to 1 d.p.

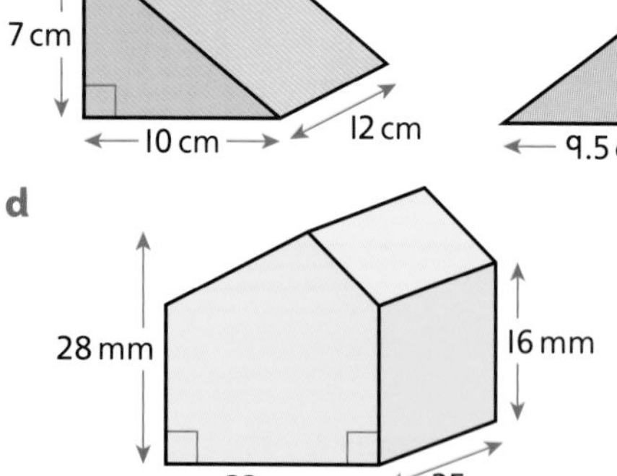

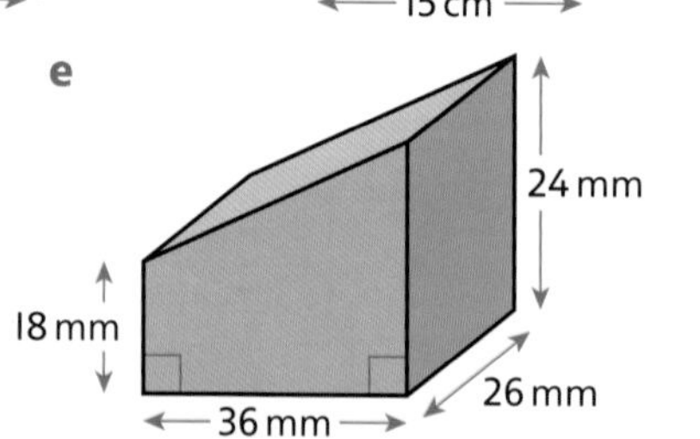

Learn this

A prism is a 3-D shape that has the same cross-section along its length.

2 A prism is 12 cm long.
Its ends are shaped as shown in the diagram, with a semicircular curved section.
Find its volume.

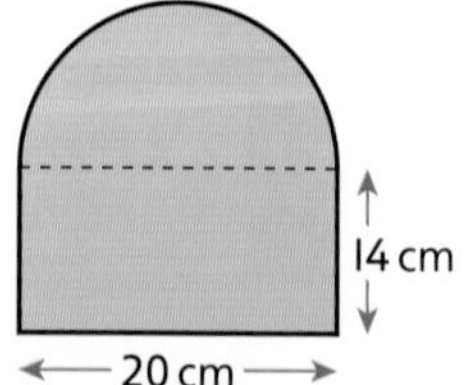

cross-section cuboid cylinder

Level 7

7b I can calculate the surface area of cylinders

3 Find the surface area of each of these prisms.

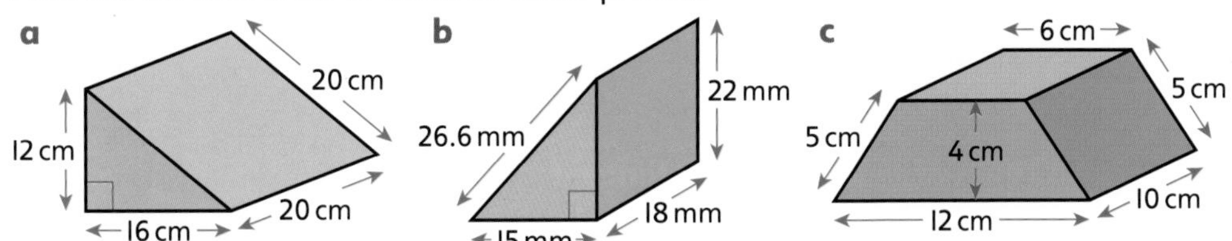

7a I can calculate the length and surface area of a prism given the volume

4 The area of the cross-section of a triangular prism is 220 cm^2. The volume is 860 cm^3. Work out the length of the prism.

5 The volume of a triangular prism is 120 cm^3. Its cross-section is a right-angled triangle with base 6 cm and height 8 cm. Work out its surface area.

7a I can calculate the volumes of cylinders

6 Calculate the volume of each of these cylinders.

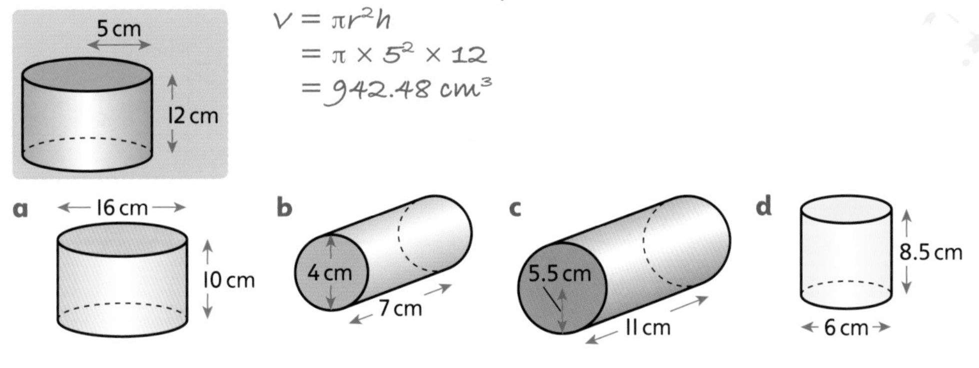

7 An oil drum is 1.2 m high and has diameter 0.6 m. Work out the volume of the oil drum.

Level 8

8b I can calculate the surface area of a cylinder

8 Work out the surface area of each of these closed cylinders.

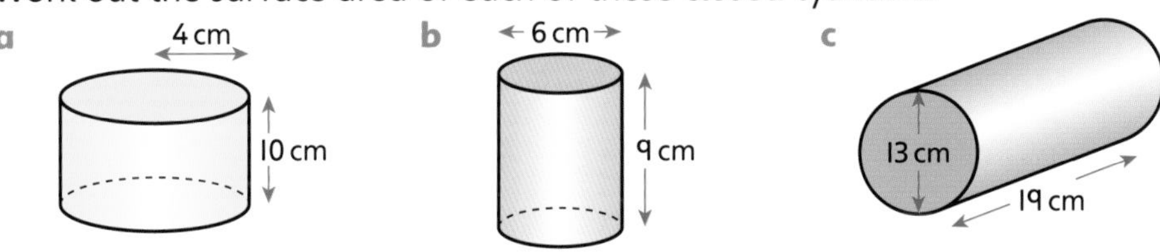

8a I can calculate the length or radius of a cylinder given the volume

9 The volume of a cylinder is 500 mm^3. The radius is 5 mm. Work out the height of the cylinder.

10 The volume of a cylinder is 850 m^3. The height is 11.5 m. Work out the radius of the cylinder.

Now try this!

A Card match

Work with a partner. You need 18 blank cards. On one card, draw a prism and mark its dimensions. On a second card write the volume of the prism and on a third, write the surface area of the prism. Repeat this for five different prisms, so that you have a total of six prisms.
Shuffle the cards and challenge another pair to sort them into sets of three.

B Investigating cylinders

A cylindrical can has volume 500 cm^3. What dimensions might it have?

Learn this

You need to learn the formula for the volume of a cylinder.

prism surface area volume

14.5 Using trigonometry 2

- Understand the trigonometric ratios of sine, cosine and tangent
- Use sine, cosine and tangent to find the lengths of sides and sizes of angles in right-angled triangles
- Identify the right-angled triangle and the ratio you need to use to solve the problem

Why learn this?

Trigonomtry is used to manoevre the robotic arm on the International Space Station.

→ In any right-angled triangle,

$$\sin\theta = \frac{\textbf{opposite}}{\textbf{hypotenuse}} \qquad \cos\theta = \frac{\textbf{adjacent}}{\text{hypotenuse}} \qquad \tan\theta = \frac{\text{opposite}}{\text{adjacent}}$$

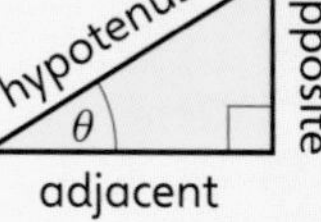

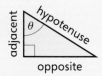

You can rearrange these trigonometric ratios to find the lengths of sides and the sizes of angles in right-angled triangles. **Level 8**

Practice, practice, practice!

Level 8

8C I can use trigonometry to find the lengths of unknown sides in a right-angled triangle, using straightforward algebraic manipulation

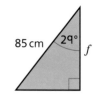

1 Calculate the lettered lengths in these triangles. Give your answers to 1 d.p.

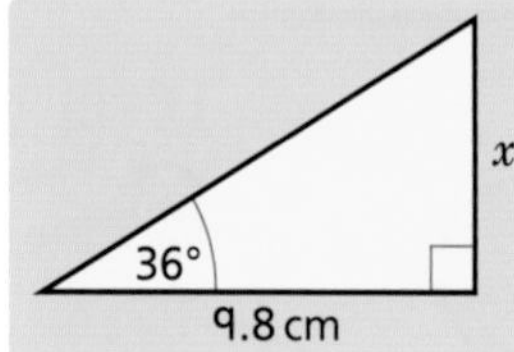

The lettered side is opposite the given angle.
The 9.8 cm side is adjacent to the given angle.
So the ratio to use is tangent.

$\tan 36° = \frac{\text{opp}}{\text{adj}} = \frac{x}{9.8}$

So $x = 9.8 \times \tan 36° = 7.1$ cm (to 1 d.p.)

Watch out! Check that your calculator is in degree mode before you start.

a

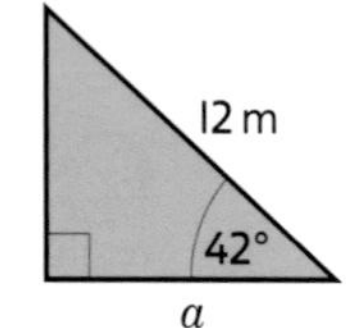

b

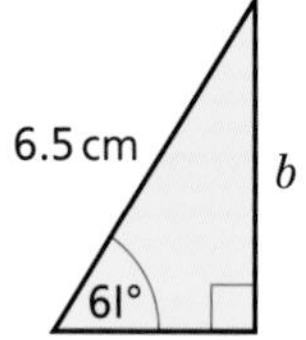

c

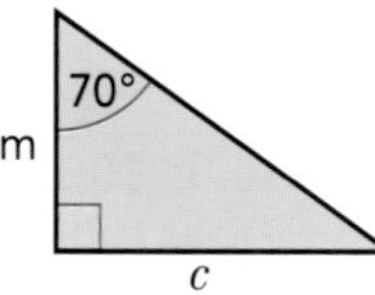

d

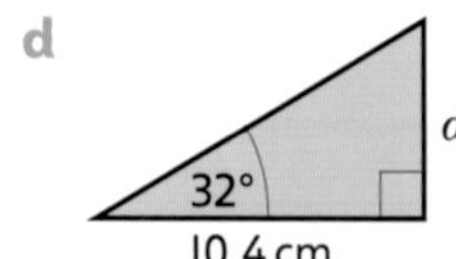

e

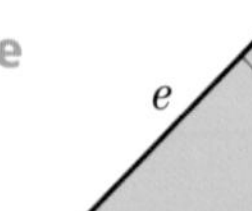

f

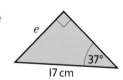

85 cm, 29°, f

2 Calculate the lettered lengths in these triangles. Give your answers to 1 d.p.

a

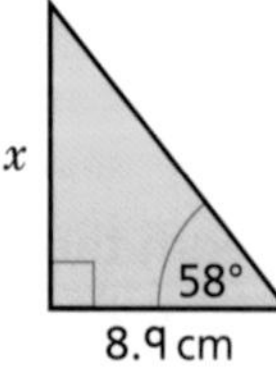

b

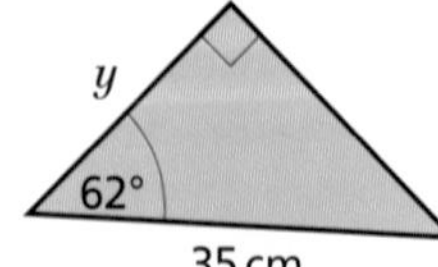

c

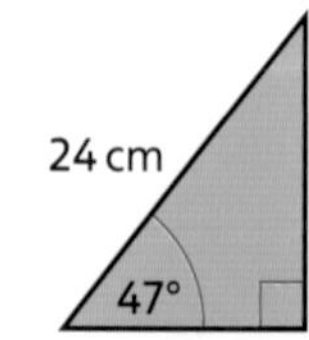

Watch out! The triangles are not drawn to scale.

adjacent cosine (cos) hypotenuse

3 Calculate the lettered lengths in these triangles to 3 s.f.

a
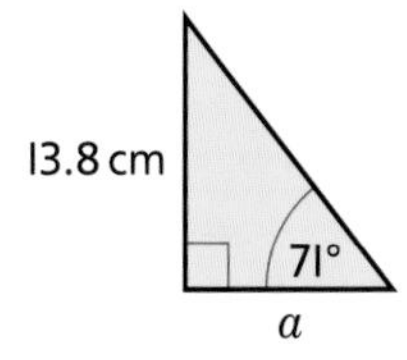

b
7.5 cm
b
28°

c
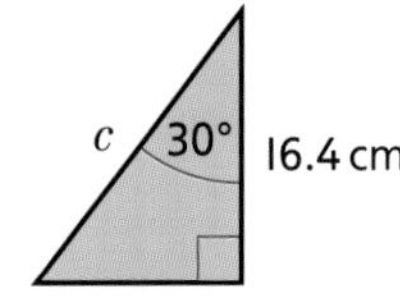

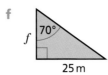

d
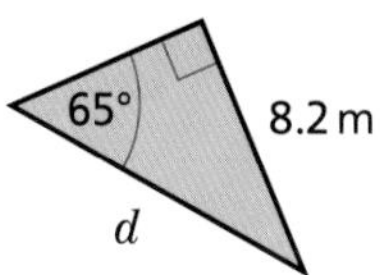

e
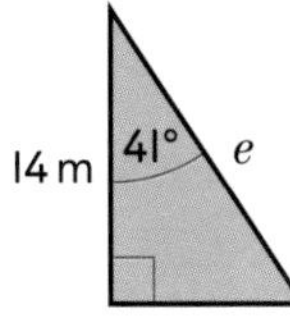

f
f
70°
25 m

4 Find the size of each of the lettered angles to 3 s.f.

a
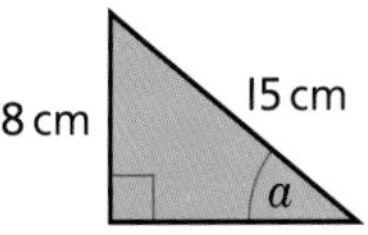

b
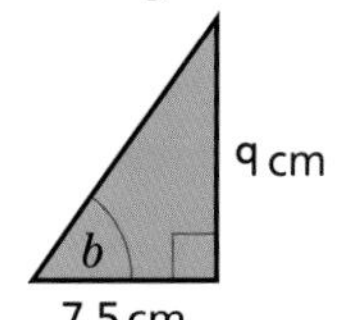

c
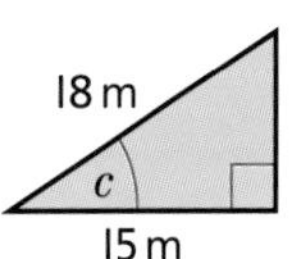

d
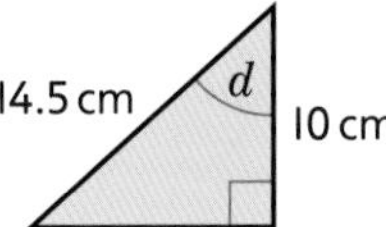

e
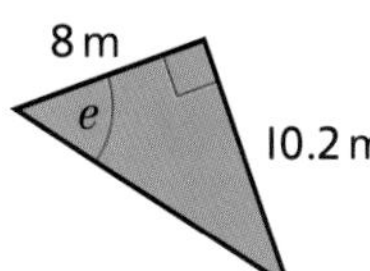

f
f
165 mm
110 mm

5 Work out the size of each lettered angle to 3 s.f.

a
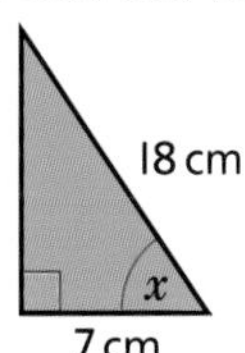

b
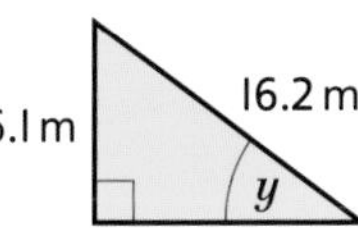

c
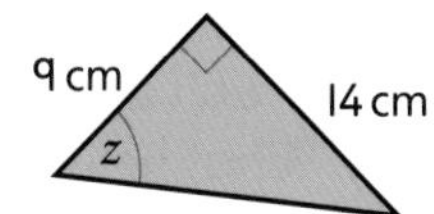

6 Find the size of the lettered angle in each of these triangles.
Give your answers to 3 s.f.

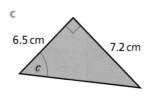

a
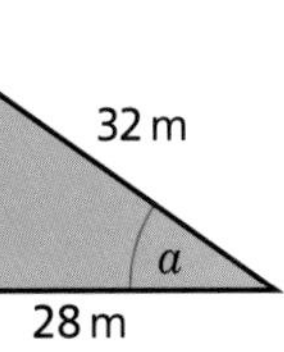

b
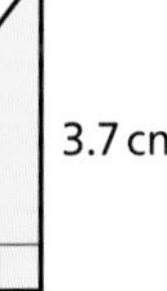

c
6.5 cm
7.2 cm
c

d
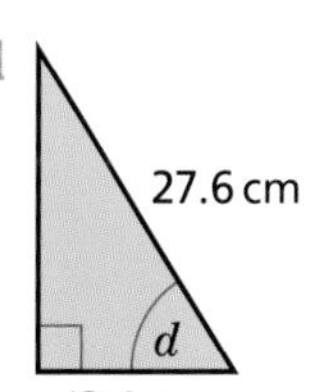

Level 8

8b I can use trigonometry to find the unknown lengths of sides in a right-angled triangle, using more complex algebraic manipulation

8a I can use the trigonometric ratios to find the size of an angle in a right-angled triangle

Now try this!

A Trigonometry questions

Make up three questions about triangles with a mystery angle or side for your partner to solve. Swap questions then mark them – how well do you do?

B Investigating tangents

a Use your calculator to find the values of tan 0°, tan 10°, tan 20° etc. Write all values up to tan 180° to 2 d.p.

b Plot a graph of the tangent ratio. What do you notice at 90°? Why is this?

Did you know?

Before calculators, people had to look up the values of sine, cosine and tangent in tables and then do the calculations by hand.

opposite sine (sin) tangent (tan)

14.6 Solving problems in trigonometry

- Identify the right-angled triangle and the ratio you need to use to solve the problem
- Use trigonometry to solve problems in two dimensions

Why learn this?

Astronomers use trigonometry to work out how far away stars are from Earth.

What's the BIG idea?

→ You can use the trigonometric ratios of **sine**, **cosine** and **tangent** to solve problems. Level 8

Learn this

You need to learn the trigonometric ratios.

$\sin\theta = \dfrac{\text{opposite}}{\text{hypotenuse}}$ $\cos\theta = \dfrac{\text{adjacent}}{\text{hypotenuse}}$

$\tan\theta = \dfrac{\text{opposite}}{\text{adjacent}}$

Practice, practice, practice!

Level 8

8b I can use trigonometry to solve a range of problems.

1 The length of the diagonal of a square is 17 m. Find the side length of the square.

Tip Draw a diagram to help you.

2 A ladder leans against a wall at an angle of 75° to the ground. The ladder is 4 m long. What height does it reach up the wall?

8a I can use trigonometric ratios to find a length, or angle, in a right-angled triangle.

3 Work out the missing length or angle in each triangle. Give all your answers to 3 s.f.

a

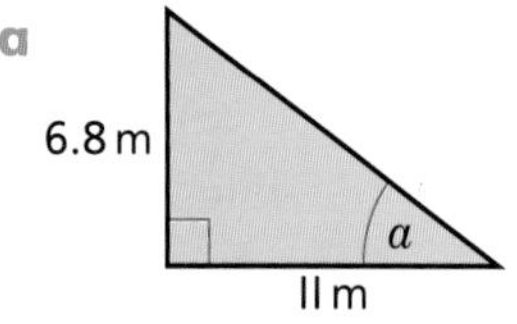

b

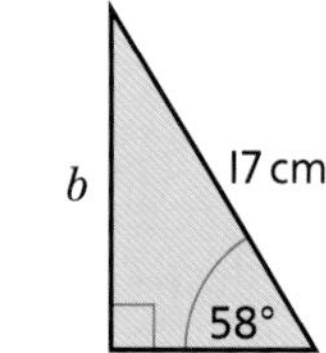

c

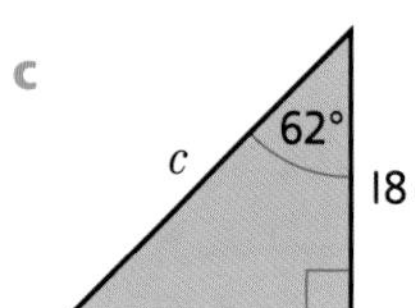

d

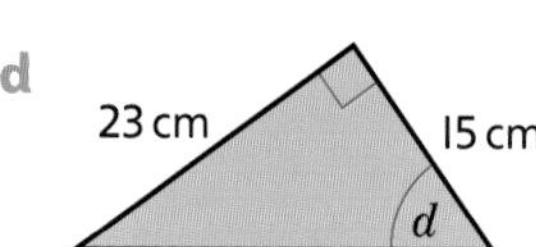

e

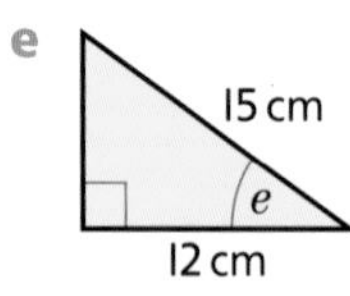

f

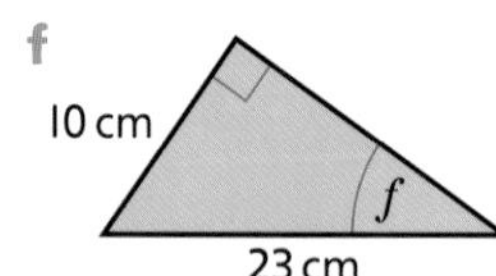

8a I can use trigonometry to solve a range of problems.

4 Work out the size of the angle at the corner A of this kite.

Hint: Find the length of the blue line.

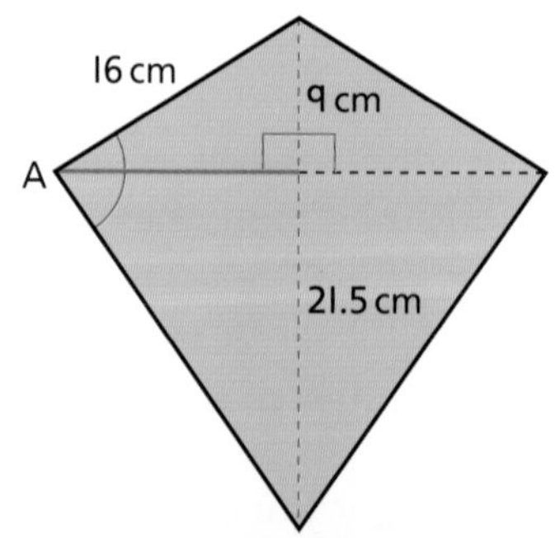

cosine sine

5 A 3 m high scramble net has two supporting wires in line with each other as shown.

Find the distance between the pegs that fasten the wires into the ground.

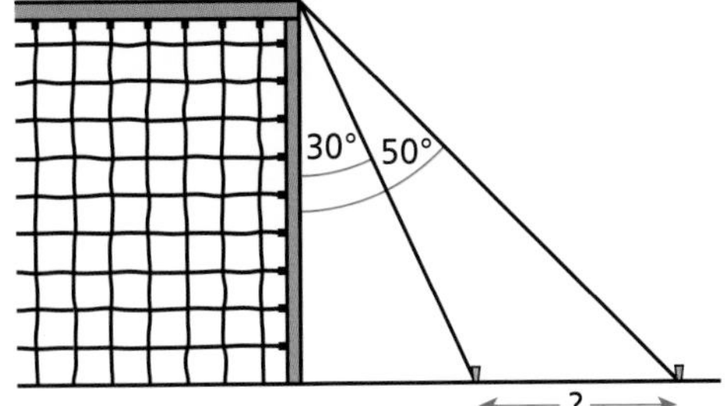

Level 8

8a I can use trigonometry to solve a range of problems.

6 A ship is 2 km off the coast and due south of a lighthouse.

The ship sails due east.
Some time later, the bearing from the ship to the lighthouse is 317°.

How far has the ship sailed?

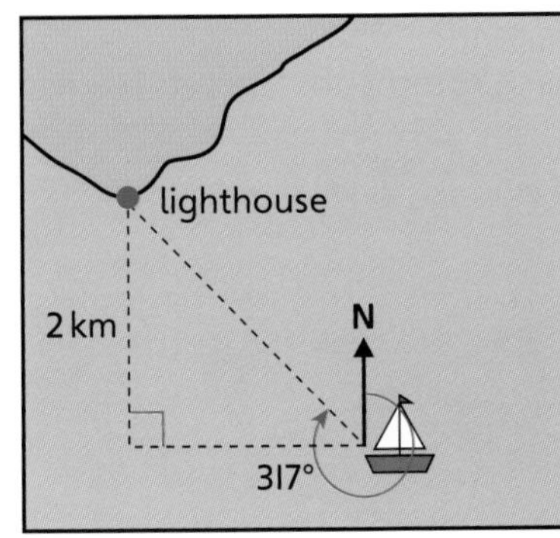

Super fact!

The word 'sine' comes from the Latin 'sinus' meaning 'bay' or 'inlet'. It was probably a 12th century mis-translation of an arabic term in an algebra text.

7 Find the area of each of these triangles.

a

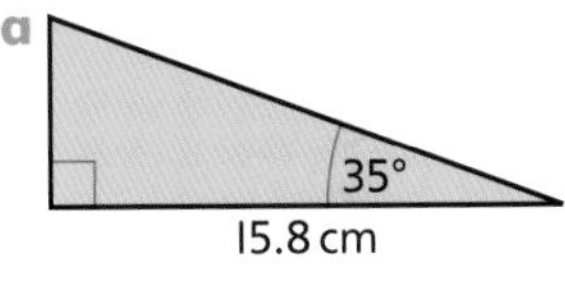

b

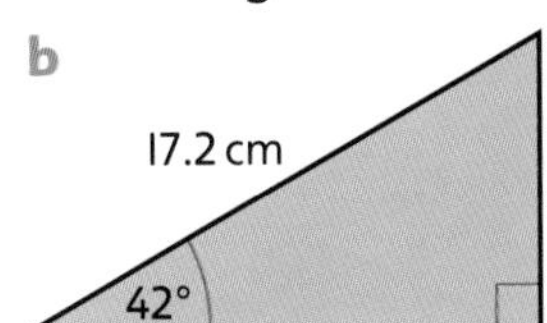

c

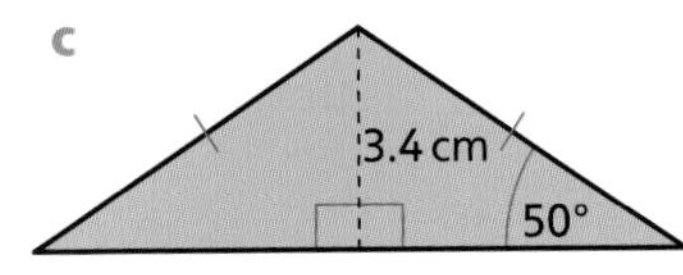

8 Here is a regular hexagon.

a Work out the height of the orange triangle.

b Work out the area of the orange triangle.

c Work out the area of the regular hexagon.

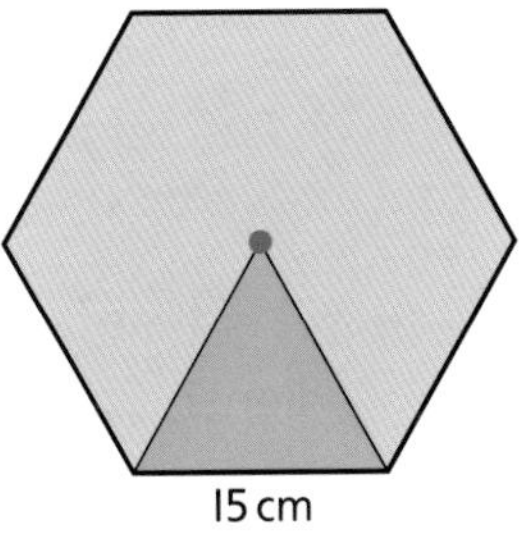

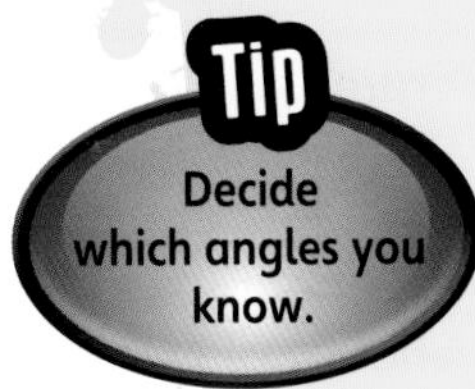

9 A regular octagon has side length 12 cm.
Use a similar method to the one in Q8 to find its area.

Now try this!

A Height of a tree

How can you use trigonometry to find the height of a tree?

Hint: You may need to add on your own height to get the height of the tree.

B Area of a decagon

Use a spreadsheet and the method in Q8 to find the area of a regular decagon with any length side.
Adapt your spreadsheet to find the area of an n-sided polygon with side length s.

tangent

Accessibility problems

All public places in the UK must be accessible for wheelchair users. This can be achieved using wheelchair ramps. These ramps might be permanent fixtures, allowing access to a public building, or they might be automatic, such as the boarding ramps fitted to buses and coaches.

1 The diagram shows a permanent ramp constructed from two triangular sections.
What are the sizes of angles a and b? **Level 6**

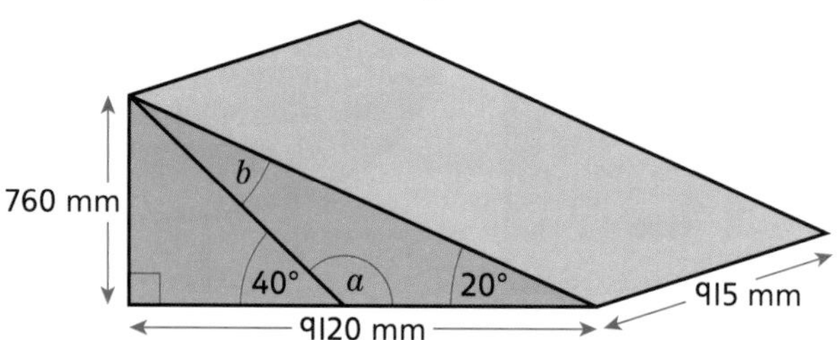

2 The ramp shown in Q1 is made of concrete. What volume of concrete will be required?
Give your answer in cubic metres. **Level 7**

3 What is the sloping length of this ramp?
Give your answer in metres. **Level 7**

20 cm
2 m

4 A ramp of sloping length 10 m has an angle of elevation of 8°.
What is the horizontal distance covered by this ramp? **Level 8**

5 Most people who use wheelchairs can manage a slope of 1 in 16.
What angle would give this gradient? **Level 8**

6 Look at the diagram of a vehicle boarding ramp.

a What is the maximum horizontal distance of zone A?

b What is the minimum horizontal distance? **Level 8**

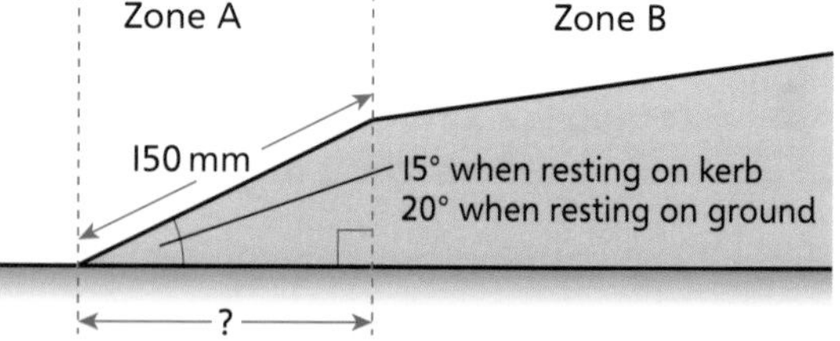

7 A wheelchair ramp is to be built to go up three steps.
The ramp will have an angle of elevation of 4.8°.
The steps are 18 cm high and have a depth of 25 cm.
Calculate distance x to find out how far away from the steps the ramp should start. **Level 8**

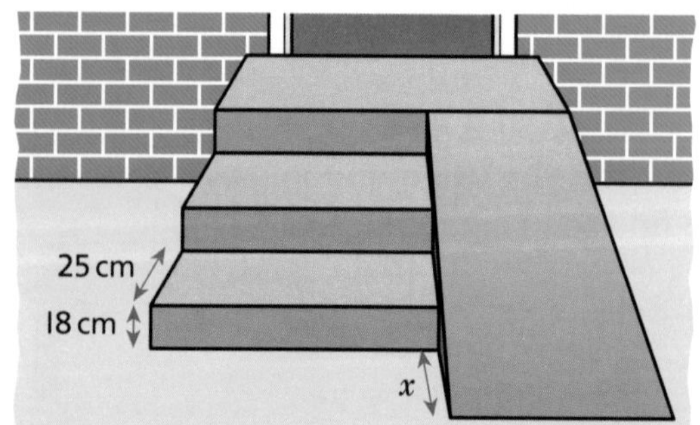

The BIG ideas

→ You can use your knowledge of the properties of triangles and **quadrilaterals** to solve problems. **Level 6**

→ To calculate the **volume** of a **prism**, find the area of its **cross-section** and then multiply by the length.
V = area of cross-section × length **Level 7**

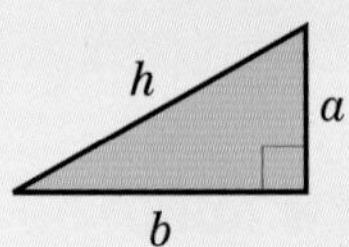

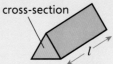

→ **Pythagoras' theorem** is $h^2 = a^2 + b^2$, where h is the length of the hypotenuse and a and b are the lengths of the shorter sides. **Level 7**

→ You can use Pythagoras' theorem to solve problems. **Level 7**

→ In any right-angled triangle,

- $\sin \theta = \dfrac{\text{opposite}}{\text{hypotenuse}}$
- $\cos \theta = \dfrac{\text{adjacent}}{\text{hypotenuse}}$
- $\tan \theta = \dfrac{\text{opposite}}{\text{adjacent}}$

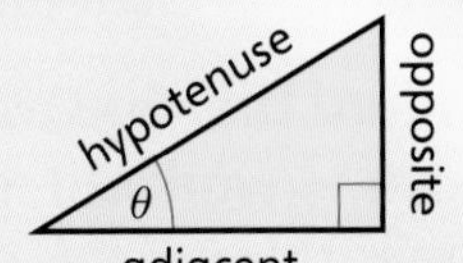

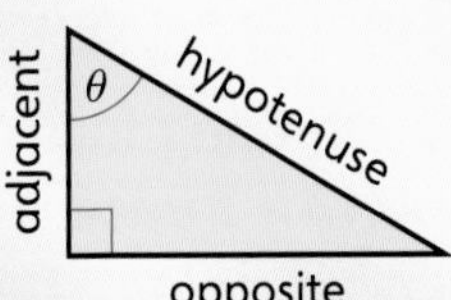

You can rearrange these formulae to find the lengths of sides and the sizes of angles in right-angled triangles. **Level 8**

Find your level

Level 6

Q1 The volume of this cuboid is 200 cm^3. What is the width, w, of the cuboid?

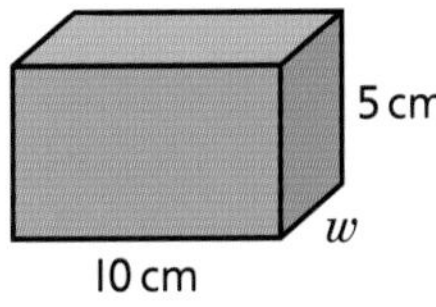

Q2 **a** Work out the size of angle a.

b What is the value of b?

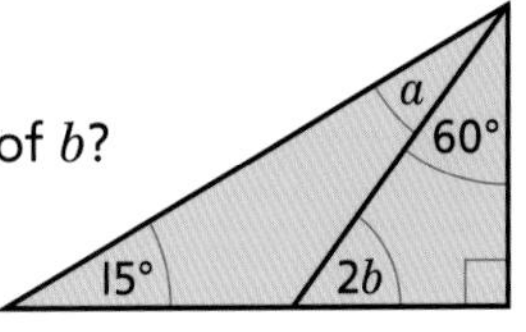

Level 7

Q3 Jacob walks 100 m north and then 100 m east.
How far is he from his starting position?

Q4 Which of these triangles will contain a right angle?
You must show your workings.

a

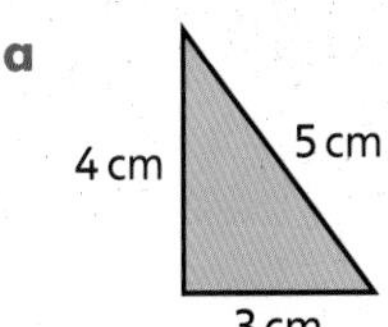

b

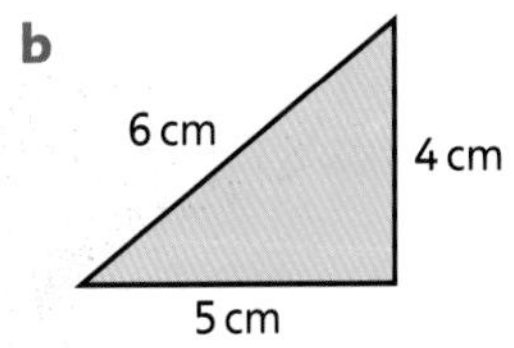

Q5 One end of a rope is tied to the top of a 12 m flagpole. The rope is pulled tight and the other end is tied to the ground.
The rope is 16 m long.
How far is the end of the rope from the base of the flagpole?
Give your answer to 1 decimal place.

Level 8

Q6 Calculate the perimeter and area of this triangle.
Give your answers to 2 decimal places.

15 cm
20°

Q7 A 5 m ladder leans against a wall.
The foot of the ladder is 1 m from the base of the wall.
Calculate the angle between the ladder and the ground.

Q8 A boat sails 50 km on a bearing of 070°.

a How far east does the boat travel?

b How far north does the boat travel?

Give your answers to 1 decimal place.

15 A likely story

This unit is about how to apply probability theory in real-life situations.

Insurance companies consider probabilities when they are working out how much people have to pay for their premiums. If the probability of someone making a claim is high, the companies charge more for their insurance.

Anne Vorley is an underwriter working for Standard Life. She assesses people's applications for health insurance. 'I have to work out the probability of someone making a claim,' she says. 'Also, if they were to make a claim I have to look at the probable cost of the claim – it has to be within the bounds of what the company can afford.

'Medical history is important. For example, if someone has already had a heart attack they are more likely to claim again as the heart is likely to have sustained damage during the first attack.' She concludes, 'No two people's medical histories are ever the same – that is the human factor that makes each case a little bit different.'

Activities

A To estimate the probability of an applicant having a heart attack, insurance companies collect data like this.

Category	Number of people in category	People in category who had a heart attack in one year
non-smoker, healthy weight	50 000	500
smoker	15 000	450
overweight	30 000	600

- Estimate the probability of having a heart attack for each category.
- Match the health insurance premiums to the people.

Mr Green: 6 kg overweight	£40 per month
Mr Brown: has never smoked, healthy weight	£60 per month
Mr White: smokes, 5 kg overweight	£80 per month
Mr Grey: smokes	£100 per month

B Young drivers need to pay more for their car insurance than older drivers. Why do you think this is? Do you think this is fair?

Before you start this unit...

1 **a** Write all the letters of your name on squares of paper. Select a letter at random. Note whether it is a consonant or a vowel, then replace it. Repeat 30 times.

b What is the relative frequency of getting a vowel?

page 172

2 A bag contains 8 red pencils and 12 green pencils. A pencil is selected at random, the colour recorded, and then it is replaced. This is repeated.

a Draw a tree diagram to represent this.

b List all the mutually exclusive outcomes.

c What is the probability of selecting 2 red pencils?

page 170

3 Here are results from a class survey of mobile phones: 10 Yodaphone; 5 Satsuma; 8 ChatChat; 4 H_2; and 5 pupils didn't own a mobile.

a What is the relative frequency of a pupil owning a ChatChat?

b In a school of 500 pupils, how many are likely to own a mobile?

page 172

Did you know?

If a driver has no accidents in their car for several years, they are awarded a No Claims Bonus. This makes the insurance much cheaper. The insurance companies do this because the probability of a driver with an accident-free record making a claim is far lower than the probability of a driver with a record of prangs making a claim.

15.1 This will probably be familiar!

15.2 Growing trees

15.3 Experiment!

15.4 What's the problem?

15.5 Back to the future

maths! The birthday problem

Unit plenary: Understanding probabilities

Plus digital resources

15.1 This will probably be familiar!

- Use the vocabulary of probability in interpreting results involving uncertainty and prediction
- Find all the possible mutually exclusive outcomes for single and two successive events and record them systematically using diagrams and tables
- Identify all the mutually exclusive outcomes of an experiment
- Know that the sum of the probabilities of all mutually exclusive outcomes is 1 and use this when solving problems
- Estimate probabilities from experimental data
- Understand relative frequency as an estimate of probability and use this to compare results of experiments

Why learn this?

Statistics are used to compare the effectiveness of different medicines and treatments.

What's the BIG idea?

- You can compare probabilities using fractions. Find a common **denominator** to compare two fractions such as $\frac{3}{5}$ and $\frac{1}{4}$. Level 6
- There are nine possible combined **outcomes** for two successive events, when each event has three possible outcomes. Level 6
- The sum of the probabilities of all **mutually exclusive** outcomes is 1. So if the probability of an outcome is p, the probability of it *not* happening is $1 - p$. Level 6
- You can use the results from experiments to predict the probability that an outcome will occur. Level 6
- If you know the probabilities of all but one outcome of a set of mutually exclusive outcomes, you can calculate the probability of the remaining outcome. Level 7
- The **relative frequency** of an event is the estimated probability derived from experimental data. Level 7

Super fact!

In the 1800s, the death rate amongst women giving birth was often as high as 40%. Nowadays, it is around 13 in every 100 000 – only 0.013%.

Practice, practice, practice!

1 Explain why these statements cannot be quite true.

  a There is a 50% chance that the next baby born in our town will be male.

  b My train has been delayed three days in a row. It can't *possibly* be late again today.

  c To play 'Beetle' you need to roll a 6 to start. But I can *never* roll a 6.

2 Jade is getting ready for a bonfire night party. She is making three main courses and three puddings.

  a List all the possible different combinations of main course and pudding you could choose.

  b If Jade made an extra pudding, how would that affect the number of different combinations you could choose?

Spicy sausages
Veggie hotpot
Bonfire burgers

Jelly sparkle
Pumpkin pie
Baked apple

Level 6

6c I can use the language of probability to interpret results involving uncertainty and prediction

6c I can list all possible outcomes for two successive events, with three possibilities at each stage

denominator likelihood mutually exclusive

3 Work out the probability that a person chosen at random from each of these groups of internet users did not log on during the past week.

a In a group of 36 people under the age of 30, 28 people did log on.

b In a group of 750 people between the ages of 30 and 75, 575 did log on.

c In a group of over-75s, 72 out of 240 people did log on.

4 In a clinical trial, medicine A was compared with medicine B. Patients with a certain illness were given either medicine A or medicine B at random.
20 out of the 60 patients given medicine A showed a noticeable improvement.
20 out of the 70 patients given medicine B showed a noticeable improvement.

a Which medicine seems to be the more effective? Explain your answer.

b In a second trial, only 50 patients were given medicine B, and 20 of these patients again showed a noticeable improvement.
Which seems to be the more effective medicine now?

5 Doctors wanted to see which method worked better for giving up smoking: nicotine patches (P) or nicotine chewing gum (G). Of 360 people given patches, 240 managed to stop smoking by the end of a month. Of 500 people given gum, 320 managed to stop smoking by the end of a month.

a Write the probability of success with each method.

P(P) = ____ P(G) = ____ Give your answers in their simplest terms.

b Compare the probabilities and say which method was more effective.

Level 6

6c If I know the probability of an event happening, I can work out the probability of it not happening

6b I can compare probabilities using equivalent fractions

6 In a survey into painkillers for toothache, $\frac{3}{5}$ of patients preferred ibuprofen, $\frac{1}{5}$ said that paracetamol worked best and 17% found that aspirin was most effective. The rest of the 100 people surveyed had never had toothache.
A patient is chosen at random.

a What is the probability of a patient not having had toothache?

b In a village of 870 people, how many are likely to have never had toothache?

7 Max the magician recorded the colours in a pack of 100 silk handkerchiefs. The tally chart shows his results.

a Calculate the relative frequency of each colour.

b If Max bought 10 packs, how many handkerchiefs might he expect to be red?

Colour	Tally
Red	卌 卌 卌 卌 卌 卌 卌 ||
White	卌 卌 卌 ||
Yellow	卌 卌 卌 卌 卌 卌 |
Blue	卌 卌 卌

Level 7

7c I can use the fact that the probabilities of mutually exclusive outcomes total 1 to solve problems

7b I can understand that the relative frequency of an event is the estimated probability derived from experimental data

Watch out!
Relative frequency is only an estimate of probability.

Now try this!

A It's a shoe-in

Working in groups of 5 or 6, each write your shoe size on a piece of paper, then put the pieces of paper in a bag. Each draw a piece of paper out of the bag, read it and then replace it. Do you draw out your own shoe size? What is the probability of this happening? How might the probability change if everyone in the school put their shoe size in?

B Spinners

Work with a partner. Make two hexagonal spinners, and label each from 1 to 6. Take turns to spin both and multiply the numbers to obtain a score. Do the results show a bias towards particular scores?

outcome relative frequency

15.2 Growing trees

- Draw tree diagrams for complex independent events
- Draw tree diagrams for events where the probability for a second event depends on the outcome of the first event

What's the BIG idea?

- **Tree diagrams** can be used to represent sequences of **independent** events and their **probabilities**. Level 8 & Level 8+
- Tree diagrams can also be used when the probability of one event is **dependent** on a previous event, for example, if chocolates are taken from a box one by one and eaten.This is called conditional probability. Level 8 & Level 8+
- To find the probability that two possible outcomes both occur, multiply the probabilities along the branches. For example, a box has four hard-centre and eight soft-centre chocolates. The tree diagram represents two chocolates taken at **random** and eaten.

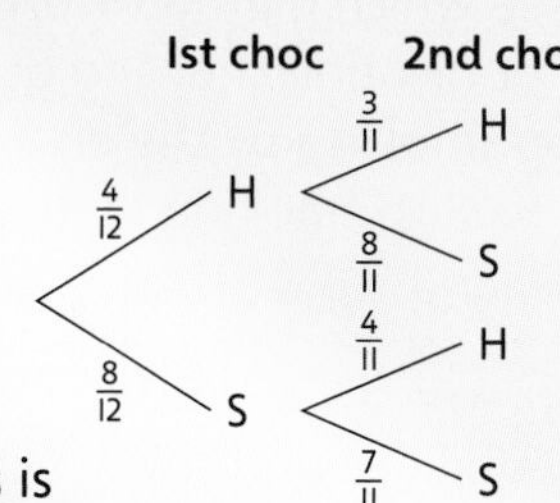

The probability of randomly taking two soft centres is $\frac{8}{12} \times \frac{7}{11} = \frac{2}{3} \times \frac{7}{11} = \frac{14}{33}$ Level 8 & Level 8+

Why learn this?

Tree diagrams can help you to visualise more complicated probability problems, especially when the probabilities of the different outcomes aren't equal.

Practice, practice, practice!

1 Imagine this experiment with an ordinary pack of playing cards.
You pick one card at random. It is an ace, a picture card or a number.
You put it back and then pick another card at random.

a Draw a tree diagram to represent the possible outcomes and probabilities in this experiment.

b What is the probability of picking two aces?

c What is the probability of picking two picture cards?

2 a Draw a tree diagram to represent the information in this report.

Omega 3 and exercise

Research has shown that taking Omega 3 and taking part in moderate amounts of exercise stops the development of excess body fat. The study examined a large sample of people over a three-month period. Half the sample were given Omega 3 and the other half were given sunflower oil. Each group was then sub-divided into those doing 45 minutes of moderate exercise and those who did no exercise, in the ratio 2 : 1.

After the trial the group who took Omega 3 and exercise had lost an average of 4.5 lb. None of the other groups had lost any weight at all.

The conclusion was clear: daily doses of Omega 3 can boost the rate at which the body burns calories when taking exercise.

b Out of 90 people taking part in the study, how many should lose weight?

Level 8

8C I can draw tree diagrams to represent outcomes of two independent events

8C I can draw and use a tree diagram to solve a problem

Learn this

Probabilities are always multiplied along the branches.

dependent independent

3 A bag of coloured counters contains eight blue counters (B) and four red (R). Alison picks one at random, then she puts it back. So does Brian, and then Carlo.

- **a** Copy and complete this tree diagram to represent the possible outcomes and their probabilities.
- **b** What is the probability of them all picking a red counter?
- **c** What is the probability of them all picking a blue counter?
- **d** What is the probability of at least one of them picking a red counter?

Level 8

8a I can draw tree diagrams to represent outcomes of more than two independent events

4 Keri has a bag of sweets. There are four strawberry (S), six orange (O) and two lime (L).

- **a** Copy and complete this tree diagram to represent Keri randomly picking and eating two sweets.
- **b** What is the probability that she randomly picks two limes?
- **c** What is the probability that she randomly picks two sweets of the same flavour?

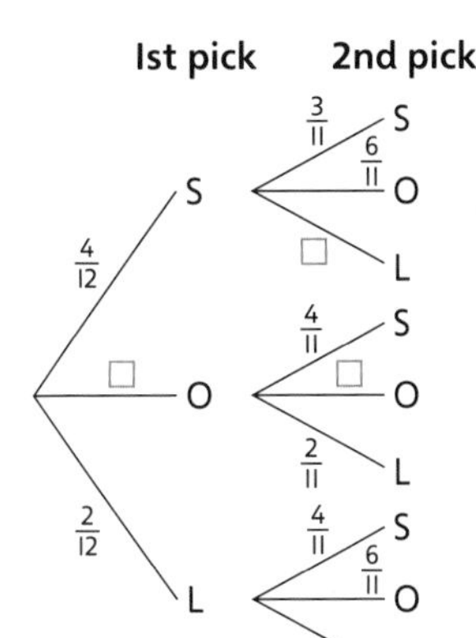

8+ I can draw tree diagrams to represent two events with dependent probabilities

Tip
Check that each group of probability 'branches' totals 1.

5 As in Q1, imagine an experiment with an ordinary pack of cards. This time you don't put your first card (an ace, a picture or a number) back. You keep it and then pick another card at random.

- **a** Draw a tree diagram to represent the possible outcomes and probabilities in this experiment.
- **b** What is the probability of picking two aces?
- **c** What is the probability of picking two picture cards?
- **d** Compare these probabilities with those in Q1.

6 There are 48 cars in a car park: 20 silver, 12 red and 16 blue. They are all equally likely to leave at any time, and no more cars can arrive in the car park.

- **a** Draw a tree diagram to represent the colours of the first three cars leaving the car park.
- **b** What is the probability that all three cars are the same colour?

8+ I can draw tree diagrams to represent complex events with dependent probabilities

Now try this!

A Which colour?

Work in pairs. Make 12 coloured squares of paper using three or four different colours. Draw a tree diagram to represent picking two coloured squares at random. What is the probability of selecting two squares of the same colour?

B Lucky dip

Work in pairs. A lucky dip has 30 prizes. There are 12 packets of sweets, 10 packs of cards and 8 sets of crayons. Draw a tree diagram to represent picking three prizes at random. What is the probability of getting at least one packet of crayons? Make little cards to represent the prizes and work out an estimate of the experimental probability.

probability random tree diagram

15.3 Experiment!

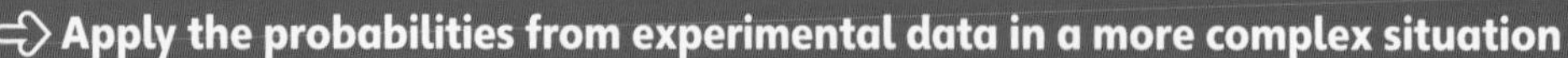

- Apply the probabilities from experimental data in a more complex situation
- Identify conditions for a fair game
- Recognise that with repeated trials experimental probability tends to a limit
- Compare experimental and theoretical probabilities in a range of contexts
- Recognise that an estimate of probability from experiment may be the only way of estimating probability when outcomes are not equally likely
- Plot and use relative frequency diagrams

What's the BIG idea?

- A game is **fair** if all players have the same chance of winning. Level 6
- Imagine two **surveys**. Survey A finds out people's favourite sandwich filling. Survey B finds out their favourite drink. You could then use the results of both surveys to find out the probability of people buying each sandwich–drink combination. Level 8
- The more times you repeat an experiment, or the greater the number of people in a survey, the more accurate the results are likely to be. Level 8
- Diagrams are often a good way of visualising **relative frequencies**. Level 8

Why learn this?

Supermarkets use experimental data (buying trends) to predict how much of each item they need to stock.

Practice, practice, practice!

Level 6

6b I can identify conditions for a fair game

6a I can work out if a complex game is fair

1 Two friends make a hexagonal spinner labelled 1 to 6. Each adds up their scores from two spins. Make up three pairs of rules which give them both the same chance of winning.

2 A game has nine targets, labelled 1 to 9. Each player adds the scores for their two goes. Players choose one of these rules for how to win.

Rule 1: The total score is 18.
Rule 2: The total score is less than 5.
Rule 3: The total score is more than 10.
Rule 4: The total score is even.

a Which rule will give you the most chance of winning? What is the probability of winning with this rule?

b Which rule will give you the least chance of winning? What is the probability of winning with this rule?

c Which rule is fairest? What is the probability of winning with this rule?

d What is the probability of winning with rule 2?

Tip

In a fair game the proportion of wins for each player should be equal.

Level 7

7b I can use relative frequency as an estimate of probability

3 100 people were asked which of four washing powders they would choose.

The tally chart shows the results.

A shop sells 5000 boxes of washing powder each week. How many boxes of the most popular powder should it expect to sell?

Powder	Tally
Parsil	卌 卌 卌 卌 卌 卌
Tyde	卌 卌 卌 卌 卌 I
Emu	卌 卌 卌 IIII
Diz	卌 卌 卌 卌 卌

experimental probability fair frequency diagram

4 A supermarket carried out two surveys, each with a representative sample of 100 people.

Here are the results of Survey A, which asked about people's favourite sandwiches.

32 cheese   16 egg   28 chicken   24 prawn

Here are the results of Survey B, which asked about people's favourite drinks.

24 cola   16 diet cola   28 orange   18 lemonade   14 water

The surveys showed that people's choice of sandwich did not affect their choice of drink.

a If the supermarket sells 600 sandwiches a day, how many of each type should it stock?

b The supermarket has a special offer. In this offer what is the probability of someone buying

Buy any sandwich and any drink for £2.

i a cheese sandwich and an orange drink?

ii a chicken sandwich and some water?

c If 300 people buy a sandwich and a drink, how many would you expect to buy a prawn sandwich and a cola?

5 Make a circular spinner and divide it into three sections. Colour 180° red, 120° blue and 60° purple.

a Spin your spinner 10 times. Record the number (N) for each colour in a table like this one.

Now work out the relative frequency (R) for each colour and write it in the table.

	Number of spins							
	10		20		30		100	
	N	R	N	R	N	R	N	R
red								
blue								
purple								

b Repeat for another 10 spins. Work out the relative frequencies based on 20 spins.

c Repeat, recording results every 10 spins up to 100.

d Copy and complete this relative frequency diagram to compare the relative frequency of red with its thoeretical probability (shown by the horizontal red line).

e Sketch relative frequency diagrams for blue and purple.

f Imagine the experiment in Q4 was repeated 2000 times. Using your experimental results, how often would you expect to get

i a red   **ii** a blue?

Explain your answers.

Level 8

8b I can undertand relative frequency as an estimate of probability and know when to add or multiply probabilities

8a I can plot and use relative frequency diagrams and recognise that with repeated trials experimental probability tends to a limit

Learn this

The more times you carry out an experiment, the more the experimental results will tend towards theoretical ones.

Watch out!

Remember that relative frequency is only an estimate of probability.

Now try this!

A Playing cards

Shuffle an ordinary pack of cards. Pick a card at random. Record whether it is a number card, a picture card or an ace. Replace the card. Repeat 99 more times. Use the results of your experiment to find an estimate of the relative frequency of picking each type of card.

B Which way up?

Roll a small eraser (or similar small object) a number of times to determine the way it is most likely to land. Use your results to devise a fair game.

relative frequency   survey   theoretical probability

15.4 What's the problem?

➩ Solve problems involving probability

Why learn this?

Solving problems involving probability can help you to work out the chances of winning a game.

What's the BIG idea?

- Probability of outcome A or outcome B happening: P(A or B) = P(A) + P(B) **Level 6**
- Two events are **independent** if the outcome of the first has no effect on the outcome of the second, for example when you toss two coins. **Level 7**
- Two event are **dependent** if the probability of the second depends on the outcome of the first, for example, when a sweet is picked from a bag of assorted sweets and not replaced before a second sweet is picked. **Level 8 & Level 8+**
- Probability of outcome A and outcome B happening: P(A and B) = P(A) × P(B) **Level 8 & Level 8+**
- You can use different techniques, such as drawing **tree diagrams** or working out probabilities of consecutive events, to solve problems involving probability. **Level 8 & Level 8+**

Practice, practice, practice!

Level 6

1 The sweets in one bag (A) of pick'n'mix are equal numbers of sour grapes, flying saucers, strawberry laces and sherbet lemons. In another bag (B) there are equal numbers of flying saucers, sour grapes and sherbet lemons. Jonas says that there is more chance of picking a sherbet lemon from bag A than bag B. Is he right? Explain your answer.

6c I can use the language of probability to compare probabilities

2 During one year, 0.7% of a hospital's admissions were for broken legs. 70% of these people were male and 30% were female. 0.4% of the hospital's admissions were for broken arms. These were divided equally between men and women. The hospital has at total of 10 000 admissions annually

- **a** How many are for women with a broken leg?
- **b** How many are for men with a broken arm?
- **c** How many are for women with a broken arm and a broken leg?

Explain your answers.

6a I can apply probabilities from experimental data

3 Ozzie and Sharon play a game. A fair, six-sided spinner is marked with the letters of Sharon's name. If it lands on a vowel, Sharon gets 3 points. If it lands on a consonant, Ozzie gets 2 points. It is spun twice and the points totalled.

- **a** Is this a fair game? Explain your answer.
- **b** If it is unfair, suggest a way of making it fair.

Tip Use a tree diagram.

6a I can identify rules for a fair game

dependent independent mutually exclusive

4 An airline calculates that the probability of a passenger being disabled is 0.01.

a Each plane carries 300 people, and four seats are allocated to disabled people. Is this likely to be sufficient? Explain your answer.

b The airline carries 2.4×10^6 passengers per year. How many of these are likely to be disabled?

5 Kate is taking a theory test and a practical test. The tests are independent. The probability of passing the theory test is 0.9. The probability of passing the practical test is 0.85.

a What is the probability

i she passes both tests ii she passes the theory and not the practical?

b The same tests are taken by 1400 students. How many of them are likely to pass both?

6 In a lucky dip, barrel A holds 18 boxes of sweets and 12 toy cars. Barrel B holds 15 boxes of sweets and 9 toy cars. There are no other prizes.

a Which barrel should you pick from to have the better chance of getting a box of sweets?

b Imagine you pick one prize from barrel A and one from barrel B. What is the probability that you get two boxes of sweets?

c Imagine you see someone pick a car from barrel A. Does this change your choice of barrel in part a? Explain your answer.

7 A fruit machine has three 'windows'. In each window you have equal probabilities of getting an orange, a lemon, a bunch of grapes or a 7.

a What is the probability of getting three 7s?

b What is the probability of getting identical items in all three windows?

c What is the probability of all the windows being different?

Tip You may want to use a tree diagram to answer this question.

7 7 7

8 In a bag of sweets there are 20 fruit sweets, 24 chocolates and 16 mints. Three friends each take a random sweet in turn.

a Draw a tree diagram to represent this.

Hint: With dependent probabilities remember to subtract the item from the total as well as from a specific category.

b Use your tree diagram to answer these questions.

i What is the probability of them all selecting chocolates?

ii What is the probability of them all selecting the same kind of sweet?

iii What is the probability of them all selecting different kinds of sweet?

Level 8

8c I can use probability to solve complex real-life problems

8b I can calculate the probability of independent and dependent events

8a I can draw and use tree diagrams to represent outcomes of more than two independent events

8+ I can calculate probabilities of dependent events

Now try this!

A Who wins?

Work in pairs. Imagine a card game where players are dealt three cards each. Create three different sets of rules so that

- every player is guaranteed their three cards would make them a winner
- no player can possibly win
- around half the players will win.

B Don't play this game!

Work in pairs. Design a fruit machine where the probability that a player will win is less than $\frac{1}{20}$.

tree diagram

15.5 Back to the future

→ Interpret results involving uncertainty and prediction

What's the BIG idea?

→ You can use mathematics to analyse real-life data and make predictions based on general trends you notice in the data. **Level 7 & Level 8**

Why learn this?

If we look at past weather trends, we can make predictions about the future.

Practice, practice, practice!

Here are seven real-life situations to investigate in groups. For some you may be able to collect the data yourself. For others you will need to access statistics from the internet or elsewhere.

Your task is to

- analyse the data
- use your findings to make predictions
- prepare and deliver a short presentation about your conclusions and predictions.

To help you analyse the data, you should draw charts, graphs and diagrams.

Your presentation should include any assumptions you made, the source of your data, how it was analysed, your charts, graphs and diagrams, and your predictions.

1 School attendance

- Access your school's statistics for the last few years.
- Decide how to group the figures – by month, by tutor group, by gender?
- Represent your findings graphically.
- Predict how many days of absence there will be from your year next week.

2 School canteen

- Look at the main meals chosen by pupils.
- Decide how to group them.
- Represent your findings graphically.
- Predict how many of each type of meal the canteen needs to supply on a given day next week.

3 Leisure activities

- Examine the numbers of people doing a range of sports at your local leisure centre or nationally.
- Which sports are the most popular.
- How do the numbers change by season?
- What other factors affect the numbers?
- Represent your findings using diagrams.
- Choose a sport. Predict how many people in your town/city are likely to do this at least once a month.

data tree diagram

4 Lunchtime and after-school clubs

- Investigate the clubs or activities offered by your school at lunchtimes or at the end of the school day.
- How many are available? Are they open to all year groups? How many pupils attend at least one club or activity?
- Do timings vary from club to club? Is there a link between timings and the numbers of pupils who attend?
- How many clubs are related to leisure activities? How many are linked to school work?
- Represent your findings using diagrams.
- Predict the total number of pupils likely to regularly attend at least one club in the same term next year.

5 Local shop

- Investigate one particular product line or compare product lines – sweets, soft drinks, fruit, newspapers, ... anything else you choose.
- Look at factors affecting sales, such as the day of the week, the time of the year, temperature and sporting events.
- Represent your findings diagrammatically.
- If you were the store manager, would you make any adjustments to the products you stock? What would you change?
- Predict sales for the coming week.

6 Your local Accident & Emergency department

- Look at data for your local A & E department.
- What percentage of people attend for follow-up appointments?
- What is the probability of waiting more than 4 hours in A & E?
- What proportion of people are admitted to a ward in the hospital after attending A & E? How long does this take?
- Draw a tree diagram to represent your findings.
- 100 people attend A & E. Predict how many of them would be coming for a follow-up appointment, would have to wait less than 4 hours and would not be admitted to the ward.

7 Weather forecasts

- Look on websites for factors affecting weather forecasts for your area at this time of year for the past few years (temperature, air pressure, rainfall etc.).
- Represent these findings diagrammatically.
- What links do you see?
- Predict the weather in your area for the next two weeks.
- Justify your predictions.

trend

maths!

The birthday problem

Probability can help you understand risk and chance, and to make informed decisions – but the answers to probability questions can often take you by surprise! For example, do any two people in your class share the same birthday? What are the chances of this happening?

The theory...

You can show that if there are at least 23 people in your class, there is a better than even chance that two of them share a birthday, like this:

For a group of n people, calculate the probability of everyone having a different birthday and take it away from 1.

When $n = 1$ P(all have different birthdays) = 1 (Certain – there is only 1 person)
P(two people share a birthday) = 1 – 1 = 0 (Impossible)

When $n = 2$ P(all have different birthdays) = $1 \times \frac{365}{366} = 0.9973$
(There are 366 possible birthdays for Person 1 but only 365 for Person 2. The probability of Person 2 having a different birthday is $\frac{365}{366}$. The probability of both having different birthdays is $\frac{366}{366} \times \frac{365}{366}$.)
P(two people share a birthday) = 1 – 0.9973 = 0.0027

When $n = 3$ P(all have different birthdays) = $1 \times \frac{365}{366} \times \frac{364}{366} = 0.9918$
P(two people share a birthday) = 1 – 0.9918 = 0.0082

- Here is a partially completed table of results.
  - **a** Calculate the missing values for groups of 4, 5 and 6 people.
  - **b** Now think carefully about how to use your calculator efficiently to calculate the probabilities for a group of 23.The correct result will show that the probability of two people sharing a birthday in a group of at least 23 people is greater than 0.5.

Hint: You could multiply the numerators and divide by the correct power of 366.

- Plot a graph with n on the x-axis and the probability of two people in a group of n sharing a birthday on the y-axis. Roughly how many people do you need for the probability of two people sharing a birthday to be greater than 95%?

Number of people (n)	Probability of different birthdays	Probability of 2 people sharing a birthday
1	1	0
2	$1 \times \frac{365}{366} = 0.9972$	0.0027
3	$1 \times \frac{365}{366} \times \frac{364}{366} = 0.9918$	0.0082
4		
5		
6		
10	0.8834	0.1166
20	0.5894	0.4106
23		
30	0.2947	0.7053
40	0.1095	0.8905
50	0.0299	0.9701
60	0.0060	0.9940

...and an experiment

If you investigated this probability experimentally by collecting data from 100 groups of 23 people, you would need 2300 people! When experiments aren't possible, safe, or cost- or time-effective, mathematicians sometimes use a simulation of the experiment.

- With a partner, simulate a group of 23 people by randomly generating 23 birth dates. For each birth date you will need to generate a month and a day.

  First flip a coin and roll a dice, and then use the left hand table to select the month.

  Then roll the dice twice and use the right hand table to select the day.

  If your roll generates a date that doesn't exist (for example 31 February), roll the dice a further two times to select a different day.

Coin

Dice	Heads	Tails
1	January	February
2	March	April
3	May	June
4	July	August
5	September	October
6	November	December

Roll 2

Roll 1	1	2	3	4	5	6
1	1	2	3	4	5	6
2	7	8	9	10	11	12
3	13	14	15	16	17	18
4	19	20	21	22	23	24
5	25	26	27	28	29	30
6	31	—	—	—	—	—

- Look at your list – did any two people in the generated group share a birthday?
- Now compare your results with the other pairs in the class – what proportion of pairs generated the same date twice?
- How does this proportion compare to the theoretical result?
- What could you do to improve your simulation?

Understanding probabilities

The risk of Coronary Heart Disease is reduced by 50% after 1 year of smoking abstinence

Stopping smoking reduces the risk of fatal heart and lung disease

Statistics and the vocabulary of probability are used on a daily basis in the media. Are they used correctly? Do you understand what they mean?

1 Imagine that a rare disease is breaking out in a town.
If nothing is done, 600 people will die.
There are two possible options.
Option A will save 200 people.
Option B will either save everyone or no one.
It has a $\frac{1}{3}$ probability of saving everyone.

a What is the probability that Option B will save no one?

b Which option would you choose?
Explain your answer. **Level 7**

2 This table shows the results of a cancer screening test for 10 000 people.

	Cancer	No cancer	Total
Test positive	85	1010	1095
Test negative	15	8890	8905
Total	100	9900	10 000

a Look at the 'Cancer' column.
If a person has cancer, what is the probability of their getting a positive result?

b Look at the 'Test positive' row.
If a person's test is positive, what is the probability that they have cancer? **Level 7**

3 There is a test for Disease A that is 98% accurate.
One out of 200 people have Disease A, and 10 000 tests are performed.

a In the sample of 10 000 people, how many are likely to have Disease A?

b How many of the people with Disease A will test positive?

c How many people without Disease A will test positive?

d Karim receives a positive test result. What is the probability that he has Disease A? **Level 8**

4 Here is some information about breast cancer screening.

- The probability that a woman aged between 40 and 50 has breast cancer is 1%.
- If the patient has breast cancer, the probability that the radiologist will correctly diagnose it is 80%.
- If the patient does not have cancer, the probability that the radiologist will incorrectly diagnose it as cancer is 10%.

100 doctors were asked 'What is the probability that a patient with a positive result actually has breast cancer?'
95 of the doctors estimated the probability to be about 75%. Were they correct?

a Draw a tree diagram to represent the problem.

b What is the probability that a patient who tests positive actually has breast cancer? **Level 8+**

The BIG ideas

- → The sum of the probabilities of all **mutually exclusive** outcomes is 1, so if the **probability** of an event is P, the probability of it *not* happening is 1 − P. Level 6
- → You can use the results from experiments to predict which outcome will occur. Level 6
- → **Tree diagrams** can be used to represent sequences of **independent events** and their probabilities. Level 8 & Level 8+
- → Tree diagrams can also be used when the probability of one event is **dependent** on a previous event. Level 8 & Level 8+
- → You can use different techniques, such as drawing tree diagrams and working out probabilities of consecutive events, to solve problems involving probability. Level 8 & Level 8+

Find your level

Level 6

Q1 The table shows the number of men and women working in two different offices.

	Office A	Office B
Men	24	15
Women	18	10

A manager is going to select a representative at random from each office.
In which office is the manager more likely to select a woman?

Q2 100 raffle tickets are sold at a fair. There is just one prize. You buy two tickets.
What is the probability that you win the prize?

Level 7

Q3 In an experiment with a four-coloured spinner, these results were recorded.

Colour	blue	red	green	orange
Successes	4	12	6	2

a Show roughly how this spinner is probably coloured.

b If it is spun twice, what is the probability of getting

i two reds **ii** no reds?

Level 8

Q4 A bag contains 4 green and 5 red counters. Ann takes a counter from the bag at random, notes its colour and then replaces it.
She then takes a second counter from the bag at random.
What is the probability that the two counters are different colours?

Q5 A bag contains 4 toffees (T), 3 mints (M) and 2 chocolates (C). Vic takes one sweet at random from the bag and eats it.
He then takes a second sweet at random.

a Copy and complete the tree diagram.

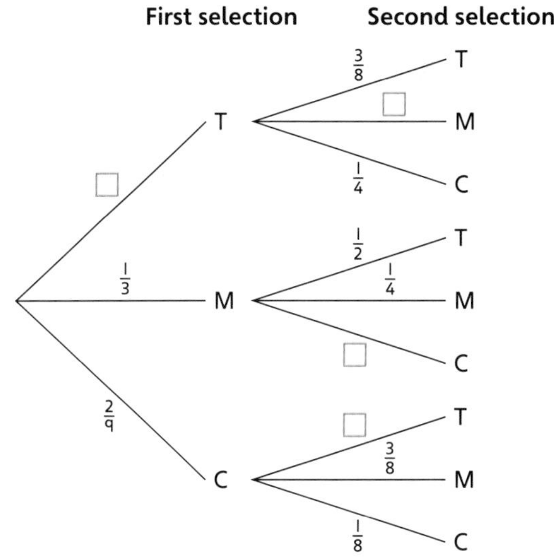

b What is the probability that Vic takes a mint followed by a toffee?

c What is the probability that Vic takes two sweets of the same type?

Revision 3

Quick Quiz

Q1 Multiply out each bracket and collect like terms.
$6(2x + 5) + 3(7 - x)$
→ *See 11.1*

Q2 Work out the size of angle x.

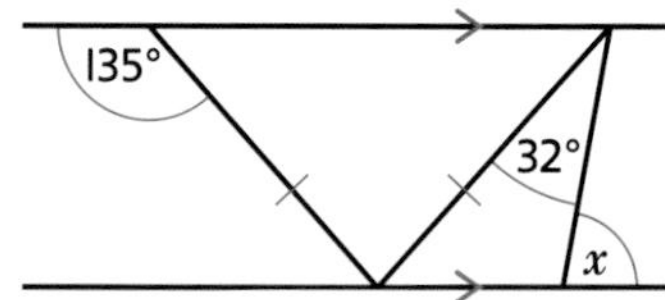

→ *See 12.4*

Q3 What type of correlation does this scatter graph show?

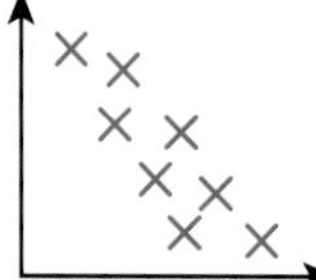

→ *See 13.4*

Q4 In a café you can buy the following coffees and cakes.

Coffees: latté, mocha or cappuccino
Cakes: fruit, orange or carrot

a List all the possible combinations of coffee and cake you could choose.
b If the café introduces a new type of cake, how many possible combinations of coffee and cake are there now?

→ *See 15.1*

Q5 Which offer is the better value?
Explain your answer fully.

Special offer! 25 black pens £8 (+VAT15%)	SALE! **25% off** 20 black pens *Normally* £11

→ *See 12.5*

Q6 The table shows the ages of 20 children at a party.

Age (years)	Frequency
$10 \leqslant \text{age} < 12$	7
$12 \leqslant \text{age} < 14$	9
$14 \leqslant \text{age} < 16$	4

Calculate an estimate of the mean age.
→ *See 13.2*

Q7 Calculate the volume of this prism.
→ *See 14.4*

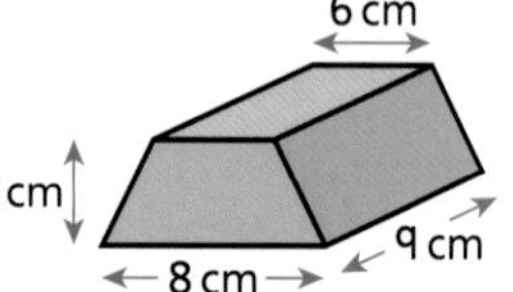

Q8 Multiply out the brackets and simplify $(x + 4)(2x - 3)$.
→ *See 11.2*

Q9

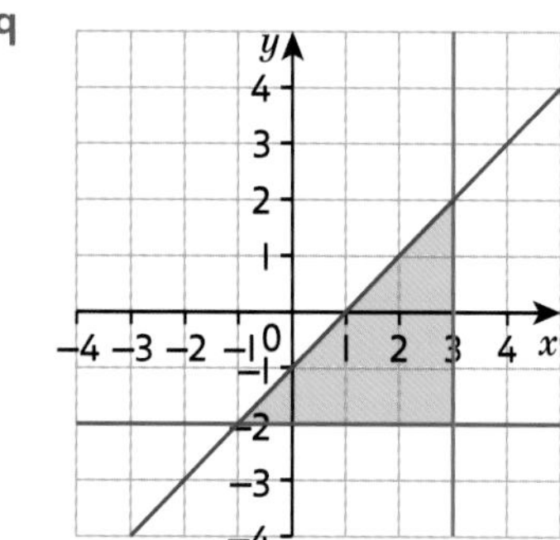

a Write the equation of each coloured line.
b Write three inequalities which are satisfied by the shaded region.

→ *See 11.6*

Q10 Calculate the length x in this triangle.
Give your answer to 2 d.p.
→ *See 14.5*

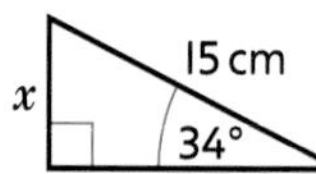

Activity

You have managed to get a Saturday job working in a café.

LEVEL 6

Q1 In the café, water is heated for tea and coffee in a cylindrical urn of height 80 cm and radius 21 cm.
The volume of a cylinder is given by the formula $V = \pi r^2 h$.

The tea is made in a spherical teapot of radius 10 cm.
The volume of a sphere is given by the formula $V = \frac{4}{3}\pi r^3$.
The maximum volume of water used in the urn or the teapot in one go is $\frac{3}{4}$ the volume of the urn or teapot.
How many pots of tea can be made from one urn of hot water?

Q2 Soup and a bread roll is the most popular Saturday lunch. This part of the menu shows the soups and bread rolls available.

Soups: tomato, minestrone, chicken, vegetable
Bread rolls: white, brown, granary

a List all the possible combinations of soup and bread roll.

b If all the combinations are equally likely, what is the probability of a customer ordering chicken soup and a brown roll.

c On a previous Saturday, 40% of all customers chose tomato soup and 45% of those customers chose a white roll. There are likely to be 200 customers this Saturday. How many are likely to choose tomato soup and a white roll?

LEVEL 7

Q3 Which of these offers would be better for customers? Explain your answer fully.

Soup: £1.85 Roll: 55p
Saturday special
Buy a soup and a roll and get a 30% discount

Soup and a roll: £2.20
Saturday special!
Buy a soup and a roll and pay $\frac{3}{4}$ the normal price!

Q4 The cafe owner keeps a record of the amount spent by each customer. This table shows the results for one Saturday.

Amount spent (£)	Frequency	Amount spent (£)	Frequency
$0 < \text{amount} \leq 5$	22	$15 < \text{amount} \leq 20$	52
$5 < \text{amount} \leq 10$	29	$20 < \text{amount} \leq 25$	34
$10 < \text{amount} \leq 15$	48	$25 < \text{amount} \leq 30$	15

Calculate an estimate of the mean amount spent per customer.

LEVEL 8

Q5 Estimate the median amount spent by customers in Q4.

Find your level

LEVEL 6

Q1 The formula to work out the area of any triangle with sides of length a, b and c is:

Area = $\sqrt{s(s-a)(s-b)(s-c)}$

where $s = \dfrac{a+b+c}{2}$

Work out the area of the triangle with side lengths $a = 4$ cm, $b = 6$ cm and $c = 8$ cm.

Q2 During one year, 65% of visitors to a theme park were children, the rest were adults. 42% of the children were girls, and 58% of the adults were female.
The theme park estimates that next year it will have 120 000 visitors.

a How many are likely to be girls?

b How many are likely to be men?

LEVEL 7

Q3 A pupil investigated whether students who revise more get better test scores. The scatter graph shows her results. The line of best fit is also shown.

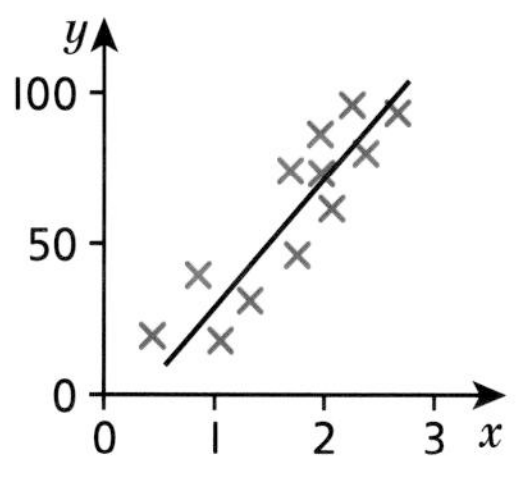

a What type of correlation does the graph show?

b The pupil says that the equation of the line of best fit is $y = x + 50$.
Explain how you can tell that this equation is wrong.

Q4 Calculate the diameter, d, of this semicircle.
Give your answer to 1 d.p.

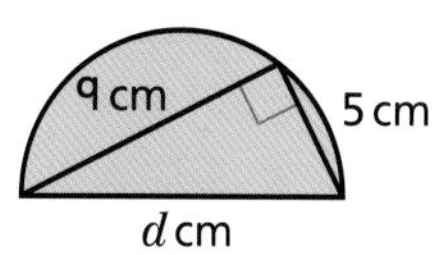

Q5 Work out the value of x in this triangle.

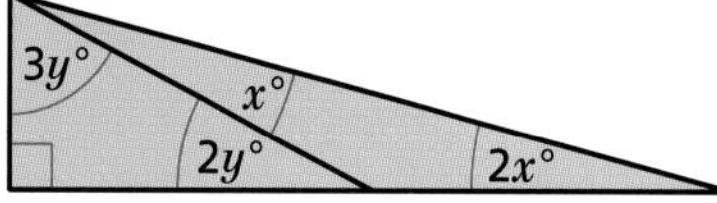

LEVEL 8

Q6 Multiply out the brackets in this expression and simplify the answer. $(k + 3)(k + 5)$

Q7 **a** Calculate the length y.

b Calculate the size of angle x.

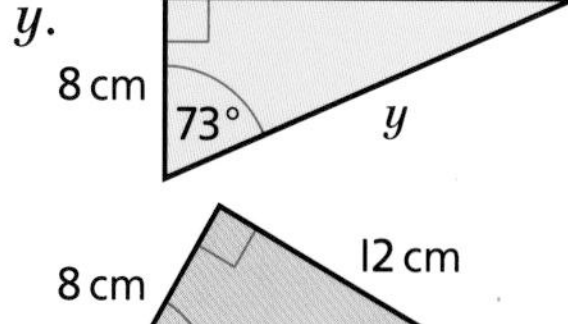

Q8 **a** The population of the world is approximately 6 707 000 000.
Write this number in standard form.

b The population of China is approximately $\frac{1}{5}$ the population of the world.
Approximately what is the population of China?
Write your answer in standard form.

16 Dramatic mathematics

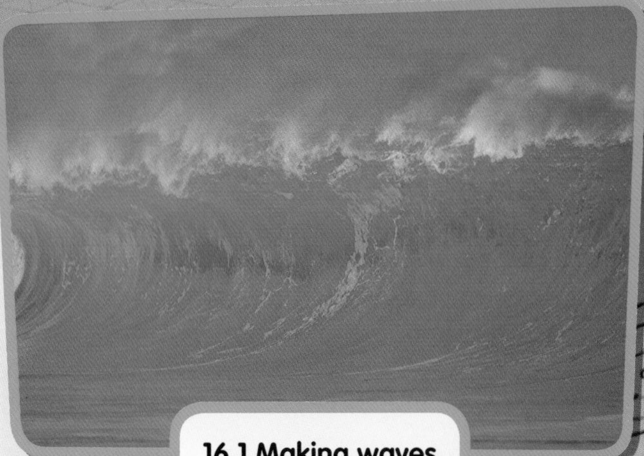

Although you might not be aware of it, mathematics is at work in almost every area of your life. Wherever there are quantities, shapes or connections, mathematics can be used to model what is happening and give a better understanding of the situation.

This unit shows you six of the thousands of ways that mathematics is used today - from tidal waves to earthquakes, from healthcare to manufacturing, and from distant stars to snowflakes. Each topic will use some of the mathematics you have already learned in this book, as well as introducing some new ideas and tools.

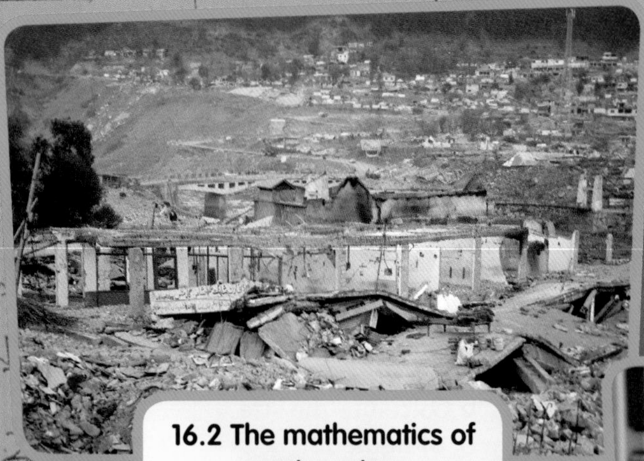

16.1 Making waves

16.2 The mathematics of earthquakes

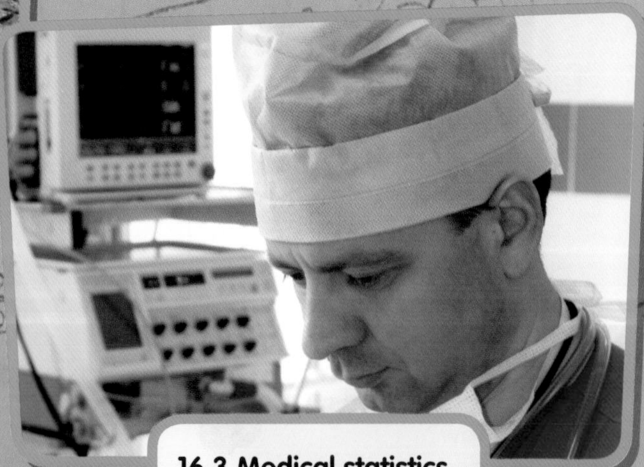

16.3 Medical statistics

Unit plenary:
Mathematics at work

16.6 Fractals

Mathematics counts – especially when it comes to your wages! Research suggests that pupils who achieve an A-level in mathematics will earn about 10% more on average than pupils with otherwise similar qualifications.

Curves of pursuit

16.5 Stellar mathematics

16.4 Made to measure

16.1 Making waves

- Sketch the graph of $y = \sin x$ and variations of it
- Explore the effect of changing values in related functions
- Understand how waves can model real-life events

What's the BIG idea?

→ The graph of the **trigonometric function** $y = \sin x$ produces a simple, or harmonic, wave. This shape, and variations of it, can be used to represent a range of real-life situations.

→ A mathematical **model** is a simplified representation of a real-world situation, using only the important mathematical elements and properties.

Why learn this?

Waves are all around you – from the sounds you make to the heartbeat inside your body.

Practice, practice, practice!

1 **a** Copy this table, then use a calculator to complete it.

x	0°	30°	60°	90°	120°	150°	180°	210°	240°	270°	300°	330°	360°
$\sin x$	0												
$2 \sin x$	0												
$3 \sin x$	0												

b Draw a set of axes from 0° to 360° for the x-axis and −5 to 5 for the y-axis, like this

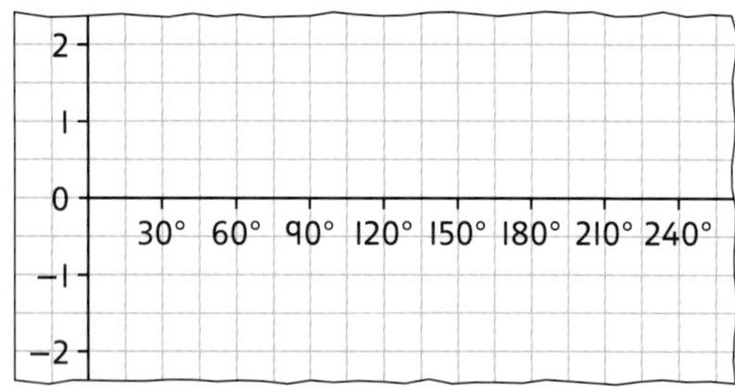

Using the results from your table, sketch the graphs of $y = \sin x$, $y = 2 \sin x$ and $y = 3 \sin x$ on the same set of axes.

c Write down the maximum and minimum values of each graph. What do you notice?

d In the equation $y = A \sin x$, the value A is called the amplitude. Without calculating any further points, sketch the graph of the wave with amplitude 5. Label the graph $y = 5 \sin x$.

Watch out!

Make sure your calculator is in degree mode.

Learn this

Amplitude is the maximum displacement from a reference level, in either a positive or negative direction.

2 **a** Copy this table, then use a calculator to complete it.

x	0°	30°	60°	90°	120°	150°	180°	210°	240°	270°	300°	330°	360°
$\sin x$	0												
$1 + \sin x$	1												
$2 + \sin x$	2												

b Draw a set of axes from 0° to 360° for the x-axis and −3 to +3 for the y-axis. Using the results from your table, sketch the graphs of $y = \sin x$, $y = 1 + \sin x$ and $y = 2 + \sin x$.

c Without calculating any further points, sketch the graph of $y = -1 + \sin x$.

amplitude maximum value minimum value model

3 A buoy is floating at a fixed point in the sea.
As the tide comes in and out, the height of the buoy in metres above the seabed is given by the equation $y = 15 + 8 \sin x$.

a Write down the maximum and minimum values of $8 \sin x$.

b Hence, work out the buoy's height above the seabed
i at high tide **ii** at low tide.

4 The height of a carriage on a ferris wheel is given by an equation involving $\sin x$. This is the graph of height plotted against time.

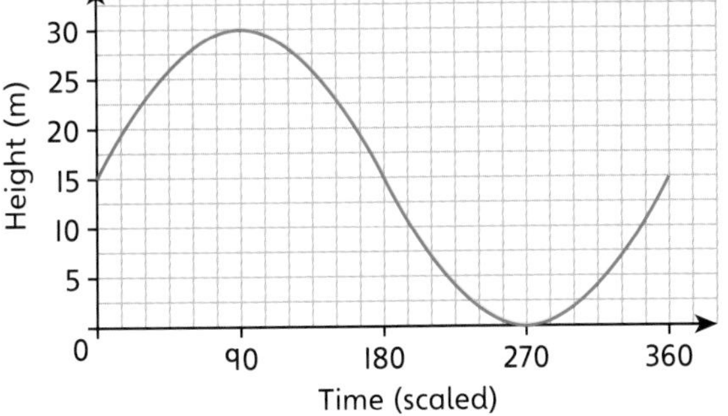

a What is the height of the ferris wheel?

b What is the amplitude of the wave?

c Write down an equation for the height of the carriage, h.

5 More complicated waveforms can be made by adding two or more simple waves together. Musical instruments all sound different because they produce different combinations of waves with different wavelengths.

a The wavelength of a wave with equation $y = \sin Bx$ is given by $\frac{360°}{B}$.
Write down the wavelengths for $y = \sin 2x$ and $y = \sin 3x$.

b Copy this table, then use a calculator to complete it.

x	0°	30°	60°	90°	120°	150°	180°	210°	240°	270°	300°	330°	360°
$\sin 2x$	0												
$\sin 3x$	0												
$\sin 2x + \sin 3x$	0												

c Use your results to draw the graph of $y = \sin 2x + \sin 3x$.

6 Use ICT to explore what happens when you put different combinations of simple waves together.

Learn this

The wavelength is the distance between two corresponding points on a repeating wave.

Super fact!

The human heartbeat is made up of a combination of waveforms. Different waves are formed as valves open and close in the cardiac cycle, and these add together to make the complicated wave shape of the human pulse.

Count on maths!

The mathematics of waves can be used to model and understand tsunamis. The wave equations used are more complicated and involve more variables because the height of the wave depends on time, distance and the depth profile of the seabed. Mathematicians use tools called partial differential equations to model the motion of the wave across the sea. The heights of the waves build up, as faster waves pile up behind slower waves, until eventually the waves break near the shore and the water moves in like an extremely strong tide.

sine (sin) trigonometric function wavelength

16.2 The mathematics of earthquakes

- Use algebraic formulae to model real-life situations
- Use loci and construction skills to find the epicentre of an earthquake
- Investigate whether different sets of variables are correlated

Why learn this?

Seismologists can use mathematics to understand the seismic waves of earthquakes and to make some predictions of the damage they might cause.

What's the BIG idea?

- A **locus** is a set of points that follow a rule or a set of conditions called **constraints**.
  For example, you can find the epicentre of an earthquake by measuring its strength in different places and then drawing loci of possible locations.
- You can use **scatter graphs** to investigate whether two different variables are **correlated**.
  For example, you can investigate whether the magnitude of an earthquake is correlated to the number of deaths it causes.

Did you know?

The strength, or magnitude, of earthquakes is sometimes measured using the Richter scale. Every whole number on the Richter scale represents a tenfold increase in magnitude, so an earthquake measuring 6.0 is ten times greater than one of 5.0 and 100 times greater than one of 4.0.

Practice, practice, practice!

1 Three towns experience a small earthquake. A scientist suggests that the measured power, P, of the earthquake at different points and the distance in miles, D, to the epicentre are related by the formula $D = \frac{1}{2}P + 18$.

a Using this formula, what is the distance to the epicentre if the power is
i 20 units **ii** 50 units?

b Explain why this formula cannot be the rule that describes real-life earthquakes.

Did you know?

An earthquake's epicentre is the point on the Earth's surface immediately over where it happened.

2 A better rule is given by the formula $D = \frac{400}{P^2}$.
Using this formula, what is the distance to the epicentre if the power is

a 20 units **b** 10 units **c** 8 units?

3 The three towns are positioned as shown.

Biggleston — Little Hampton: 10 miles
Biggleston — Missenden: 7 miles
Little Hampton — Missenden: 4 miles

a Use a ruler and compasses to construct a scale diagram using the scale 1 cm : 1 mile.

b The towns record these power levels.

Missenden	20 units
Little Hampton	10 units
Biggleston	8 units

Use your answers from Q2 to construct three loci to show where the epicentre might be. Hence, find the actual location of the epicentre.

constraint construct correlation hypothesis line of best fit

4 Resource sheet 16.2 contains data for some earthquakes. Max and Kathy want to use the data to predict the number of deaths from future earthquakes.

a Max and Kathy decide not to use the data from the four earthquakes in China. Why do you think they made this decision?

b Max thinks that an earthquake with a larger magnitude will result in a larger number of deaths.

i Use ICT to draw a scatter graph (with magnitude from 6 to 9 on the x-axis and deaths from 0 to 30 000 on the y-axis) to see if Max's hypothesis is correct.

ii What kind of correlation does your graph in part **i** show? Is it a strong correlation or a weak correlation?

iii Draw a line of best fit on your graph. Would you be able to make an accurate prediction from this line?

c Kathy thinks that more people will have died in earlier earthquakes, as buildings are now built to a higher standard of safety.

i Use ICT to draw a scatter graph (with years from 1900 to 1990 on the x-axis and deaths from 0 to 30 000 on the y-axis) to see if Kathy's hypothesis is correct.

ii What kind of correlation does your graph in part **i** show? Is it strong correlation or weak correlation?

iii Draw a line of best fit on your graph. Would you be able to make an accurate prediction from this line?

d **i** What other factors might influence the number of deaths from an earthquake?

ii Can you suggest a better way in which Max and Kathy could build a predictive model?

Learn this

An observation that is far removed from the rest of the data is called an outlier. Outliers are usually removed from the data as they can distort any calculations.

Resource

16.2 Table showing deaths from earthquakes

Country	Year	Magnitude	Deaths
Afghanistan	1956	7.7	2000
Alaska	1964	8.4	131
Algeria	1920	7.3	4500
Armenia	1988	6.9	
Bosnia	1963	6.0	
Chile	1906		
Chile			

Count on maths!

Mathematics can be used to explore and understand earthquakes. Engineers also use mathematics to design buildings which are more resistant to earthquakes.
Scientists send artificial seismic waves into the Earth and use mathematics to analyse the results and work out what materials lie hidden under the ground.

locus model outlier scatter graph

16.3 Medical statistics

- Consider the statistical design of medical experiments
- Recognise how to work with probabilities associated with more complex events
- Understand the ideas of symmetry and skew in continuous data distributions

Why learn this?

Mathematics plays many important roles in the world of medicine – from testing out new treatments to understanding the statistics of the human body.

What's the BIG idea?

→ **Probability** ideas can be used to help assess new treatments and medical tests, but you have to be careful when designing the experiments.

→ When describing **distributions** of **continuous** data, you can say whether they are roughly **symmetrical** or **skewed** to one side. The amount of skew sometimes tells you something about the physical situation you are investigating.

Practice, practice, practice!

1 A new treatment is being given to patients with a certain illness.
After three weeks, 42 patients claimed to have experienced an improvement, while 98 did not.

a Using these figures, estimate the probability that someone undergoing the treatment will feel better after three weeks.

b Is the probability that someone undergoing the treatment will feel better after three weeks the same as the probability that the treatment will cause them to feel better after three weeks? Explain your answer.

2 Sometimes people experience a psychological phenomenon known as the placebo effect.
The table shows the results of a trial where three different treatments and a placebo were given out at random to patients with a certain illness.

	Placebo	Treatment A	Treatment B	Treatment C
Improved after three weeks	38	42	96	81
No improvement	101	98	44	61

a Work out the probability that someone will feel better after three weeks for each of the four treatments.

b Do you think treatment A is successful? Explain your answer.

c Which treatment seems to be the most successful? Explain your answer.

d Why do you think it was important that the treatments and the placebo were given out at random?

e How could this trial be improved?

Did you know?

A placebo treatment in a medical experiment is one that has no actual medical value. A person's symptoms are reduced simply because they believe they are taking an effective medicine or because they would have got better anyway.

bias continuous data distribution frequency diagram

3 Doctors are trying out two new methods of testing for the Bigedd virus. They test each patient with test A or new test B and then check their results using an expensive older method that is known to be accurate. The table shows the results of the tests.

	Test method A		Test method B	
	Patient has virus	Patient does not have virus	Patient has virus	Patient does not have virus
Test is positive	183	5	172	1
Test is negative	2	495	3	508

a A 'false positive' occurs when a patient is told they have the virus when they do not. Work out the probability that a positive result is a false positive with
i test method A **ii** test method B.

b A 'false negative' occurs when a patient is told they do not have the virus when they do. Work out the probability that a negative result is a false negative with
i test method A **ii** test method B.

c Which test method do you think is the better? Explain your answer.

Did you know?

More accurate medical tests are often more expensive, so doctors often start by screening the population using cheaper tests which produce many false positives and then check the positive results by using the more expensive tests.

4

Weight (g)	Births (%)
$1000 \leqslant w < 1500$	1
$1500 \leqslant w < 2000$	3
$2000 \leqslant w < 2500$	8
$2500 \leqslant w < 3000$	11
$3000 \leqslant w < 3500$	28
$3500 \leqslant w < 4000$	27
$4000 \leqslant w < 4500$	13
$4500 \leqslant w < 5000$	7
$5000 \leqslant w < 5500$	2

The table gives information about the birth weights of babies born last year.

a Draw a frequency diagram to display this data.

b What is the modal class of the data?

Hint: Where would the 50% point be?

c In which class would the median lie?

d Is the shape of the distribution roughly symmetrical or skewed to one side? Write a sentence to explain what this tells you about the expected weight of a newborn baby.

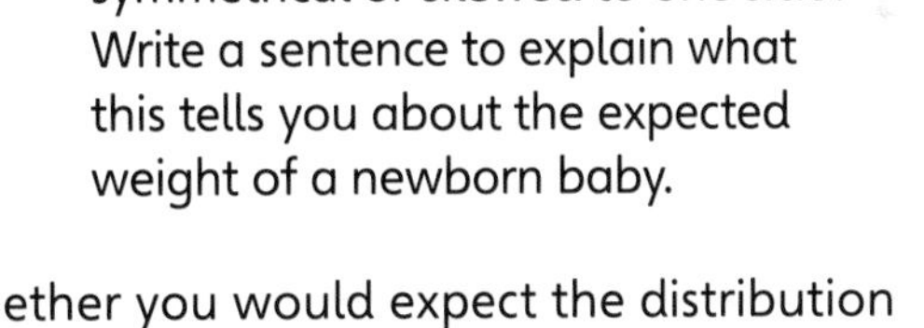

Learn this

If the data is biased towards the right-hand side of a diagram, it is said to have a negative skew. If the data is biased towards the left-hand side of a diagram, it has a positive skew.

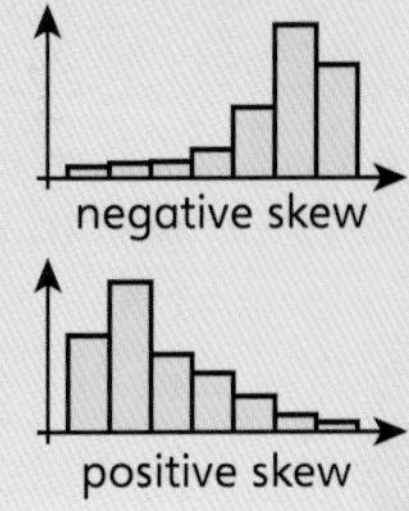

negative skew

positive skew

5 For each of these sets of data, say whether you would expect the distribution to be roughly symmetrical, negatively skewed or positively skewed.

a The distribution of adult heights in the UK.

b The distribution of adult weights in a country experiencing food shortages.

c The distribution of adult weights in a country with a high rate of obesity.

d The distribution of IQ test scores in the UK.

Count on maths!

The mathematics of false positives and false negatives is also used in other types of testing, such as iris recognition software and 'spam filtering' of emails.

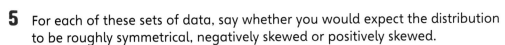

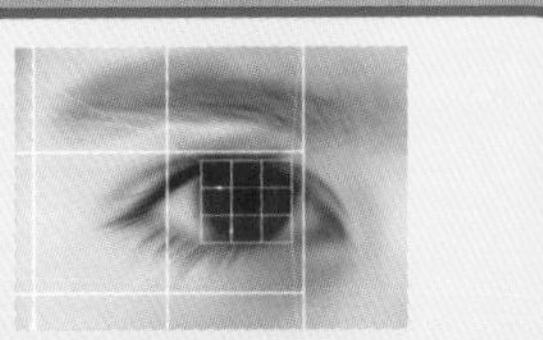

modal class placebo probability skew symmetry

16.4 Made to measure

- Convert recurring decimals into equivalent fractions
- Know the difference between rational and irrational numbers
- Calculate the volume of cylinders and cones
- Maintain appropriate accuracy through a multi-step calculation

Why learn this?

The use of computers enables manufacturers to make products with high levels of accuracy and precision – as long as they can input accurate data to begin with.

What's the BIG idea?

- People who use mathematics in real situations often want to calculate results to a very high degree of accuracy. Results can be ruined if you **round** too early in calculations.
- **Rational numbers** are numbers that can be written as fractions. For example, 3.25 is rational as it can be rewritten as $\frac{13}{4}$.
- **Irrational numbers** can only be written in number form as a never-ending, non-repeating decimal. For example, $\sqrt{2}$ and π are irrational as they have no equivalent fractions.
- $\sqrt{2}$ is an example of a **surd**. A surd is an irrational root of a whole number.
- If a calculation involves an irrational number, you sometimes leave it in your final answer to make it as exact as possible.

Did you know?

Although π cannot be written as a single fraction, you can make it by adding an infinite number of fractions together. Leibniz's formula for π tells us that $\frac{\pi}{4} = 1 - \frac{1}{3} + \frac{1}{5} - \frac{1}{7} + \frac{1}{9} - \frac{1}{11} + \frac{1}{13} - \ldots$

Practice, practice, practice!

1 Adam is manufacturing a component that needs to be two thirds of the height of a previous component. He sets the computer to multiply the previous height by 0.66.

a Will this give Adam an accurate result? Explain your answer.

b What mathematical operations should Adam have used?

2 Convert these decimals into fractions and simplify them.

$0.\dot{1}\dot{7}$ Call the number x:

$$100x = 17.17171717\ldots$$
$$x = 0.17171717\ldots$$
$$99x = 17$$
$$x = \frac{17}{99}$$

a $0.\dot{3}\dot{6}$ **b** $0.\dot{1}\dot{5}$ **c** $0.\dot{5}\dot{7}$ **d** $0.\dot{1}2\dot{3}$

3 Sort these numbers into rational numbers and irrational numbers.

7π | $\frac{43}{50}$ | $\sqrt{3}$ | $\frac{1}{3}$ | $\sqrt{5}$ | $3\sqrt{2}$ | $\sqrt{7}$ | $\sqrt{9}$ | $0.\dot{1}\dot{3}$

Did you know?

Although recurring notation can represent the exact value of a fraction with decimals, it is very difficult to input into a computer.

infinite irrational number Pythagoras' theorem rational number

4 Work out the volume of each shape. Leave your answers in terms of π.

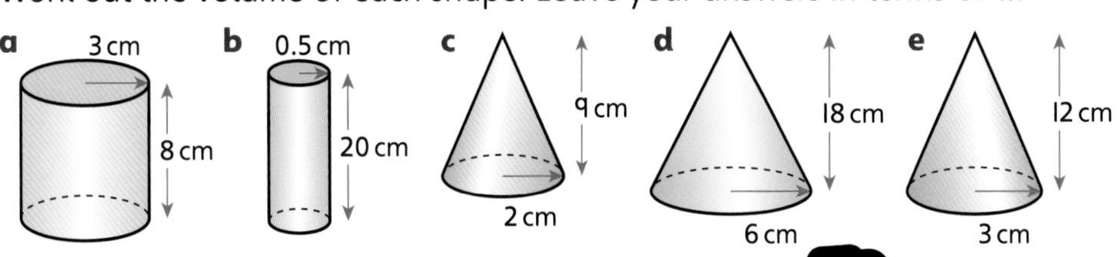

A cone takes up exactly one-third of the space of a cylinder with the same height and radius, so the formula for the volume of a cone is $V = \frac{1}{3}\pi r^2 h$.

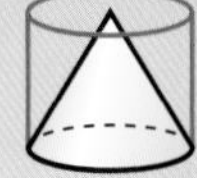

5 Work out the shaded area of each shape. Leave your answers in terms of π.

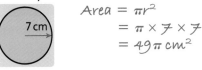

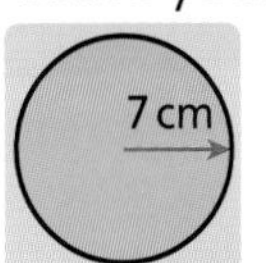

$$\text{Area} = \pi r^2$$
$$= \pi \times 7 \times 7$$
$$= 49\pi \text{ cm}^2$$

Tip

π is an irrational number so mathematicians often leave their answers in terms of π to make sure they are exact.

a

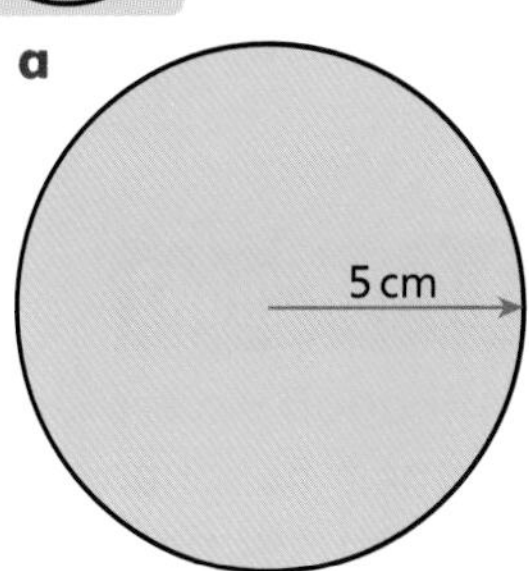

b

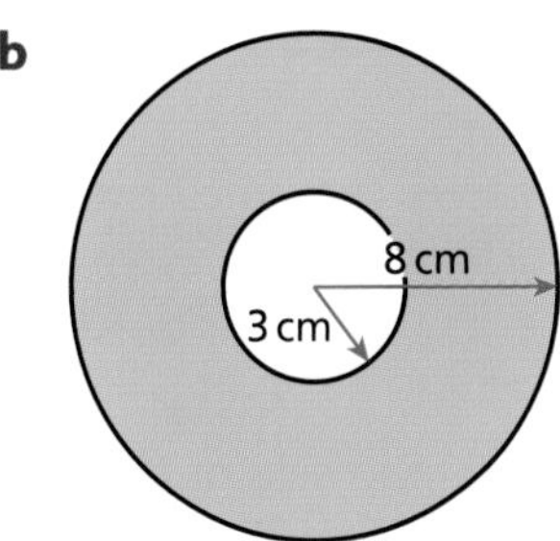

c

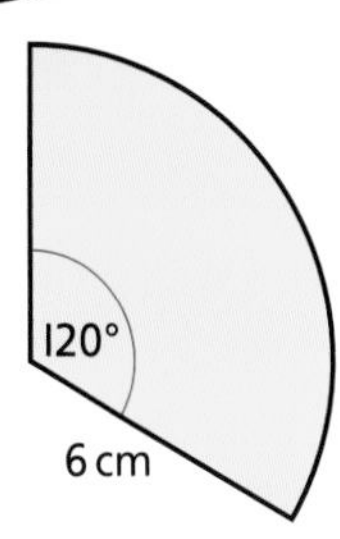

6 The diagram shows a shape made up of three right-angled triangles.

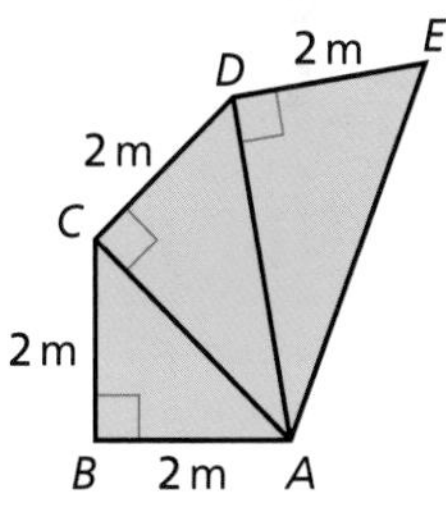

a Method 1

i Using Pythagoras' theorem and a calculator, work out the length of AC to one decimal place.

ii Using your answer from part **i**, work out the length of AD to one decimal place.

iii Using your answer from part **ii**, work out the length of AE to one decimal place.

b Method 2

i Without using a calculator, work out the length of AC. Leave your answer as a square root.

ii Using your answer from part **i**, work out the length of AD. Leave your answer as a square root.

iii Using your answer from part **ii**, work out the length of AE. Evaluate your answer.

c Is there a difference between the two values you calculated for the length of AE?

7 Without using a calculator, work out the area of this rectangle.

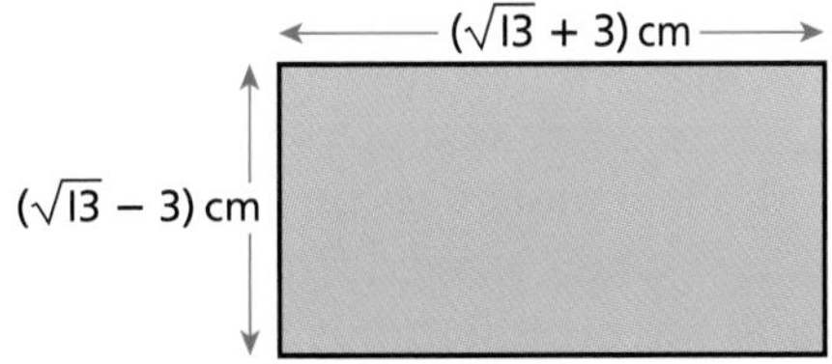

Count on maths!

Manufacturing industries are concerned with both accuracy and precision. A manufacturing process is precise if it can consistently reproduce the same or very similar results. Statistical analysis and the use of graphs such as quality control charts can help factories monitor their output and produce the best possible quality of product.

recurring decimal **round** **surd** **terminating decimal**

16.5 Stellar mathematics

- Consider arc lengths and sector areas in an astronomical context
- Understand and manipulate numbers in standard form
- Calculate the volume and surface area of a sphere
- Substitute values into complex astronomical formulae and interpret the results

Why learn this?

The tools of mathematics can help astronomers examine and explore the universe around us – without having to go there!

What's the BIG idea?

→ By modelling stars and planets as spheres we can calculate good approximations for their volumes and surface areas. We often write astronomical measurements using standard form as they can be very large.

→ If we can estimate the surface temperature and size of a star, we can use formulae and a Hertzsprung–Russell diagram to find out what kind of star it is.

Did you know?

Although we think of stars and planets as spheres, they're actually not quite spherical. The Earth is closer to being an oblate spheroid – a rounded shape with a slight bulge around the equator.

Practice, practice, practice!

1 The planet Earth orbits the Sun in an approximate circle with a radius of 1.496×10^{11} m. A complete orbit takes one year.

a Calculate the distance that the Earth travels in one year.

b There are 52 weeks in a year. Calculate the length of the arc that the Earth travels along in one week.

2 Most planets travel around the Sun in elliptical orbits. Kepler's second law is a piece of mathematics that shows that a planet will always take equal amounts of time to trace out equal areas. Look at the diagram. At which part of the orbit is the planet travelling

a the fastest **b** the slowest?

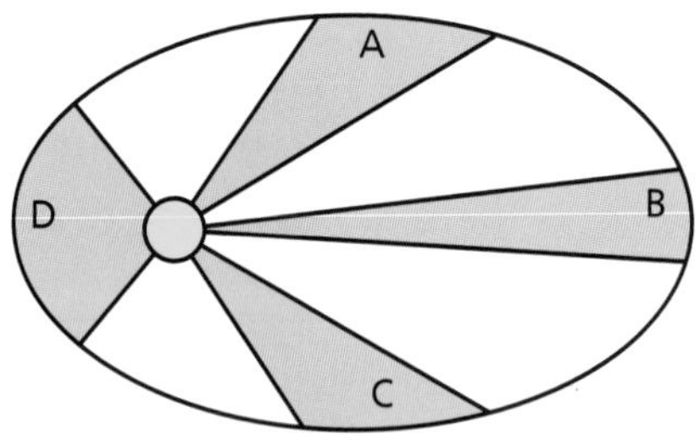

A

B

C

D

3 The diameter of the Sun makes an angle of approximately 0.533° when viewed from the Earth. Using this fact, and the orbital radius of the Earth, find the radius of the Sun to three significant figures.

1.496×10^{11} m

0.533°

Sun

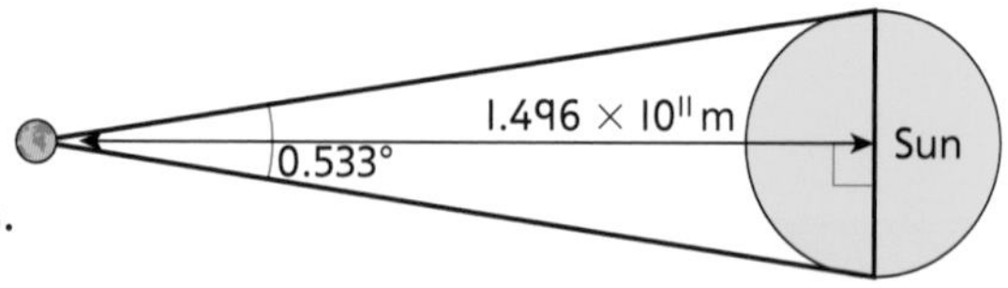

4 A sphere of radius r has a volume of $\frac{4}{3}\pi r^3$ and a surface area of $4\pi r^2$. Using your answer from Q3, find

a the volume of the Sun

b the surface area of the Sun.

Give your answers to three significant figures.

Tip

Use a calculator for these questions! Make sure you know how to enter numbers in standard form into your calculator.

arc Hertzsprung–Russell diagram luminosity

5 The total power that a star radiates is called its 'luminosity'.
The luminosity, L, of a star is calculated in watts using the formula:

$$L = 4\pi r^2 \sigma T^4$$

r is the star's radius in metres, T is its surface temperature in kelvins and σ is a special constant equal to 5.67×10^{-8}. Using the value of the radius from Q3, and given that the surface temperature of the Sun is 5778 K, calculate its luminosity to three significant figures.

6 The luminosity of a star and its surface temperature tell us what type of star it is. The vast majority of stars belong to the 'main sequence' of stars, but stars can also be 'white dwarfs', 'giants' or 'supergiants'. A Hertzsprung–Russell diagram is a special diagram which tells us what group a star belongs in.

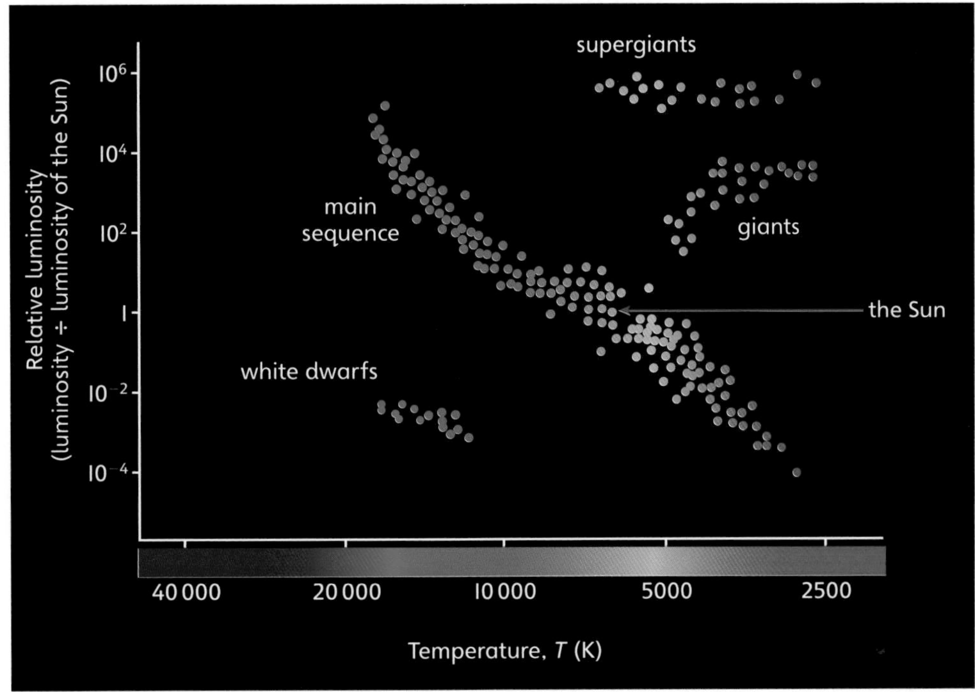

The luminosity is scaled relative to the Sun, so to use the Hertzsprung–Russell diagram we first have to divide a star's luminosity by the luminosity of the Sun, which you found in Q5.

Use the formula in Q5 to copy and complete this table, and find out what type of star each one is.

Name	Radius (m)	Temperature (K)	Luminosity (W)	Luminosity ÷ luminosity of the Sun	Type of star
Sun		5778		1	main sequence
Betelgeuse	4.38×10^{11}	3500			
40 Eridani B	9.74×10^{6}	16 500			
Alpha Centauri A	8.54×10^{8}	5790			

Count on maths!

Maths is used throughout astronomy, for example, in planning the path of a rocket to the Moon. This requires many calculations, since the Earth, the rocket and the Moon are all moving in different ways.

surface area volume

maths!

Curves of pursuit

A curve of pursuit is a curve formed when A is chasing B by following a curve that is always pointing towards B, and A and B are both moving at constant speeds.

Such a curve might describe the path of a bird of prey as it swoops to catch a moving target – or a patrol boat as it travels to intercept a suspect vessel.

Pet pursuit

To plot a simple example of a curve of pursuit, imagine a dog chasing a car. The dog moves towards wherever the car is, but the car is also moving at three times the speed of the dog.

Plot the starting positions of a dog and a car.

- Move the dog 1 cm towards the car. Now move the car forwards 3 cm in a horizontal line.
- The dog now changes direction because the car has moved. Move the dog another 1 cm and the car another 3 cm.
- Repeat this as many times as you can. Describe the shape of the path of the dog.
- What happens if the car only moves twice as fast as the dog?
- What happens if you change the starting positions of the dog and the car?

Mutual pursuit

Imagine that there are three dogs at the vertices of an equilateral triangle. Each dog is chasing the dog on its right.

- Construct an equilateral triangle with side length 10 cm.
- Move each dog 1 cm towards the dog it is chasing.
- Each dog now changes direction because the other dogs have moved. Move each dog another 1 cm, along its new direction.
- Repeat this until the dogs meet. Describe the shape of the dogs' paths.
- What happens if you start with four dogs at the vertices of a square, or six dogs at the vertices of a regular hexagon?
- What do you think would happen if the dogs started at the vertices of a scalene triangle?
  Make a prediction and test it.
  Were you right?

Dog and duck

A duck is swimming around the edge of a circular pond, at the centre of which lurks a dog in some shallow water. If the dog swims in a curve of pursuit to catch the duck, and swims at twice the speed of the duck, what shape will the curve be?

16.6 Fractals

- Explore the geometry of fractals
- Find the nth terms for more complicated sequences and find their limits as n tends to infinity

Why learn this?

Fractal geometry can help us understand some amazing shapes from the world around us, such as this Romanesco broccoli.

What's the BIG idea?

- **Fractals** are shapes that exhibit **self-similarity**. This means that every part of the shape is a reduced copy of the original shape.
- Fractal shapes lead to some interesting mathematical results. They can also be used to model some real-world shapes or to construct copies of those shapes in a simple way.

Did you know?

The word 'fractal' comes from the Latin word 'fractus', which means 'broken' or 'fractured'.

Practice, practice, practice!

1 A famous example of a fractal is the Koch snowflake.
Begin with an equilateral triangle. At each stage divide every line into three equal sections.
Using the middle sections as the bases, construct new smaller equilateral triangles.
Finally, rub out the bases of the new triangles.

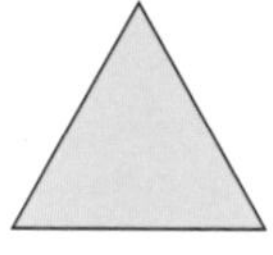

Stage 0

Stage 1

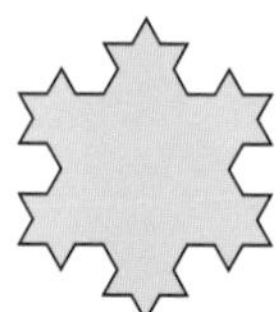

Stage 2

Stage 3

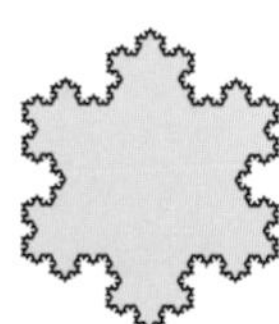

Stage 4

…

The true Koch snowflake is formed after an infinite number of stages.

a Use a ruler and compasses to construct Stage 2 of the Koch snowflake. Start with an equilateral triangle of side length 6 cm.

b The area of the true Koch snowflake is given by the formula $\frac{2\sqrt{3} \times s^2}{5}$, where s is the original side length. Calculate the area of your Koch snowflake.

c **i** Write down the perimeter of your Koch snowflake at Stage 0, Stage 1 and Stage 2.
ii What fraction are you multiplying by each time?
iii What is happening to the length of the perimeter each time?

d **i** Write down an expression for the length of the perimeter at Stage n.
ii What happens to the length of the perimeter as n heads to infinity?

Tip

In the Koch snowflake three sections are being replaced by four, so what fraction are you multiplying the perimeter length by each time?

area fractal infinite limit nth term perimeter

2 Another famous example of a fractal is the Sierpinski triangle.
It begins with a coloured equilateral triangle. At each stage an inverted triangle is taken away, of half the height and width.

Stage 0

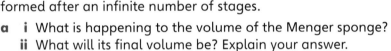

Stage 1

Stage 2

…

Stage 6

The true Sierpinski triangle is formed after an infinite number of stages.

a Draw an equilateral triangle of side length 16 cm on isometric paper.
Construct the Sierpinski triangle up to Stage 3.

b **i** What is happening to the area of the Sierpinski triangle that is coloured?
ii What will its final area be? Explain your answer.

c **i** What is happening to the perimeter of the Sierpinski triangle? (Include the edges on the inside.)
ii What will the final perimeter be? Explain your answer.

3 The Menger sponge is an example of a 3-D fractal.
Begin with a cube. At each stage remove a smaller cube with a side length one-third as large as before from the centre of every cube and from each of its faces. The true Menger sponge is formed after an infinite number of stages.

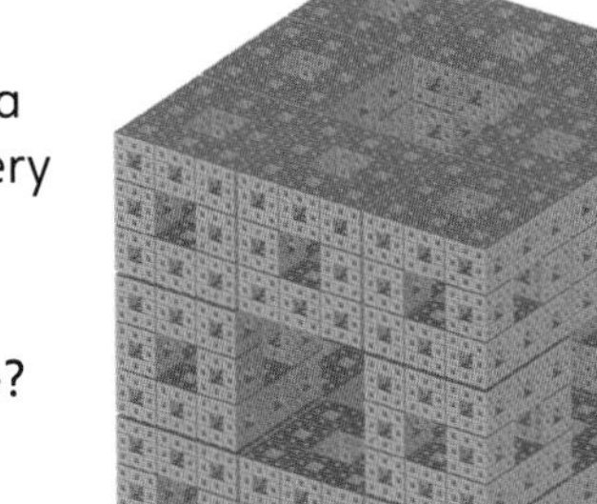

a **i** What is happening to the volume of the Menger sponge?
ii What will its final volume be? Explain your answer.

b **i** What is happening to its surface area?
ii What will its final surface area be? Explain your answer.

c The tubes in a pair of human lungs split into smaller and smaller tubes in a similar way to a fractal. Explain why the human body might benefit from a lung shape which has a very large surface area and a relatively small volume.

4 A 'fractal tree' can be generated like this.
Begin with one branch. At each stage add on two branches to the end of every new branch.

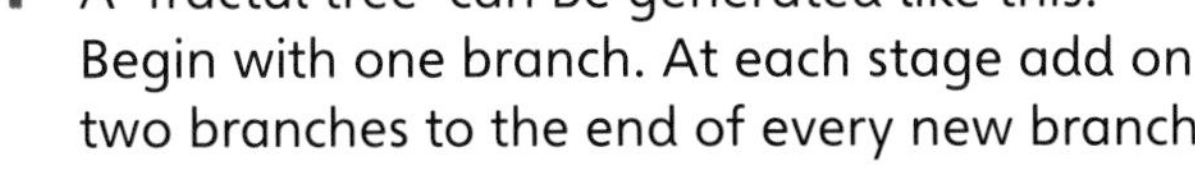

Stage 1 Stage 2 Stage 3 …

a Draw your own fractal tree up to Stage 5.

b Copy and complete this table.

Stage	1	2	3	4	5
Number of branches	1	3			

c **i** Describe the sequence that the number of branches is following.
ii Hence, or otherwise, write down the number of branches in a Stage n fractal tree.

Count on maths!

The geometry of fractals has been used to help scientists study snowflakes, blood vessels, lightning and even the shape of the entire galaxy.
Fractals are also connected to the mathematical area of complex numbers, which has applications in many areas of science and engineering.

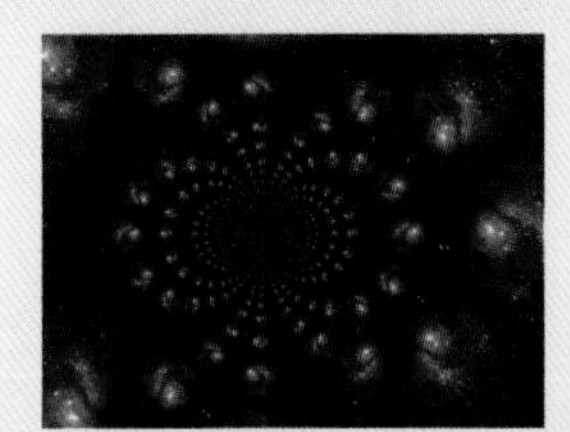

self-similar sequence surface area volume

Mathematics at work

This unit has touched on just six ways that mathematics is used in the world around us – but there are thousands more.

Choose one of the subject areas below and research the role of mathematics in that area. Prepare a short presentation for your classmates.

Cryptography

Cryptography is the science of writing codes.
Clever mathematical methods are responsible for keeping data safe and private – even your private e-mail!
Many areas of mathematics are used to develop increasingly complicated codes, such as prime factor decomposition and some special curves called 'elliptic curves'.

Key research questions

- What is frequency analysis and how can it help crack a code?
- What is public-key cryptography and how does it relate to prime numbers?

Knot theory

As strange as it may sound, knots have their own branch of mathematics. This can help to explain why some magicians seem to tie a knot which disappears when pulled and to tell whether two knots that look different are actually the same. It also has some surprising practical uses: in the last few decades, scientists have begun to use it to study knotting in DNA and polymers.

Key research questions

- What is an 'unknot'?
- What are the Borromean rings?

Econometrics

Econometrics uses mathematical and statistical techniques to try to explain patterns of growth and recession in the economy. This helps economists test their ideas against real data and predict what might happen next.
This area of mathematics uses simultaneous equations, as well as a range of advanced statistical methods.

Key research questions

- How is econometrics used to test economic theories?
- What kinds of graphs might be used in econometrics?

Game theory

Game theory is a relatively new area of study that uses mathematics to model competitive behaviour. It was initially developed to predict the best strategy to use in certain games. It also has applications in biology, politics and the business world.

Key research questions

- What is a zero-sum game?
- What is the Prisoner's Dilemma?

Chaos theory and weather forecasting

Many real-life systems are very complicated. Changing one variable slightly will, over time, result in a huge difference in outcome. These systems are known as 'chaotic' and are the subject of chaos theory. One common example is the weather. However, chaos theory also helps us understand the orbits in the solar system and the turbulence experienced by a plane.

Key research questions

- What is the butterfly effect?
- Why can't we predict the weather exactly?

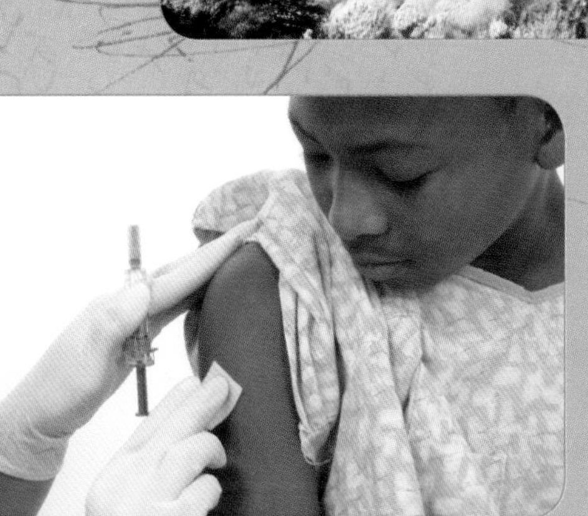

Mathematical epidemiology

Mathematics can be used to study and explain the spread of disease. Mathematicians can analyse statistical data and use differential equations to create models that predict the speed and direction of the spread of a disease. The models can help find out critical values too, such as what percentage of the population needs to be vaccinated to prevent future outbreaks of a disease.

Key research questions

- Why is vaccination important?
- What is the SIR model?

Index

a

adding fractions 40
alternate angles 62, 222, 252
ancient Egyptians 41
ancient Greeks 73, 256
angles 62–7, 72
approximating 56, 128
arcs (circles) 108
areas 104, 112, 114, 196
arithmetic sequences 6
arrowheads (shape) 62
assumed means 88
astronauts 102
astronomy 294–5
autostereograms 192
average speeds 106
averages (*see* mean, median, modal classes)
axes (graphs) 242

b

Babylonians 31, 196–7
Benford's law 164–5
biased samples 84–5, 234
bisectors (angles) 72
black holes 210
brackets 20, 24, 198–9
  double brackets 200–1

c

calculators 126–7, 130–1, 134–5
  trigonometry 188
capacities 104
capture and recapture method 217
chaos theory 301
chords (circles) 76
cicadas 140
circles 76, 108, 112
circumferences 198
class intervals (data) 86, 96
codes 300
common denominators 40
complex numbers 299
compound interest 46, 224
compound measures 106
compound shapes 112
congruent shapes 180–2
constructions 72–5
conventions 62
conversion graphs 48
correlation 240, 288
corresponding angles 62, 222, 252
cosine ratio (cos) 188–91, 222, 260–2
cross-sections (prisms) 258
cryptography 300
cube roots 52
cubic expressions 29
cubic functions 154
cuboids 114, 116
curves of pursuit 296–7
cyclic quadrilaterals 77
cylinders 116, 258

d

data handling cycle 216
decimals 54, 56, 124, 132
decreasing sequences 7
definitions 62
denominators 38, 40, 44, 54, 124, 144
densities 106
dependent events 270, 274
derived properties 62
diagonals 62, 223
diagrams (algebra) 200
diameters 108, 112
direct proportion 48, 106, 224
diseases 301
distance–time graphs 14, 158
distributions (data) 244–5, 290
dividing
  decimals 56, 132
  fractions 44
  powers 146, 148
double brackets 200–1
du Sautoy, Marcus 140

e

earthquakes 288–9
econometrics 300
Einstein, Albert 210
eliminating (equations) 24, 32
enlargements 50, 186
epidemiology 301
equations 20–33, 220
equilateral triangles 66
equivalent calculations 132
equivalent equations 24, 30, 32
equivalent formulae 206
equivalent fractions 38, 40, 56

estimates 56, 124
events (probabilities) 268
expanding brackets 20, 198–9, 202
explicit facts 22, 226
expressions 20
exterior angles 64

f

factorising (algebra) 198, 202–3
factors 52, 142, 144
false positives and negatives 291
Fibonacci sequence 9
first differences (sequences) 4, 6–7
Formula 1 racing 214–15, 230
formulae 158, 204–7
  spreadsheets 226
four colour problem 219
fractals 298–9
fractional powers 148
fractions 38–45, 54, 124, 144
  calculators 126
frequency diagrams 96
frequency polygons 96, 236, 244
frequency tables 86, 96
front elevations 256
function machines 12
functions 12

g

game theory 301
generalising 160, 204
Goldbach conjecture 228
googolplexes 130
googols 130
gradients 150, 152, 226
graphs
  data 216
  inequalities 208
  linear functions 12, 150, 226
  quadratic functions 12, 154
  simultaneous equations 30–1
  trigonometric functions 191, 286
grouped data 236

h

halfway test 29
Hertzsprung–Russell diagrams 294
highest common factors (HCFs) 142
  algebraic expressions 198
hypotenuses 68, 188–9, 222
hypotheses 216

i

identities 20, 200
implicit facts 22, 226
improper fractions 40
independent events 170, 270, 274
indices 146–9
inequalities 38, 208–9
inscribed polygons 76
intercepts (graphs) 150, 226
interior angles 64
interpreting data 216
inverse functions 12
inverse proportion 48
irrational numbers 292
isosceles trapeziums 62
isosceles triangles 66

k

kites (shape) 62
knot theory 300

l

like terms 198
line graphs 94
linear expressions 6
linear functions 12, 150, 226
lines of best fit 240
loci (*singular* locus) 288
lower bounds 104, 128, 218–19
lowest common multiples (LCMs) 142, 144

m

mapping diagrams 12
mappings 12
mathematical models 286
maxima (graphs) 154
mean (data) 88–9, 236–7, 244
measurements 104, 292–3
median 88–9, 236–7, 244
medical statistics 290–1
Mersenne primes 162
metric units 104
mid-points (lines) 186
Millennium Prize Problems 219
minima (graphs) 154
misleading graphs (data) 242
mixed numbers 40, 44, 126
modal classes (data) 88–9
mosaics 110–11
multiples 142
multipliers (percentages) 46–7, 224

multiplying
  decimals 56, 132
  fractions 44
  powers 146, 148
mutually exclusive events 168

n

negative correlation 240
negative powers 148
Nightingale, Florence 82
number lines 208
numerators 38, 44, 54, 124

o

oblate spheroids 294
odometers 108
order of operations 5, 52, 126, 134, 204
ordering fractions 38
oscillating sequences 135
outcomes 168, 268

p

parabolas 156–7
parallel lines 30
  graphs 150, 152
parallelograms 62, 114
patterns 10, 18, 160, 178–9
pentagons 253
percentages 46, 54, 224
peripheral drift illusion 178
perpendicular lines 72, 76
  graphs 152
pie charts 90
plan views 256
planes of symmetry 256
planets 294
polygons 64
populations (statistics) 217
position-to-term rules 4
positive correlation 240
powers 52, 130, 146, 148
pressures 106
primary data 84
prime factor decompositions 142
prime numbers 140, 142
principal (interest) 46
prisms 116, 258
probabilities 166, 268, 290
  dependent events 270, 274
  independent events 170, 270, 274
  mutually exclusive events 168
problem solving 136, 214–30
proofs 180–81
proportion 48, 224
Pythagoras 69, 71, 254–5
Pythagoras' theorem 68–71, 222–3, 248, 255
  proof 254
Pythagorean triples 70

q

quadratic expressions 202–3
quadratic functions 12, 154
quadratic sequences 4, 8
quadrilaterals 62, 66, 250
questionnaires 234

r

radii (*singular* radius) 108, 112
ranges (data) 88–9, 244
ranges (measurements) 104, 106
rational numbers 292
ratios 50
reaction times 217
real-life data 276–7
rearranging formulae 206
reciprocals 44, 126–7, 148
rectangles 62
recurring decimals 124–5
reflections 182
regular polygons 64
relative frequencies 172–3, 268–9, 272
renewable energy sources 92–3
resistors (electricity) 42–3
rhombuses 62, 223
right-angled triangles 68, 70, 74, 188, 190, 222, 260
rotations 182
rounding 56, 124, 129, 134, 218, 292

s

samples (data) 84, 234, 242
scale drawings 256
scale factors 50
scatter graphs 240, 288
Schwarzschild radius 210
second differences (sequences) 8
secondary data 84, 234
sectors (circles) 112
self-inverse functions 12
self-similarity 298
sequences 4–11, 160
series 10
side elevations 256
significant figures 56, 128
similar shapes 186–7

simultaneous equations 30–3, 220
ine ratio (sin) 188–91, 222, 260–3, 286
kewed data 290
ound waves 191
patial sequences 10
peeds 106
preadsheets 226
quare numbers 52, 201
square roots 52, 131, 144, 189
algebra 207
squares 62
squaring algebraic expressions 200
standard form 130, 218–9
stem-and-leaf diagrams 88–9
Stonehenge 60–1
straight-line graphs 150–3
subjects (formulae) 204, 206
substitution (algebra) 32, 204–5, 220
subtracting fractions 40
surds 52, 144, 222
surface areas 114, 116, 258
surveys (data) 235, 272

t

tangent ratio (tan) 188–91, 222, 260–2
tangents (circles) 76
term-to-term rules 4
tessellations 65, 251
time dilation 211
time-series graphs 94
tolerance 42
translations 182
trapeziums 114
tree diagrams (probabilities) 170–1, 270–1, 274
trial and improvement 28, 136, 144
triangles 62, 66, 74, 250
area 114
congruent triangles 180
trigonometric functions 190
trigonometry 188–91, 260–3
two-way tables 86

u

unitary method (percentages) 46
units (measurement) 104, 158, 227
unknowns (equations) 30
upper bounds 104, 128, 218–19

v

variables 20, 224
VAT (value added tax) 205
volumes 104, 114, 116, 258

w

waves 286–7
weather forecasting 301